DESIGN THROUGH TIME

DESIGN THROUGH TIME

Evolving Landscapes, from Alcatraz to Prospect Park

Mark H. Hough

University of Virginia Press
CHARLOTTESVILLE AND LONDON

The University of Virginia Press is situated on the traditional lands of the Monacan Nation, and the Commonwealth of Virginia was and is home to many other Indigenous people. We pay our respect to all of them, past and present. We also honor the enslaved African and African American people who built the University of Virginia, and we recognize their descendants. We commit to fostering voices from these communities through our publications and to deepening our collective understanding of their histories and contributions.

University of Virginia Press

Printed in the United States of America on acid-free paper

First published 2024

9 8 7 6 5 4 3 2 1

Library of Congress Cataloging-in-Publication Data

Names: Hough, Mark H., author.
Title: Design through time : evolving landscapes, from Alcatraz to Prospect Park / Mark H. Hough.
Description: Charlottesville : University of Virginia Press, 2024. | Includes bibliographical references and index.
Identifiers: LCCN 2024002732 (print) | LCCN 2024002733 (ebook) | ISBN 9780813951652 (hardcover) | ISBN 9780813951669 (paperback) | ISBN 9780813951676 (ebook)
Subjects: LCSH: Landscape architecture—United States—History. | Landscape design—United States—History.
Classification: LCC SB469.33 .H68 2024 (print) | LCC SB469.33 (ebook) | DDC 712/.60973—dc23/eng/20240329
LC record available at https://lccn.loc.gov/2024002732
LC ebook record available at https://lccn.loc.gov/2024002733

Cover art: Historic and contemporary views of front facade of Casino, Vizcaya. (Photograph by F. E. Geisler, Vizcaya Museum and Gardens Archives, Miami, Florida, 1925; photograph by the author, 2019)
Cover design: David Fassett

In memory of my mother, Harriet Houston Hough,
who always encouraged me to write a book

A sense of the stream of time is more valuable and more poignant and engaging than a formal knowledge of remote periods. New things must be created, and others allowed to be forgotten.

—Kevin Lynch, urban planner and author, 1972

It is the bane, or perhaps the beauty, of many designed landscapes that significant and ongoing care is required not only to preserve day-to-day appearances, but to ensure that the design's structure evolves fully.

—Dan Kiley, landscape architect, 1999

When I became a landscape architect in the late 1950s there was an underlying belief, among most of us, that landscapes, once built, were forever.

—Stuart Dawson, landscape architect, 2002

CONTENTS

ILLUSTRATIONS

ACKNOWLEDGMENTS

This book is the result of two decades of professional practice during which I have been fortunate to collaborate with and get to know many talented landscape architects. Both directly and through a process of intellectual osmosis, they have educated and inspired me as a designer and a writer. Through lengthy conversations over drinks and dinners, interviews, work sessions, joint presentations, and charettes, they have helped me better understand design, designed places, and the cultural importance of our profession. Among them, Thomas Woltz, Christy Ten Eyck, Michael Vergason, Linda Jewell, Thaisa Way, and Eric Kramer have all influenced how I see and experience landscapes. The same is true for Glenn Allen and Peter Schaudt, who sadly are no longer with us. Laurie Olin and Warren Byrd earn a special nod not only for their influence on me but also for their consistent and good-natured support over the past twenty years. Above all, my sincerest thanks to Gary Hilderbrand for his unwavering friendship, encouragement, advice, and mentorship. His constant support instilled in me the confidence I needed to tackle this book.

I also want to acknowledge professionals in roles like mine who have inspired my work at Duke University and the writing of this book. Foremost is Mary Hughes, who until recently was my counterpart as the university landscape architect at the University of Virginia. While there, she was a paragon within our quiet professional niche, a humble and passionate steward of one of the most significant, complex, and dynamic cultural landscapes in the country. I have been trying to write a story about the evolving campus landscape at UVA—even a planned chapter in this book—for years, but for reasons I can't fully determine, I have not been able to do it justice.

David Barnett of Mount Auburn Cemetery, Cathy Blake at Stanford University, Michael Boland at Presidio Trust, Richard Bumstead at the University of Chicago, Paul Dolinsky and Maureen Joseph of the National Park Service, Russell Fredericks and David Henderson at Sea Pines Resort, Shelagh Fritz of the Golden Gate National Parks Conservancy, Liza Gilbert of the Dumbarton Oaks Park Conservancy, Chris Nolan and Doug Blonsky of the Central Park Conservancy, Ian Simpkins at Vizcaya, Preston Stockton of Reynolda Gardens (another case study that I had to set aside because of the book's growing length), and Christian Zimmerman of Prospect Park Alliance all earn my thanks. Each of these like-minded professionals has directly or indirectly informed this book; all exemplify the practice of landscape stewardship.

Sincere thanks to the historians, librarians, and archivists who offered their expertise and assistance. Rebekah Burgess (New York City Parks), Michele Clark (Frederick Law Olmsted National Historic Site), Julie Cain and Josh Schneider (Stanford University), Dana Chandler (Tuskegee University), Andrew Colligan (Missouri Botanical Garden), Amy Peck (Prospect Park), Anatole Tchikine (Dumbarton Oaks), Karen Urbec (Vizcaya), and Meg Winslow (Mount Auburn) were all generous with their time and provided me with invaluable support and resources.

Special thanks to Boyd Zenner, who died unexpectedly while I was finishing the manuscript. Boyd was an acquisitions editor at the University of Virginia Press who several years ago asked me, "If you're not working on a book project, might you consider one?" This book obviously would not have happened without her, so thanks to her for having confidence that I could do this. Many people miss her humor and talent. Thanks also to her successor, Mark Mones, for his patience and guidance as I made my way through the process.

I also want to thank my father, Royce Hough, whose ribbing about missed deadlines and persistent stress helped keep things in perspective. Finally, I want to thank my husband, David Roberts, for his support and encouragement. Much of this process occurred during the worst of the coronavirus pandemic, when stress was abundant and somewhat debilitating for much of the world. As it has since 1989, his calming presence kept me in check.

Thanks to you all and to others who have contributed to this book.

DESIGN THROUGH TIME

INTRODUCTION

There is an apparent inclination among landscape scholars, writers, and practitioners to consider designers, their patrons, and those involved in the creation of gardens, campuses, cemeteries, neighborhoods, and other designed landscapes the most important players in the history of such places. Everyone, it seems, is captivated by the design process, especially when it is tied to a well-known site or person. This inclination is evident from the many books, academic journals, and design magazines packed with stories celebrating historic American landscapes and the visionaries who helped create them. Although a fondness for historic landscapes' origin stories is not bad, it does underwrite a limited narrative in which a landscape's early history—often referred to as its *period of significance*, when it is subjectively deemed to be at its best—is presented as the definitive chapter in the story of a maturing landscape. Places created decades or centuries earlier by people living in societies dealing with far different cultural realities and aesthetic preferences often enjoy rarefied status, to the point that proposed changes to them frequently are questioned, challenged, and sometimes protested. Again, this viewpoint has some commendable points, but it nurtures a blindness to inevitable change. The true histories of designed landscapes are not rooted in a specific period but rather span an open-ended length of time during which the landscapes continuously change.

While books on the evolution of specific landscapes and published accounts documenting the degradation and subsequent restoration of some high-profile sites certainly exist, the process of how such places are managed, maintained, and periodically adapted in the context of ongoing change remains a relatively unexplored topic. This matters because all landscapes change in ways that dictate some process of alteration or renovation, much of which will deviate from an original vision. A nineteenth-century park, for instance, will inevitably need to be modified to succeed in a twenty-first-century city regardless of how much people love its initial design. The resulting process, in which owners and managers determine how best to balance historical integrity and contemporary realities, can be at least as important in the life of a landscape as the original designer's vision, especially if one considers the environmental effects of climate change and constantly evolving cultural influences. History is essential, but it is part of a continuous narrative rather than a standalone story.

The subject of evolving landscapes has been central to my professional practice and writing for a couple of decades. One of my first jobs in landscape architecture was as a designer at the Central Park Conservancy about twenty years into its mission to lead the restorative transformation of the legendary New York City park. I already knew something of the park's history and that of its original designers, Calvert Vaux and Frederick Law Olmsted, the latter of whom, well over a century after his death, is still the world's most famous landscape architect. While working at the conservancy, I continued to learn about and appreciate the beautifully crafted intricacies of the Greensward Plan, Olmsted and Vaux's original design, which laid out the landscape and its features when construction began in 1858. I was captivated by the park's progression over time and how it evolved from a picturesque nineteenth-century escape within an industrializing Manhattan into an intensively used, contemporary urban park sporting decades' worth of intriguing additions, intrusions, and adaptations.

While in New York, I immersed myself in the park's landscape and its history and sought to understand the key figures associated with its ongoing evolution. One of the most prominent is Elizabeth Barlow Rogers, the conservancy's cofounder and first president, who was no longer working there by the time I started. Also important was the well-documented influence of Robert Moses, New York City's twentieth-century parks commissioner and notorious "Master Builder," who built a spate of new playgrounds and utilitarian facilities that added activities to the park while partially eroding Olmsted and Vaux's vision of a pastoral and naturalistic landscape. I delved into Central Park's less told history, shaped by generations of unheralded people who have in various ways helped sustain the park's significant legacy. Many of them have played important roles in the broader history of New York City yet have been relatively forgotten in accounts of Central Park, which tend to emphasize the park's original design and its decline and subsequent rescue by the conservancy.

When I returned to my North Carolina roots in 2000 and embarked on a lengthy tenure as the first university landscape architect at Duke University, I applied what I had learned at Central Park to Duke, which has its own Olmsted tradition. Much of the original campus was sited, laid out, and designed on five thousand acres of Piedmont woodlands and abandoned farmland by the Olmsted Brothers firm, run by the second generation of landscape architects, who learned and honed their craft under the tutelage of the senior Olmsted. Unlike with Central Park, however, the history of the campus landscape was absent from the school's documented design legacy, which focused more on the Duke family and the architects with whom the Olmsted Brothers firm quietly collaborated.[1]

In my more than two decades at Duke, I have uncovered and documented the Olmsted Brothers firm's forty-year legacy on the campus while leading the restoration of historic quadrangles, advising on the management of natural resources, creating innovative and ecologically progressive landscapes, and overseeing the contemporary design associated with a long and ambitious campus expansion. Through this work, it has become evident to me that however important Duke's origin story and design legacy are, they are only waypoints along an extended and divergent continuum packed with countless events and decisions that affect the campus's ongoing quality and relevance. In other words, like other mature designed landscapes, Duke's campus is more accurately understood as a product of evolution than as a singular work of design.

This seemingly simple idea provided the impetus for me to write this book. I chose to focus on the influence of a broader group of people—administrators, advocates, horticulturalists, institutional leaders, elected officials—whose collective efforts allow landscapes to mature in ways that simultaneously respect established legacies and respond to shifting cultural, ecological, and economic realities. Rather than taking a nostalgic approach that celebrates landscapes as they were initially intended to be, I strive to understand them as the places they have become, and how they got there.

Preservation Paradox

To understand the history of a particular landscape, it is important to acknowledge that it exists in a state of constant change. While change is a universal constant, it is strikingly so for landscapes, from undeveloped forests and coastal wetlands to urban parks and residential gardens. As Stanley W. Abbott, a regional landscape architect for the National Park Service (NPS) who is credited with the design of much of the Blue Ridge Parkway, put it in 1950:

> Strong forces are against us; and nature itself is one! For nature is a matter of growth and decay, not of time stopped. The hardest granite in the statue spalls, the historic shore line builds or it recedes, the insects bore as well at the puncheon floor in the house of an early president as in any other, and the hungry forest reaches again for the open green where a glorious battalion held its field of fire.[2]

Although Abbott is referring primarily to preserving historic sites rather than designed landscapes, his meaning works for both. Natural processes dictate that trees and plants grow, spread, and die. Over time, and almost without fail, concrete spalls, bricks crumble, wood rots, and fountains leak. Floods, fires, droughts, and powerful storms inflict damage at a rate that will likely escalate along with rising global temperatures. At the same time that such deterioration and destruction take place, cultural shifts alter how people experience and are permitted to experience public spaces, while economic uncertainties threaten everything from maintenance and restoration budgets to staffing and resource prioritization. None of this is news, particularly to those designers and scholars who study landscapes. To this point, the esteemed landscape architect, author, and educator Laurie Olin stressed in a 1996 essay, "It is a cliché that landscapes are dynamic, not static."[3] Even if this dynamism is obvious to many, the relative paucity of documentation on the reasons behind it and responses to it makes it a viable and crucial focus of study.

Another important (and perhaps even more obvious) concept at the base of this topic of change is how differently buildings and landscapes evolve. Both are affected by such forces as natural disasters, yet the relative protection provided by walls and roofs allows works of architecture to be maintained and repaired while remaining relatively unchanged for centuries—we need only think of Thomas Jefferson's Monticello or Independence Hall in Philadelphia, whose late history embodies the desire to preserve significant buildings with reverential accuracy. Of course, windows break and roofs need replacing, but such maladies can typically be repaired in kind to look as if nothing had happened. Such apparent stasis is barely possible for designed landscapes. Although deteriorated paving, walls, and structures can be rebuilt, less straightforward complexities inherent in landscapes—dynamic ecologies and ecosystems, ephemeral trees and plants, and changing edge conditions and cultural contexts—make it more difficult to keep them intact. Similarly, modifying programs or uses of a landscape is more challenging than for a building, which can mask significant alterations behind stationary walls. For instance, an interior elevator or lift can provide accessibility without necessitating any alteration of a historic façade. In a landscape, however, replacing steps with an exterior ramp to achieve the same goal will fundamentally alter both design intent and spatial organization.

Nonetheless, there remain persistent and passionate cries to salvage, recreate, and preserve the artistic integrity and design heritage of certain historic landscapes. The argument seems to be: *If architects' work can be deemed sacred enough to preserve intact, why is the same not true for designed landscapes?* There is no definitive response to that question, and attempting to find one would lead one into a warren of subjective issues, conflicting opinions, and passionate arguments, with no reasonable route of escape. Even if there were consensus that designed landscapes *should* be preserved, making it happen would be exceedingly difficult. Another challenge is that landscape architecture is centuries behind architecture as a widely recognized and respected discipline, leaving landscapes generally undervalued as works of design. This lack of appreciation, combined with landscapes' inherent fragility, makes landscapes exponentially more prone to redesign efforts, redevelopment, or even destruction.

There is no predetermined path that landscapes follow toward maturity. Those that manage to survive and thrive while retaining or recovering some degree of authenticity still face ongoing and unpredictable threats. For every celebrated or beloved place that garners support thanks to a significant design provenance or cultural legacy—Central Park and the campus at Duke University, for instance—there are other places that struggle and risk disappearing forever. Included in this category are landscapes that are either aesthetically unconventional, meaning not traditionally expected or accepted, or products of less appreciated or understood design periods, such as modernism or postmodernism. Whether (or how) to "save" such places is less straightforward than in the case of a beloved campus or garden. Either way, a process occurs in which a landscape and the community in which it is located evolve, while people—owners, managers, advocates—are forced to determine how to balance a desire to preserve certain elements with the need to facilitate change.

City Hall Plaza, Boston

A unique example of the challenges associated with landscape preservation is the long-disparaged brutalist-era City Hall Plaza in downtown Boston, which, after surviving decades of threats and criticism, experienced a long overdue reimagining in 2022. The complicated process leading up to the plaza's redo is central to its relatively short history and provides insight into the issues and decision-making that accompany the management of landscapes as they age. Unlike the examples explored later in the book, where processes of restoration or adaptation took hold thanks to effective advocacy and the work of impassioned individuals, City Hall Plaza's history is rooted in apathy rather than in action. It was never a beloved place, yet its central downtown location and considerable civic potential instilled in it a gravitas that belied its questionable aesthetic value. What to do with such landscapes as they age is an equally important question and surely a more challenging problem than what to do with places people and communities cherish.

The plaza was designed by I. M. Pei and Partners and opened in 1968. Envisioned as a vast, seven-acre sea of red brick surrounding the equally maligned City Hall building, the plaza embodied a version of civic landscapes distinct from familiar urban parks.[4] Instead of the familiar revisioning to infuse nature into the city, it was seen as a grand stage for public engagement. The original design adheres to the tenets of brutalism (from *béton brut,* meaning raw concrete), with a simplicity in form and material, a lack of ornamentation, and a utilitarian nature—all characteristics better adapted to a building dedicated to work and transaction than to a public open space meant for use and enjoyment. The building and plaza were constructed prior to the passage of the Americans with Disabilities Act, and the thirty-foot grade change across the site made universal accessibility nearly impossible.

From the start, the plaza was periodically filled with programmed events such as marches, protests, and celebrations. At other times it sat relatively empty, drawing public disdain as it quickly came to symbolize failed urban placemaking. The space became a hollow downtown void that sharply contrasted with the verdant lawns and constantly bustling walkways found in the nearby and generally beloved fifty-acre Boston Common, which has been only modestly modified since its seventeenth-century beginnings as one of America's first public open spaces. The connection between the two spaces is intentional, as Pei's original plan even proposed a large grassy lawn, "sort of a mini-version of Boston Common," according to the architect Henry Cobb of Pei, Cobb, Freed & Partners, who believed the greener version would have been more embraced by the public.[5] Some people did appreciate the plaza's stark aesthetic, including the architecture critic Wolf Von Eckardt, who called it a "great urban plaza" in a critique that celebrated the project's "guts and conviction" rather than its success at placemaking.[6] Thomas Boylston Adams, the Boston writer, academician, and politician, wrote optimistically in a 1969 *Boston Globe* opinion piece that there is "nothing like it in America, nothing half so good," and offered that the plaza "will be copied and repeated in old cities seeking new life and in new cities rising."[7] While Adams's assessment could easily apply to the early days of Boston Common, it never proved true for City Hall Plaza.

In 1973, with more time to experience the space, Robert Campbell, the longtime architecture critic for the *Boston Globe,* presented a different opinion, writing in his column that the plaza was "always too big, too empty, too grand" and looked like a "parking lot temporarily cleared of cars."[8] He blamed the emptiness partly on the city, which prohibited such commercial uses as cafés and coffee houses that could enliven the space. That policy soon changed, however, and by 1976, several outdoor restaurants and as many as ten sidewalk food vendors were competing for business . . . at least during Boston's few warm months. While this helped somewhat, particularly during lunchtime on weekdays, the space remained largely unoccupied much of the time, prompting another *Globe* columnist to write a decade later, "Isn't it time to admit that the Plaza, from a people view, is the pits—unused, unloved, a brick barren."[9]

It is difficult to discern in hindsight who, if anyone, in Boston's city government—the owner—was advocating on behalf of the plaza during this time. Its continued unpopularity led officials to shift maintenance resources to other parks and landscapes in the city. Unsurprisingly, brick pavers crumbled, concrete spalled, drains clogged, and trees suffered and died. A prominent fountain that had served as one of the few humanizing elements in the space was shut off in 1977 and sat derelict for many years. Even so, with relatively few things to go wrong, the landscape was easier to keep in apparent order than either Boston Common or the adjacent, twenty-four-acre Public Garden, which was created by a group of horticulturalists beginning in 1839 with the goal of filling it with expected garden elements—plants, lawns, water, and trees. These landscapes were not only more popular than the plaza, they were also more susceptible to the forces of time owing to their relative intricacy and reliance on living organisms. The visual impact of dead and invasive plants, trodden lawns, and eroded features made their decline more apparent than the stark City Hall Plaza's, and when their condition declined dramatically during the 1960s, they were readily embraced by engaged citizens.

It is not surprising that Boston Common and the Public Garden garnered more public support than the plaza since they present accessible images of what designed landscapes are "supposed" to look like. In 1970, a group of concerned volunteers organized the Friends of the Public Garden, one of the first grassroots advocacy and stewardship groups in the country focused on cultural and designed landscapes. That same year, Jane Holtz Kay, an architecture and urban design critic, publicized the state of the historical landscapes in a *Boston Globe* column titled "Common, Public Garden Are a Mess." In it, she described eroded lawns, broken walkways, vandalized areas, tipped-over benches, and trees "poised on the brink of destruction from a decade of abuse."[10] Although poverty, pollution, public abuse, vandalism, Dutch elm disease, and a proliferation of hippies are addressed as contributing factors, Kay quoted Jim Falck, a landscape architect with the Boston Redevelopment Authority and a proclaimed "chief Garden and Common booster," who blamed the decline on "just years and years of apathy."[11]

The Friends group's initial goal was to rescue and help restore the Public Garden. Subsequently, it adopted the Common and, later, the adjacent, thirty-two-acre Commonwealth Avenue Mall, which was designed in 1856 by the architect Arthur Gilman as a connective green space. Owing in large part to this advocacy and support, the City of Boston was able to reverse the systemic neglect and embark on a process of restoration. This ongoing work, which coincided with a renaissance for Olmsted's reputation and the founding of similar groups at both Central Park and Prospect Park in New York (see chapter 5), was at the leading edge of a national trend toward salvaging, protecting, and stewarding cultural landscapes. Buoyed by increasing public support, individuals and organizations devoted to landscape preservation took the fight to those wanting to destroy or redo cherished places. With a stated charge to preserve the nation's history, the NPS joined the efforts and expanded its leadership role to include designed sites (see chapter 2). With support from historians and landscape architects, the NPS produced volumes of guidelines and methods that facilitated the research, documentation, and management of places at risk of being lost forever. Over the next few decades, dozens of notable historic landscapes were rescued and restored—at least temporarily.

While progress spread across the country, Boston's City Hall sat relatively untouched, a menacing Cold War–era relic plagued by slow and persistent decline and stymied by municipal paralysis. Consensus seemed to favor change, as several studies and master plans for the site proposed solutions of various magnitudes and viability. Among the most dramatic of these involved demolishing City Hall and relocating it and the associated open space to another part of the city. Ideas from a series of mayors for improving the plaza included converting the space for drive-in movies, public vegetable gardens, a baseball diamond dedicated to Babe Ruth, European-style gardens à la the Public Garden, and a 150-foot-tall wind turbine. None of the proposals went very far.

By 1994, the plaza was hosting three hundred programmed events each year. At other times, however, it failed to exhibit much life. Then mayor Thomas Menino, an established proponent of the public realm who referred to City Hall Plaza both as "Boston's symbolic front yard" and as "a vast wasteland," directed city staff to spruce up the plaza in time for the summer tourists.[12] Owners of a coffee house and a sandwich shop operating in the space agreed to provide tables and chairs by the still dry fountain, which had been filled with thousands of flowers. Sixteen dead trees were replaced. Maintenance crews used "new state-of-the-art steam machines" to wash away graffiti.[13] Spending exorbitant amounts of taxpayer money on the plaza was always controversial, yet these small moves signified a changing perception about the space: even if it was far from perfect, it was worth investing some money to make it more hospitable.

Paralleling this clean-up effort was a public competition, the Revitalizing City Hall Plaza Ideas Competition, which was initiated by Menino to help plot an appropriate future for the space. After considering nearly two hundred, wildly different entries, the eight-person jury recommended modifying the plaza with a "series of small interventions to create lovable, sacred spaces" and advised against "re-historicizing" the plaza.[14] Out of this set of recommendations emerged a new planning effort populated by urban designers, architects, landscape architects, city staff, and invested local constituencies. The process was led by Trust for City Hall, a nonprofit formed by Menino to consider public-private opportunities for City Hall. Unlike the Friends of the Public Garden, which organized concerned citizen advocates, the trust was populated by influential businesspeople and developers. As such, transforming the space to accommodate development appears to have been a more important driver than preserving any perceived design sanctity of the landscape. Among the plan's proposals was softening the plaza with a large bosque of trees and adding a civic-scaled lawn and an interactive fountain. More controversial was the plan's proposal to sell off more than a third of the property to private developers. The scale of the space—its most recognizable attribute—was to be compromised with the insertion of a 350-room hotel, a 700-space underground parking garage, and other rent-capturing amenities marketed as a way to fund upkeep of the landscape. It, like plans that came before and after, garnered more criticism than support and went virtually nowhere.

Reaction against the proposal to privatize the plaza seems to have had less to do with preserving a particular design legacy or the plaza's status as a cultural landscape than with a fundamental opposition to ceding public open space to developers.[15] Such desire to keep public land in public hands is integral to the American democratic ethos and has motivated people to save parks and other places for many years. In Central Park, for instance, advocates have argued against encroachment from unwanted intrusion since its beginning, as noted in a 1919 piece in the *New York Herald:*

> Friends of Central Park must be ever vigilant. Many times they have saved it from uses not coinciding with the views of those who think of a park as a park. Its friends have successfully resisted efforts to convert it into a race track, athletic field, fair grounds, site for tenements, military parade ground; they have beaten off those who longed to fill it with Coney Island delights, popular priced hotels, music halls, beer gardens.[16]

Bostonians may not have liked the plaza in the same way New Yorkers valued Central Park, but neither were they eager to sell off the public realm to developers.

While City Hall Plaza never received the same support as Boston Common or the Public Garden, it did attract the attention of the Cultural Landscape Foundation (TCLF), which had been founded in 1998 by the landscape architect Charles Birnbaum. Through a fifteen-year stint as the coordinator of the NPS's Historic Landscape Initiative, Birnbaum helped to redefine how people think about cultural and designed landscapes. Among his most meaningful contributions in that position was a brief he wrote for the NPS titled "Protecting Cultural Landscapes: Planning, Treatment and Management of Historic Landscapes," which became the seminal document directing the emerging field of landscape preservation.[17] After removing himself from the prescriptive rules and cumbersome bureaucracy at the NPS, he became a leading advocate for "saving" threatened landscapes, with a focus on those reflecting the mid-century modernist style. Whereas groups such as the Friends of the Public Garden and the Central Park Conservancy worked on behalf of one landscape, TCLF eagerly championed landscapes across the country.

In 2008, TCLF included City Hall Plaza on its list of twelve "Marvels of Modernism"—period landscapes deemed notable and in danger of being lost

Fig. 1. As designed, Boston's City Hall Plaza was a large, unfriendly, and frequently empty open space intentionally designed to accommodate major civic events, celebrations, and protests. (Photo by author)

because of neglect, redevelopment, or a combination of these and myriad other factors. That same year, a group calling itself Citizens for City Hall formed to fight a proposal by Menino to demolish the now iconic brutalist building, which they considered architecturally significant. While this group joined a growing list of scholars, architects, and preservationists arguing loudly to save the building, TCLF was part of a less vocal contingent speaking out on behalf of the landscape. William Thompson, who was editor of *Landscape Architecture Magazine* at the time, wrote against taking a preservationist approach, stating: "Some modernist-era landscapes . . . merit preservation, but many are cold, inhuman expressions of architectural arrogance—of which the 'windswept plazas' of City Hall Plaza is perhaps the epitome."[18]

While the massive City Hall building garnered support, relative apathy signaled an uncertain future for the plaza. TCLF did take a stand, endorsing a compromise approach that accepted updates and modifications while attempting to hold on to some of the original design intent. "By acknowledging, understanding and respecting the heritage of the plaza, its essence can be preserved and interpreted while it is simultaneously enriched through intelligent change that will heighten the experience of all its users," the foundation commented in a published statement.[19] A similar degree of tepid support for the plaza came from the Boston Preservation Alliance, a nonprofit advocacy organization charged with supporting Boston's historic buildings, landscapes, and neighborhoods. Its board of directors issued a statement in 2006 stressing the need to preserve and rehabilitate the building while favoring a seemingly unrestricted redesign approach for the landscape. Despite the ongoing conversation, inaction again prevailed.

In 2010, the year Menino announced he was giving up on plans to relocate City Hall, Boston became one of five American cities selected to take part in an initiative by the Environmental Protection Agency (EPA) called Greening America's Capitals. The program, which prompted cities to become more environmentally sustainable, gave Boston money to hire urban planners and landscape architects to develop realistic proposals for improving the area in and around City Hall Plaza. This program coincided with an emerging renaissance for urban landscapes associated with a growing list of well-publicized projects that transformed abandoned infrastructure into twenty-first-century parks. Brooklyn Bridge Park and the High Line in New York City, Millennium Park in Chicago, and the Rose Kennedy Greenway in downtown Boston are prominent examples of this significant trend. These influential projects, along with others, drew accolades and press attention around the world and prompted politicians, planners, and citizens to consider opportunities for their own urban landscapes. As happened in the 1990s when the NPS began focusing on landscape preservation, the prospects of many cultural and designed landscapes improved with this new awareness and appreciation.

When candidate Martin Walsh was running for Boston mayor in 2013, he reiterated Menino's proposal for selling City Hall and the plaza to developers. Once elected, however, his administration took a different approach by embarking on a plan to enliven the space with, among other events, concerts, celebrations, and World Cup viewing parties. It even considered building an outdoor skating rink on the site. In 2015, Walsh announced yet another planning process to determine the future of the space, saying the plaza could be Boston's version of the High Line or Brooklyn Bridge Park.[20] Later that year, a team led by the Boston architectural firm Utile, in collaboration with a long list of consultants, including the Cambridge-based Reed Hilderbrand landscape architects, embarked on a master planning process paid for by the Greening America's Capitals program. Walsh said the purpose of the plan was to "re-imagine City Hall Plaza as the thriving, healthy, innovative space that it should be."[21]

The plan, which was released in 2017, succinctly describes the landscape this way: "The Plaza lacks the infrastructure for successful programming and events, has suffered from deteriorated conditions over time due to weathering, limited maintenance, and poor treatment by event organizers, and does not provide for comfortable day-to-day use." Proposed improvements included adding outdoor dining venues, breaking down the vast scale, accommodating universal accessibility, planting significantly more trees to provide shade and mitigate the urban heat island effect, and incorporating green infrastructure with sustainable solutions for managing stormwater. Although there was a consultant on the team who focused on history, the plan barely mentioned respecting or responding to the design legacy of the plaza. Such an omission suggests how subjective and even arbitrary decisions can be when determining which aspects of historic landscapes are worthy of preservation over others.

The master plan proved to be the impetus for transformation, and a design process was undertaken to finally improve City Hall Plaza. Sasaki, a Boston-based multidisciplinary firm, was hired to lead the redesign and renovation of the space. Adhering closely to the master plan, the firm's work focused on sustainability, accessibility, and creating a flexible and welcoming space, none of which was part of the original iteration. At the same time, respecting the space's design legacy seems to have been an afterthought at best. Sixty percent of the new surface is permeable, with extensive stormwater catchment and engineered filtration reducing negative impacts downstream. The plaza went from being virtually inaccessible throughout to being nearly entirely accessible thanks to a series of sloping paths replacing what once were steps. Around one hundred trees were planted to add shade and comfort to the formally severe space, and a large and active playground for children was added.[22] While the introduction of similar playgrounds makes sense in public parks—such as Central Park, where the new playgrounds were initially criticized—the bustling space

Fig. 2. A recent renovation to City Hall Plaza significantly transformed the landscape with trees, activities, and accessible walkways. An active playground, while popular, challenges historical precedent and the plaza's civic intent. (Photo by author)

packed with water features and whimsical play equipment seems intrusive in a civic landscape initially conceived as being grand and simple.

Quibbles aside, the plaza's renovation is an unquestionable win for the people of Boston. It is, however, unclear how differently the lengthy process might have played out had a respected landscape architect been responsible for its initial design rather than an architectural firm. With TCLF leading the charge, the landscape architecture community has become a vital champion for maintaining the integrity of designed landscapes, especially when the subject is a work created by a fellow designer. Advocacy is not always successful, however.

In Charlotte, North Carolina, for instance, Thomas Polk Park, a small downtown postmodern landscape designed by the little-known Bulgarian-born landscape architect Angela Danadjieva, was demolished in 2023 despite pleas for salvation by

TCLF and others in the profession. Upon hearing of the city's plan to tear down the park, a cohort of landscape architecture faculty and scholars from North Carolina State University joined TCLF in protest by penning an opinion piece in the leading Charlotte paper arguing for the repurposing or relocation of the landscape they called a "masterwork by one of our greatest female landscape architects."[23] Using a familiar refrain, they accused the city of failing to adequately maintain the park while attributing safety concerns to an inability to address a growing homelessness problem. Unswayed, Charlotte officials explained that the park as designed no longer met the needs of the community and that it had limited space for public use, was not compatible with adjacent buildings, and posed safety challenges. Plus, they said, it would be prohibitively expensive to adapt the space as advocates had proposed. Despite continued outcries from the landscape architecture community, and without an effective public engagement process, the plaza was destroyed.

City Hall Plaza and Thomas Polk Park present the challenges associated with "preserving" public landscapes. Neither is a wholly representative example of the issues involved, yet combined, their stories make evident that the trajectory of every individual place or space follows a unique process specific to its context and community. Whether they are considered examples to follow or pose cautionary tales to avoid, the opposing fates of the two sites provide insight into the evolution of designed landscapes.

Points of Inflection

Change is inevitable. Decisions will always need to be made as to how to address it. Individual renovations such as the dramatic transformation of City Hall Plaza should never be regarded as the final goal because, no matter how successful they may be, they will not remain unchanged for long. All physical interventions, regardless of when they occur during the life of a landscape, become periods of pause in an interminable cycle. Even so-called "landscape preservationists" seem to understand this and, as is the case with TCLF, typically advocate responsible adaptation that allows for change while respecting a specific design history.

The challenge involves defining the level of sanctity assigned to an original or influential designer or patron (e.g., an "Olmsted landscape") versus the value of subsequent modifications made in response to contemporary drivers. Every mayor who sought to improve (or raze) Boston's City Hall Plaza made decisions that affected its evolution. Turning off the fountain instead of fixing it and moving maintenance resources to other places in the city are examples of such decisions, as is the hiring of the consulting firms Reed Hilderbrand and Sasaki. Even seemingly minor decisions can spur deviation. Moreover, deciding *not* to do something can also have long-term consequences as periods of neglect often generate obstacles that are difficult to overcome. Across landscape types, change is dictated by such decisions and by a myriad of events that may include the death of a patron, destruction or threats from natural disasters, government mandates, amended management policies, philanthropic windfalls, economic downturns, and changes in ownership, leadership, or mission. These become temporal markers that trigger change or adaptation and can be understood as *points of inflection.*

My time at Duke University has provided me with unique insight into how places evolve and the aggregate impact of inflection points over time. In addition to the many hours I have spent researching and analyzing Duke's history, my work as a practitioner addressing the needs of historical, period, and contemporary spaces has exposed me to many issues associated with managing a mature landscape. In the course of thinking and writing about Duke and other designed landscapes, I have come to realize that their transformative change—cultural, aesthetic, and ecological—is as unpredictable as it is inevitable. Left on their own, these carefully crafted places would deteriorate and eventually revert to some version of their natural state, essentially moving backward in time. Obviously, this process would take longer at a strictly urban place like City Hall Plaza than at a lush and traditional park or garden. Complete deterioration, whether materially or in overall character, would likely mean a landscape had been completely abandoned. This is relatively rare, however, because unless designed landscapes are destroyed and replaced with something different, a person or agency typically intervenes, at least temporarily, to stop or reverse the deterioration. Such involvement can be as simple as performing routine maintenance, assuming it is sufficiently prioritized and funded.

Inevitably, decisions and actions will propel a landscape's physical form and condition in a positive or negative direction, either closer to or further from its original design intent. At Central Park, for instance, municipal elections, the appointments of new parks commissioners, budgetary and staffing decisions, and economic recessions have all catalyzed some level of transformation. One obvious example is the 1980 founding of the Central Park Conservancy, which immediately began resuscitating the landscape and restoring elements of the Greensward Plan, along with making certain alterations to Olmsted and Vaux's vision that were deemed valuable.

While there are countless examples of how these inflection points can affect a landscape, certain momentous events can affect them all in some way. An example is the Great Depression, a cataclysm so severe it disrupted the lives of even the wealthiest Americans. In different ways, the coronavirus pandemic of 2020 shook the world so completely that essentially no place was left unaltered. Whether universal or specific to a locale, such major events trigger periods of neglect, restoration, or redesign that frequently overlap as parallel processes. How effectively they are managed helps determine which landscapes survive and retain their relevancy over time. Such effective management almost always results from invested and impassioned leadership.

The Process of Change

It is important to remember that, depending on the particular landscape, points of inflection can occur simultaneously and frequently overlap one another. This is particularly true for landscapes created and expanded across multiple design eras, when responding to historical precedent is balanced against the pressures of changing programs and conditions. At Duke, for instance, a recent period saw three such processes occurring at once. One was a multi-billion-dollar construction boom that rapidly and dramatically transformed the campus's contemporary landscape by turning parking lots into building sites, transforming numerous outdoor spaces into high-functioning plazas and courtyards, and, sadly, removing significant stands of remnant woodlands to accommodate new development. At the same time, the two historic quadrangles designed by the Olmsted Brothers firm were on opposing trajectories. A well-funded initiative restored and enhanced one of them in response to historical precedent while the other faced degradation owing to a lack of investment and a dying tree canopy. The reasons behind the different trajectories are complex and involve generations of culturally influenced decision-making and capital prioritization. While each landscape follows its own trajectory and faces its own issues, the two spaces will always be intertwined as a result of their shared governance, context, and cultural heritage. Fortunately, administrative leaders agreed to invest in restoration projects for both quadrangles in anticipation of the forthcoming university centennial celebration, which became a significant point of inflection.

Because of such complexities and how they accumulate and intersect over time, a singular chronology or timeline is an ineffective way to consider a landscape's ongoing history. Therefore, I prefer to consider the evolution of this and other designed landscapes as running on a multitrack continuum along which varying processes emerge, run, and transition from one to another. Typically, five processes—*creation, maturation, deviation, adaptation,* and *rejuvenation*—occur along this continuum in a manner unique to each individual place.

Creation

Creation may be interpreted as the process of design and construction, but it is better understood in the language of a continuum as the period during which a place's enduring legacy is first established. This, the so-called period of significance, is the story most often told in landscape histories and can involve a site's geological, ecological, and cultural histories and include the period before a designer is hired, when an individual owner, institution, or agency identifies needs, purchases property, and outlines an initial vision. In fact, the character established in such early visioning is often more integral to a place's legacy than an actual design. In public landscapes, the voices of advocates and the actions of government officials are frequently important to this process as well. The duration and complexity of this period will vary depending on the specific landscape.

Processes that are part of the creation period include the design process, in which a landscape's legacy is cemented, at least initially. The creation period typically ends once construction is complete and the landscape is made available to its users or owners. From this point forward, evolution takes hold, and, for better or worse, the place is never the same again.

Maturation

Maturation is the period during which a place evolves as originally intended by an owner or designer. Quality construction successfully fulfills design intent, trees and plants grow as anticipated, and people activate the spaces as planned. In this scenario, regular maintenance practices are adequate for successfully preserving an original vision. Thanks to the unpredictability wrought by both human interaction and environmental forces, however, this phase rarely endures for long, with specific landscapes exhibiting vastly different phase lengths. City Hall Plaza, for instance, was designed as a relatively simple open space filled with brick and concrete and only minimal plantings, allowing its maturation period to persist longer than that of Central Park, a substantially larger, greener, and ecologically more complex space. Central Park began transforming quickly as plants died and were not replaced, waterways became clogged, soil eroded, and pristine landscapes were trampled by people wandering off designated paths.

The unpredictability that accompanies landscapes as they evolve represents one of the fundamental dilemmas facing landscape architects and designers: How do you effectively envision a place that will never mature exactly as you want it to? While you can anticipate the long-term growth habit of mature trees, there is no way to predict all of the cultural and ecological curveballs that will affect a landscape by the time the trees reach maturity. Long after consulting designers have exited the project, owners and managers must balance the desire to maintain the original design integrity with the need to modify it to ensure continued success of the site.

Deviation

Once maturation proves unattainable, landscapes veer off their planned course as points of inflection interrupt intended processes while triggering a new period that I call *deviation.* Anything from a miscalculation made in the design process to changing program requirements can alter a landscape's trajectory. For example, a poorly selected plant species in a garden could fail and need to be replaced. If this plant is one among many, the result may be minimal change, even unnoticeable. However, if the plant species was used to create an important formal element, such as a structural hedge or an extended allée of trees, the effect could be dramatic. As another example, a space may be significantly altered by people not using it as designed. Unwanted paths may appear on a campus lawn because sidewalks were not provided or playgrounds may be added to serve the needs of a growing population, as happened in New York's Central Park. Because no designer is prescient, such situations are unavoidable.

Any number of points of inflection can (and will) directly trigger deviation from the original design. Deviation may manifest in a period of deterioration sparked by a funding crisis or in the devastation caused by a storm or other natural disaster. Even neglect is a trigger because neglecting something involves a decision that inevitably leads to degradation. In whatever form, these deviations are frequently considered detrimental because they challenge a specific design vision. Conversely, they may eventually prove beneficial in helping places become more resilient in the long term.

Adaptation

As deviations occur, it is up to owners, managers, and other decision-makers to weigh the relative sanctity of a design legacy or guiding vision when determining how best to respond to change. One option is to do nothing, which would be expected to propel a landscape into a period of deterioration. Another choice is to fight change and attempt to revert to a historically accurate state, which may lead to short-term success yet rarely amounts to more than a temporary solution.

An alternative I call *adaptation* involves not only accepting changes but formally incorporating them into the landscape. This could mean redesigning a portion of a garden to better adapt to changing conditions or converting makeshift paths on campus lawns into paved sidewalks. A more momentous example is provided by Central Park, where playfields and playgrounds were added in the twentieth century to accommodate the tens of thousands of people flooding the landscape in search of places to recreate or hold events. In the process, numerous spaces once envisioned as passive, picturesque-inspired meadows were transformed into recreation zones. Although one could view such change as being antagonistic to Olmsted and Vaux's Greensward Plan, these zones have become popular and cherished features deemed indispensable by both the public and supporters of the park. Once such acceptance takes place, these new or reimagined spaces become parts of a recalibrated baseline from which other changes will eventually emerge.

Rejuvenation

Rejuvenation, the final process, frequently overlaps the others. Like adaption, rejuvenation entails conscious action that instigates change. Whereas adaptation modifies form and program in response to changing conditions, rejuvenation involves improving the *condition* of a landscape to achieve something like its former state. In some cases, adaptation and rejuvenation occur in unison. For instance, when we were restoring one of the historic quadrangles at Duke, we made certain design adjustments in response to its contemporary usage while also reinforcing as many of the Olmsted Brothers' intended patterns as was feasible. History mattered, but only as much as it made sense in the culture of a twenty-first-century campus. Although Duke chose to retain the Olmsted Brothers' plan as the basis from which to work, rejuvenation does not necessarily align with an original vision.

This way of thinking about rejuvenation may remind some people of the NPS's process for treating cultural landscapes through preservation, rehabilitation, restoration, and reconstruction.[24] My interpretation, however, focuses on this intervention as one part of a continuous evolutionary cycle rather than as an end in itself. When a consultant is engaged to facilitate a specific restoration, that work may be one of many initiatives occurring sequentially or simultaneously. Overseeing this collective of work often falls to professional staff whose job it is to represent an owner's or institution's interests when addressing change.

Stewardship

As designed landscapes mature and the cycle of the five processes described above sets in, addressing changing needs and unexpected realities becomes an imperative for administrators, officials, practitioners, and others tasked with overseeing these places. Where commitment is low or resources are scarce, responding to these issues may be limited to basic or reactive maintenance. This limited response typically leads to the eventual need for significant investment to prevent the landscape from being completely lost or replaced. Conversely, when active response to change is prioritized, funded, and accomplished with careful forethought and adherence to a guiding ethic of respect and responsibility, the practice is elevated to *stewardship*. By this I mean that decisions made in response to adaptation and rejuvenation are done so in a manner that seeks to achieve a balance between protecting the integrity of the past and responding to current conditions *while also* anticipating future needs.

How effectively this balance is achieved ultimately determines how successfully a place can retain its ongoing quality and relevance. Destroying a cherished garden to build a parking lot would rarely be considered successful stewardship, nor would refusing to revise original plantings in that same garden to adapt to changing program needs or ecological and climatological conditions. Respecting a design legacy is essential, yet doing so with a disregard for the future is no more sustainable than ignoring or forgetting the history in the first place.

Although individuals can have a significant impact on a designed landscape, stewardship almost always extends beyond the efforts of one person or a single decision. When most effective, it involves a structural and cultural foundation that is built on by generations of advocates and professionals acting on behalf of a place through all phases of its evolution. In this optimal scenario, a solid and long-standing ethic becomes embedded in policymaking, fiscal prioritization, staff hiring, fundraising, and maintenance practices. This, of course, is more easily achieved with financial resources. Central Park Conservancy's endowment, for instance, surpassed $300 million in 2021. More commonly, conservancies, institutions, and agencies struggle to raise enough money to stem or manage the forces of time. Thankfully, the value of stewardship is not gauged by money spent alone. More critical is prioritizing how available resources are expended to best serve the evolving needs of a landscape.

There are endless ways for these processes and subsequent stewardship to play out over time. Although my personal experiences at Duke University and the Central Park Conservancy are meaningful and serve as effective examples, the lessons derived from them are not necessarily transferable to other places or across broader typologies. In addition to being valued, eagerly supported, and relatively well funded, both Duke's quadrangles and Central Park have reaped the benefit of oversight and advocacy by people who are passionate and knowledgeable about landscapes and history. This is not always the case, as Boston's City Hall Plaza reminds us. While gardens, parks, campuses, and other landscapes that meet traditional expectations for beauty or quality (such as Duke University's campus and Central Park) are not immune to ongoing stewardship challenges, their troubles are typically less substantial than those facing places that lack support, are considered unpleasant or outdated, or are just unloved, for whatever reason.

The following chapters present case studies of landscapes representing various types and scales from different eras and geographic regions. The ideas discussed in this introduction are embedded in the stories, even when not explicitly identified as such. In each case, generations of people facilitating and managing change take center stage, while the stories of designers and patrons help set the context. Selecting these landscapes, some of which are better known than others, involved a lengthy investigative process. Some choices reflect a personal interest on my part. For instance, I have been visiting Sea Pines my entire life, so I was already familiar with its story and with some of the people involved. Other places I had visited and found intriguing. The final list was determined both by differences and similarities between the landscapes and by how the composite presented the topics and issues I believe are important.

The variation in length and detail presented in the studies is intentional. The disparity provides the reader with a choice of quicker, targeted reads and deeper, more comprehensive explorations. Prospect Park, which receives the most extended discussion, is an exceedingly complex landscape whose densely layered history exemplifies actions and decisions directly related to the topic of managing change. The Missouri Botanical Garden, on the other hand, receives less discussion not because it is less significant than Prospect Park but because its story follows a relatively straightforward trajectory toward fulfilling the vision of its founder. Its financial and leadership stability is compelling as an example; it just lacks the variety and drama of some other case studies. Of course, none of the discussions is intended to be comprehensive. Each place reviewed has a far more complex history than what is brought forward here.

Each account begins with the establishment of a particular legacy and follows the landscape's evolution through various periods before ending in a discussion of contemporary issues. The goal is to present the myriad forces and situations involved in designed landscapes while conveying some idea of the issues, challenges, decisions, and actions that may be shared, to one degree or another, by all.

1

GARDENS

For like a story, a garden has its own developing plot, as it were, whose intrigues keep the caretaker under more or less constant pressure. The true gardener is always "the constant gardener."

—Robert Pogue Harrison, scholar, educator, and author, 2008

A *garden* is a unique landscape type. Unlike parks, campuses, and communities, all of which typically rely on some level of ongoing municipal or institutional oversight and governance structure, anyone can create and tend to their own version of a garden on even the smallest patch of land without anyone else's input. Gardens as designed spaces evolved from sustenance and food-generating landscapes into sites created for horticultural collections and displays, scientific research, and visual and experiential pleasure. Because they encompass an incredible variety of physical forms, scales, missions, and aesthetic imagery, gardens are as diverse as the communities in which they are located.

There are no guiding rules or established criteria defining what constitutes a garden, and therefore most landscapes can be referred to as gardens. Plants are generally considered needed elements, though there are examples, such as the revered Japanese ***karesansui*** or zen garden, where few or no plants may be used. Similarly, in arid regions where water is increasingly scarce, gardens may be filled with nonliving components such as gravel, stones, structures, and furniture. Nevertheless, plants, whether as horticultural design features or as scientific subjects, are integral to the purpose and aesthetic of traditional American gardens, regardless of any particular garden's provenance, style, size, climate, or ecological context. In part because they focus on plants, which inevitably morph and eventually die, gardens must adapt, often repeatedly, to persevere across decades or centuries. This need to change is made more urgent in gardens designed with trouble-prone features such as fountains or with extensive decorative details, paving, walls, and structures selected or designed more for their immediate aesthetic value than for their long-term viability. Such is particularly true for gardens created for residential use, the owners of which likely focus on the immediate quality of their landscape more than would people tasked with overseeing established public parks or higher education campuses, which require a more expansive purview.

Recognizing how daunting it would be to effectively capture a subject as relatively amorphous as garden, this chapter focuses on two specific types, *knowledge gardens* and *pleasure gardens,* while acknowledging an inevitable overlap of the two types. Either type could be labeled *botanical garden,* which is only marginally less ambiguous than *garden* alone. The American Public Gardens Association (APGA), which calls itself the leading professional organization for the field of public horticulture, offers a version of formal status as *botanical garden* to landscapes that meet the general criteria that they are open to the public, offer aesthetic or educational displays, and exhibit at least a small degree of taxonomic rigor, with records and labeling.[1] These requirements can apply equally to a former estate garden now open to the public and to a research garden committed to botanical study and the display of horticultural collections.

Knowledge gardens are distinguished by a shared mission in support of research, learning, and instruction in plant-related academic and scientific disciplines. Botany, conservation, taxonomy, horticulture, ecology, and increasingly climate science are typical areas of concentration. The origins of this landscape type date back millennia, with applicable precedent found in sixteenth-century Renaissance gardens, notably the Italian botanical garden established in Padua in 1545, which remains the oldest of its kind in the world. The Orto botanico de Padova, located around twenty-five miles outside Venice, was established with a mission to cultivate and study medicinal plants as tools for training pharmacists and physicians. At its founding, it was stated that the garden should be a setting in which "scholars and other gentlemen can come . . . at all hours in the summer, retiring in the shade with their books to discuss plants learnedly, and investigating their nature peripatetically while walking."[2] This goal of infusing access to the landscape into the processes of teaching and learning became fundamental to the design of subsequent European botanical gardens and university campuses.

The teaching garden ideal carried over to early America, where an interest in horticulture and botany was shared by the country's founding fathers. Thomas Jefferson, James Madison, and George Washington, for instance, were all plant enthusiasts who designed and cultivated elaborate gardens on their estates. The three men aspired to build a botanical garden on the National Mall in Washington, DC, and in 1820 their hopes were fulfilled when Madison, as president, signed a bill granting five acres of land for such purpose at the base of the Capitol. When funding for the garden ran out in 1837, the federal government took it over, thus triggering a period of neglect. A renewed commitment emerged in 1842 when the Wilkes Expedition (formally named the United States Exploring Expedition) returned from a four-year scientific venture circumnavigating the globe in search of exotic artifacts. Among the large number of items retrieved were plants that became the first collection of a revived version of the U.S. Botanic Garden. Out of this introduction to an expanded world of plants emerged a passion for horticulture never before seen in America.

Although it would take until 1850 for the first iteration of the U.S. Botanic Garden, the Victorian Conservancy, to open to the public, the scientific study of plants continued to gain momentum. A leader in the emerging discipline was the geologist and botanist Amos Eaton, who in 1824 cofounded the first teaching laboratory for natural sciences at the Rensselaer School in Troy, New York. Eaton was a highly respected educator whose lectures on natural history attracted thousands of students, including Asa Gray, arguably the most important American botanist of the nineteenth century.[3] Rensselaer helped science curricula evolve beyond passive lectures to include demonstration experiments that put the pupil in the place of the professor to teach the practical application of scientific principles.[4]

Knowledge gardens have since developed all over the country in a variety of scales and with differing programs. Some are born out of the opulent estates of wealthy plant lovers, while others directly support the pedagogical missions of colleges and universities or the environmental goals of scientific institutions. Though each pursues a unique path toward continued relevance, the challenges associated with simultaneously balancing scientific pursuits with the desire to create attractive display gardens remain consistent. In their purest forms, botany supports the mission while horticulture and design attract the public. Successfully providing one without undercutting the other is a frequent concern and often depends on the effectiveness of leadership over time and the availability of consistent funding. Knowledge gardens frequently developed as institutions, thus establishing an expectation of permanence and the need for a system of governance, management, and financial oversight.

Pleasure gardens, on the other hand, are distinguished in the context of this book from knowledge gardens by their purpose or mission rather than merely their appearance or program. While some scientific or taxonomic activities may take place in pleasure gardens, their primary purpose is to provide a space in which to display plants and give access to beauty and, sometimes, history. By the late eighteenth century, the practice of creating residential pleasure gardens, most filled with flowers and shrubs and situated prominently in relation to more utilitarian kitchen gardens, had permeated the upper crust of American colonists. With exotic plant species and new cultivars increasingly available thanks to the efforts of early botanists, people could supplement their kitchen gardens with more ornamental *parlor gardens* designed specifically for display and as sources for cut flowers for the home. An early American example is Middleton Place, the landscaped grounds of Henry Middleton's plantation in Charleston, South Carolina, where enslaved labor was used during the off-season for growing rice crops to carve elaborate terraces, pools, and canals alongside ornate, Renaissance-inspired formal gardens.[5] Landscapes of that magnitude, however, were limited to the wealthiest colonists.

More intimate pleasure gardens proliferated in the nineteenth century as ornamental plants became more available to the masses with the introduction of seed shops and catalogues. The horticulturalist, landscape designer, and writer Andrew

Jackson Downing became a major influence with his frequent contributions to *The Horticulturalist* and the first edition of *A Treatise on the Theory and Practice of Landscape Gardening Adapted to North America,* in which he attested, "The development of the Beautiful is the end and aim of Landscape Gardening, as it is of all other fine arts."[6] Downing encouraged emerging middle- and upper-class families to create their own rural gardens promoting the picturesque style that found its way to America via Great Britain and would soon be adopted by Frederick Law Olmsted, notably in the designs for Central Park and Prospect Park. This style favored a naturalistic aesthetic, often with rolling hills and asymmetrical compositions that contrasted with the rigid geometries and axial arrangements of formal landscapes, such as the famous gardens of Versailles, designed by André Le Nôtre.[7] Members of the upper class were surely influenced by Jackson's work as they created estate gardens on large tracts of land purchased in rural areas to escape urbanizing cities during the Country Place Era between around 1890 and 1930. These landscapes, which typically matched the grandeur of the homes to which they were attached, proliferated across the country. They were characterized by ostentation, with showy horticulture, spreading lawns, grand fountains, large terraces, grand vistas, and other landscape features. Rooted in history and exemplifying design periods such as the picturesque and the Beaux-Arts, many of these gardens were subsequently transformed into America's preeminent public or semipublic gardens.

Wealthy patrons such as Henry Francis du Pont of Winterthur in Delaware, Mabel Choate of Naumkaeg in Massachusetts, and Robert and Mildred Bliss of Dumbarton Oaks in Washington, DC, created innovative personal gardens that helped establish the careers of professional designers while building a collection of significant designed landscapes in the still young country.[8] As the original owners died and subsequent generations took over with less or greater interest, the ongoing management of these residential landscapes often followed very different paths. Perpetuating their relevance as they transitioned from private estate to publicly accessible garden was rarely seamless. Effective stewardship has typically involved family members and people in the community who value the property and want to preserve and share its history. Choate, for instance, bequeathed her Naumkaeg estate, which she designed in close collaboration with the celebrated landscape architect Fletcher Steele, to Massachusetts' Trustees of Reservations to ensure the house and gardens would be cared for and made open to the public.[9] The number of American estate gardens that have survived and flourished beyond their original design or patronage cannot be known for certain. It is likely, however, that many, even most, have been lost or altered well beyond their original design.

Despite their differences, many gardens, regardless of how they are categorized, face similar financial, operational, environmental, and practical challenges as they age. The challenges originate in part in the intricate plantings, features, and detailing common to garden design that make beautiful yet maintenance-intensive landscapes. Perhaps more influential than these tangible factors is the lack of a stable organizational structure experienced by many now public gardens that originated as private or residential landscapes. These gardens often thrived under an original owner or patron before struggling through periodic ownership and management transitions. Uncertain futures frequently compel gardens to rely on outside stewardship and advocacy groups for survival, more so than other landscape types. Because such volunteer organizations and nonprofit agencies have been so successful at rescuing individual gardens from being sold or destroyed, their efforts should be acknowledged as critical points of inflection in the lives of these landscapes.

No garden-focused organization is as well known as the Garden Club of America, which began in 1891 as the Ladies Garden Club of Athens and consisted of twelve women who met in the parlor of an Athens, Georgia, home.[10] Numerous such groups followed, including the Garden Club of Philadelphia, which Elizabeth Price Martin cofounded with Ernestine Abercrombie Goodman in 1904. The two women recognized the potential for such groups and, nearly a decade later, sought to broaden local efforts with a national organization. On May 1, 1913, they led a meeting at Stenton, a stately colonial home in Germantown, Pennsylvania, that was attended by more than twenty representatives from a dozen garden clubs in the eastern and central United States. The site was chosen in part because the home's gardens had recently been restored to period accuracy, thus reflecting an interest in historical landscapes. During the meeting, Goodman wrote down a three-pronged mission statement outlining the organization's purpose: to stimulate the knowledge and love of gardening, help protect native plants and birds, and encourage civic planting. From this the Garden Club of America (GCA) was born.[11]

Within its first year, the GCA had around one thousand members. In addition to promoting beauty in local residential gardens and seeking a standard color chart for flowering plants and arrangements, the group focused on civic issues such as enhancing the aesthetics of roadways and rail corridors. Its influence spread to the design world as well. According to a 1914 article in *American Homes and Gardens,* "Landscape architects are turning to these affiliated clubs as an authority and power not to be overlooked."[12] Among the several committees formed to tackle specific work was the Historic Gardens Committee, which was charged with promoting "the use of old-fashioned garden and plant material, notably flowers."[13] As the era of grand estates faded and the once elaborately planted gardens succumbed to neglect from loss of costly regular maintenance, the GCA took the lead in photographing and cataloguing the disappearing landscapes.

The GCA's threefold focus on history, education, and conservation spurred its members to become active participants in national politics. In 1927, an advisory council was assembled to help plan and develop the National Arboretum—a pet project of the GCA. Janet Thurston Noyes, a Washington socialite, political insider, and chair of the GCA's District of Columbia committee, was invited to represent the organization. Her influence was such that the Associated Press reported she was "among those most prominently identified with the success of legislation creating the arboretum."[14] The organization's vocal advocacy also assisted in the passage of the Highway Beautification Act (1965) and the Endangered Species Act (1973). In 1969 the organization formed the National Affairs and Legislation Committee specifically to, according to its primary founder, "bring to the attention of the [GCA] membership important legislative proposals in Washington which fall within the fields of our endeavor."[15] As the umbrella organization advocated and lobbied at the national level, local member clubs served their communities in various ways, including by giving garden tours, hosting fundraisers, promoting historic preservation, sponsoring lectures and awards, and tackling beautification and enhancement projects.

While the GCA pursued its ambitious agenda, another stewardship-focused nonprofit group, the Garden Conservancy, emerged as a potential savior for many gardens facing practical or even existential threats. Although its efforts may be less dramatic than GCA's at a national scale, the work it has done to help owners and agencies rescue and steward primarily residential gardens has helped ensure the longevity of many landscapes that otherwise might have been lost. The organization was founded by the financier and self-trained horticulturalist Frank Cabot and his wife, Anne, both of whom loved gardens and created two of their own at their former residences, Stonecrop in Cold Spring, New York, and Les Quatre Vents in Quebec, Canada. While touring the private residential garden of Ruth Bancroft in Walnut Grove, California, in 1989, the couple was inspired by Bancroft's exquisite, three-and-a-half-acre arid garden filled with an array of colors and textures of cacti, succulents, and drought-tolerant plants. Bancroft, who was eighty at the time and five years a widow, expressed concern for the future of her property, as there were no established plans for preserving the landscape she had spent nearly twenty years cultivating. Upon leaving, Cabot, not a fan of cacti yet still "overwhelmed by the experience," remarked to his wife, "We have to find some way to help this woman," to which she replied, perhaps facetiously, "Why don't you start a garden conservancy?"[16]

He did just that with the Garden Conservancy, which he founded as an independent, membership-based nonprofit organization with a fifty-person advisory committee. Cabot then set out to identify exceptional gardens in need of preservation. Bancroft's garden logically became its first project, and by 1991, the conservancy had helped establish the Friends of the Bancroft Garden, a small group of volunteers dedicated to raising money for preservation and maintenance efforts. Through the efforts of the conservancy, it became the first private garden to secure a legally binding conservation easement for the permanent protection of the property as a garden. In 1992, the same year the Cabots opened Stonecrop to the public, the Bancroft Garden opened two days a week, by appointment only.[17] Two years later, ownership was transferred from Bancroft to the newly organized nonprofit, Ruth Bancroft Garden, Inc. Facilitating such ongoing stewardship quickly became a trademark of the conservancy.

The Bancroft Garden project taught the young organization a great deal about the process of preserving private gardens for public use and enjoyment.[18] As the conservancy moved on to other landscapes, a deliberate and successful method emerged. Nominations for sponsorship filter in through several avenues, at which point an internal screening committee reviews the nominations and visits various gardens to evaluate them using established aesthetic, horticultural, and historical criteria. Even with clear direction, deeming a garden "exceptional," which is required for consideration, involves a level of subjectivity. As stated by Antonia Adezio, a former president of the conservancy, "Gardens are very personal works of art. We feel you know when you see it."[19]

Once a project committee decides which gardens are feasible to take on, plans are made to transfer ownership to a nonprofit entity and determine future needs. As was the case with the Bancroft Garden, legal options are considered as needed, such as codifying a conservation easement to protect the landscape from development. From here, a committee of professionals and individuals close to the garden maps out a long-term management plan that includes strategies for maintenance, fundraising, programming, and public engagement. Each garden has a plan tailored to its stewardship needs, as no two landscapes are alike. Networks are established to generate public interest and build financial plans, which may include an endowment or capital campaigns. The ultimate goal is for the conservancy to hand over garden management to a locally based nonprofit even as it continues to offer advice and support as needed.

Landscape preservation projects are complex endeavors that require serious commitment and can take multiple years of active involvement. Because there is no shortage of gardens in need, the conservancy expanded its mission in 1991 to include varying degrees of stewardship assistance beyond pure preservation. For instance, if a garden owner is financially able to preserve their garden into the future, the conservancy can offer its expertise in preparing an effective path forward without formally sponsoring it as a project. The same would be true in the case of a significant garden assured of remaining intact through ownership by a nonprofit institution or a public agency. If its long-term care or design integrity is at risk, staff from the conservancy may act in an advisory role, providing a needs assessment, recommendations for long-term care, and periodic evaluations of progress made.[20] Also, in cases in which a garden does not meet the conservancy's test for exceptionalism but is still deemed worthy of preservation, the conservancy can use its growing gravitas in the horticulture and design world to become an advocate, drawing attention to the garden's significance and value as a cultural resource.[21]

In 2009 the National Trust for Historic Preservation presented the Garden Conservancy with the Trustees Award for Organizational Excellence, which honors "extraordinary achievements over a long period of time."[22] By that time the organization had spent around $9 million protecting more than ninety landscapes across the country.[23] Most of the landscapes targeted by the conservancy are small residential gardens not conceived by well-known landscape architects or designers that display a variety of styles and personalities. Assisting intimate gardens crafted from the passion and perseverance of individuals ensures the survival of places that likely would otherwise have been lost. While some of these landscapes are concentrated in familiar horticultural hotspots such as New England, Charleston, and California's Bay Area, more eccentric and unexpected gardens often become gems in the conservancy's portfolio.

One such landscape is the whimsical topiary garden created by the self-taught garden artist Pearl Fryar outside his ranch-style home in the small town of Bishopville in central South Carolina. What began as a self-directed challenge to prove that an African American man could win the Yard-of-the-Month award from the local chapter of the Garden Club evolved into the creation of a garden that attracts people from around the world. Fryar, now a celebrity in the realms of horticulture and landscape design, created his garden with five hundred trees and shrubs snipped, trimmed, and pruned into an army of fantastical blobs and beasts marching across lawns and clustered in mulch beds. In 2007 the Garden Conservancy's work resulted in the creation of the Friends of Pearl Fryar Topiary Garden, securing what could have been an uncertain future.

In addition to providing direct aid to specific landscapes, the conservancy educates the public and instills a passion for gardens and landscape design. One of its most popular and effective programs is "Open Days," during which gardeners and garden enthusiasts teach and inspire each other by visiting a diverse range of landscapes. In its first twenty-five years, between 1995 and 2020, the program exposed more than 1.35 million visitors to over four thousand landscapes—creative urban rooftops, organic farms, historic estates, and innovative suburban landscapes—in forty-one states across the country.[24] These organized tours are about more than just appreciating beauty or gaining inspiration. As Adezio once remarked, "By opening these private gardens and exposing visitors to their variety, we can create the kind of informed audience that would begin to support garden preservation in this country."[25] By exposing people to otherwise hidden gardens and coaching individuals and nonprofits on the practice of long-term stewardship, the Garden Conservancy, like the Garden Club of America, has become an invaluable resource for perpetuating the effective care and management of designed landscapes.

The following case studies present one knowledge garden and one pleasure garden—a distinction meant to show how a landscape's guiding mission influences both its legacy and its ongoing evolution. The Missouri Botanical Garden in St. Louis was created as a place for science and continues to be managed within a stable institutional structure. Miami's Vizcaya Museum and Gardens, on the other hand, began as a personal, family-run estate before facing a long period of uncertainty that included decades of impactful leadership and ownership transitions. Each garden has a legacy that dates to a primary patron, yet they have followed different paths to retaining relevance and quality.

Missouri Botanical Garden, St. Louis, Missouri

The Missouri Botanical Garden in St. Louis, Missouri, is widely recognized as the first garden in the country to focus from its inception on science. A display garden at its core, the nearly eighty-acre landscape contains features familiar to those found in botanical gardens—a welcome center, gift shop, and conservatories, along with numerous individual gardens representing common design styles imported to America from such places as Japan, China, and England. It was founded by the millionaire businessman Henry Shaw, whose initial vision unfolded into a two-century commitment to translating his science-based mission into a beautifully crafted and managed landscape. The Missouri Botanical Garden grew incrementally on a piece of native midwestern prairie that Shaw purchased in 1842 and evocatively described as "uncultivated, without trees or fences, but covered with tall luxuriant grass, undulated by the gentle breeze of spring."[26] The land was the ancestral home of a confederacy of around a dozen Indigenous tribes, including the Kaskaskia, Cahokia, Peoria, Tamaroa, Moingwena, Michigamea, Chepoussa, Chinkoa, Coiracoentanon, Espeminkia, Maroa, and Tapouara. By the mid-eighteenth century, only five tribes remained in the area.

The nineteenth century was a boom period in St. Louis, which became an important hub for commerce and trade as settlers passed through the city on their trek westward in search of land or gold. Shaw, who was English by birth, moved to the city in 1819, during the era when ambitious businesspeople amassed great wealth from the industries that drove the country's westward expansion. In addition to the larger-than-life railroad tycoons who attracted history's attention, such as Leland Stanford (discussed in chapter 3), other clever entrepreneurs turned basic needs into large fortunes. Shaw was in this category. He brought to the "Gateway of the West" the family business selling cutlery and ironworks to traveling pioneers and people building homes in the growing city. When he sold his business in 1840 and retired at the age of forty, he traveled the world to, as one biographer wrote, "acquire that invaluable education which only such intelligent and observing travel can afford."[27] On his final trip to Europe in 1851, he visited rural England's Chatsworth House and its picturesque landscape designed by the landscape architect Lancelot "Capability" Brown, one of Great Britain's most revered and prolific landscape designers. Moved by the experience, Shaw reportedly asked himself, "Why may I not have a garden too?"[28]

By the time he opened "Shaw's Garden," as it became known, in 1859, gardens were becoming familiar landscapes in America. Few of them, however, shared Shaw's vision for emulating certain European gardens that focused on botanical science as much as on aesthetic beauty. Marked by a century and a half of commitment to botany and ambitious invention, his garden has survived, morphed, and thrives today as a landscape valued internationally for its substantial research and horticultural displays. With a consistent governance structure headed by a board of trustees and a series of effective directors and presidents, the Missouri Botanical Garden struggled early before growing into one of the most influential knowledge gardens in the world.

Establishing the Legacy

The garden's design legacy is directly related to Shaw, an amateur designer who planned his garden on a portion of one thousand acres of rolling fertile prairie around four miles from the growing downtown. As he explained in an 1857 letter, "I intend to have everything substantial and elegant but on a small scale," before adding that the design is "according to my own ideas gathered from horticultural works."[29] He did seek botanical and design input from Gray, as well as from William

Hooker, the director of the Royal Botanic Gardens, Kew, and the local physician, amateur botanist, and plant hunter, George Englemann. Englemann went to Europe on Shaw's behalf to study botanical garden collections, buy books for a library, and collect seeds and plant specimens for an herbarium. Shaw established a framework based on a Victorian attitude that promoted the use of an eclectic variety of exotic plants formally arranged to be viewed and studied from multiple angles, making the surrounding landscape merely the context for carpeted displays of flora. His original garden surrounded his stately home and was organized to showcase herbs, shrubs, and trees in, as described by one historian, "outdoor beds of ornamentals, a fructicetum of shrubs and fruit trees, and an arboretum of trees."[30] While such details were important to the experience of being there, Shaw had greater ambitions for the garden, expecting it to provide public education, social ennobling, and extensive scientific research. It is this focus that has always separated it from purely pleasure gardens.

In 1859, Missouri's general assembly passed an act enabling Shaw to transfer 760 acres of his property to trustees "to keep up, maintain and establish a botanic garden for the cultivation and propagation of plants, flowers, fruit and forest trees, and for the dissemination of the knowledge thereof among men."[31] The Missouri Botanical Garden opened to the public that same year as a forty-five-acre scientific display garden nestled in the once empty prairie. Frederick Law Olmsted visited in the spring of 1863, although it is not known whether he and Shaw discussed the potential for collaboration.[32] Olmsted reportedly was not impressed with the garden's design as it differed from the pastoral and picturesque aesthetic he preferred.

In 1866, Shaw gifted another three hundred acres of his property to St. Louis for the site for Tower Grove Park, a pleasure ground that, per his will, "shall be used as a park forever," under the management of an independent board of trustees and staff—the only landscape in the city to be governed as such. A year later, with his attention focused on the new park, he hired fellow Englishman and Kew Gardens–trained horticulturalist James Gurney as the Missouri Botanical Garden's first head gardener. Shaw tasked Gurney with the day-to-day operations of the garden and with helping create new landscapes. Over the next two decades, as the garden expanded in accordance with Shaw's wishes, the staff grew to thirty workers.[33] The two men collaborated closely until Shaw died in 1889, at which point his desire to be buried in the garden was fulfilled with the construction of a mausoleum near his

Fig. 3. Tower Grove House was Henry Shaw's country residence until his death in 1889. The Italianate-style house sits within a Victorian-inspired garden designed for displaying an array of colorful flowers and plants laid out in elaborately designed beds. The garden was restored in the late 1990s. (Photo by author)

house. At that point, the stewardship structure that still exists today was established. According to the conditions delineated in his will, management of the garden was transferred to a board of trustees who would maintain fiduciary responsibility specifically for the benefit of the public. Shaw made the garden his primary benefactor, deeding to the institution the vast acreage of land he had acquired in St. Louis. Trustees would be able to sell parcels as needed to fund operations and expansion. Because most of his estate was in the form of real estate, it is difficult to determine a precise figure for this endowment. After his death, the *St Louis Dispatch* estimated his wealth at $2,500,000, while a "personal friend" told the *Chicago Tribune* it was around $10,000,000.[34]

Expanding Shaw's Vision

It did not take long for the Missouri Botanical Garden to become an attraction. By the end of its first decade, around 40,000 people visited the garden annually. An 1883 report describes the garden, barely two decades old by that point, as a "chief attraction in the vicinity of the city" that "had acquired a national reputation for the taste and beauty they exhibited and the variety and value of their botanical features."[35] Gurney was among the most instrumental figures during this time; he tended the landscape and created new spaces until he retired in 1903. There were only a few, predominantly formally arranged garden areas, leaving most of the property undeveloped and awaiting its future.

Shaw anticipated his garden would last well beyond his lifetime. However, he left behind no roadmap to follow, making its future somewhat unclear. His will mandated that trustees appoint a full-time director to oversee the operations and management of the garden without interference from the board.[36] He had established the Henry Shaw School of Botany at nearby Washington University in 1885 and stipulated that the garden director simultaneously hold the university's George Englemann Professorship of Botany. He wanted the landscape tied to formal research and science and dictated that directors should, "when thereto required so to do, deliver lectures . . . upon Botany and matters connected therewith."[37] The trustees surely took this into account when they hired William Trelease, a professor of botany with a doctorate from Harvard, to be the garden's first director after Shaw's death. Trelease understood his role as a steward of the landscape and directed Gurney on incremental improvements to the grounds with the limited revenue he was given.[38] Most of Trelease's energy, however, was spent on pedagogical and scientific endeavors. While he focused on building a solid foundation for research and teaching programs,[39] the trustees were planning for the future of the physical garden with the help of the Olmsted, Olmsted, and Eliot firm (renamed Olmsted Brothers in 1895), which consulted for more than a decade, ultimately producing a general plan for development in 1905. Likely in response to Trelease's overt preference for research, the newly elected president of the board of trustees advised the director in 1911 that "we're going to have more garden and less science" from now on.[40] Trelease resigned that same year and was replaced by George Moore, another botany professor, who quickly earned the respect of the trustees for his business acumen. He led the Missouri Botanical Garden for more than forty years.

Most of the Olmsted firm's plan for the garden resembles one of the firm's famous schemes for a picturesque urban park more than it does a proposal for a botanical garden. Vast open lawns and meadows framed by naturalistic groves of trees are shown spread across the property, along with three small ponds. Everything is connected by a series of sinuous paths that add to an overall effect that mimics the Greensward Plan and countless other park designs that had become the bread-and-butter work for the firm. Little of it seems to respond to Shaw's vision for a science- and display-focused botanical garden aside from a row of rectilinear gardens designed with English, Italian, French, and Dutch influences and the circular Pagoda Garden sited next to the mausoleum in which Shaw was buried. The plan remained in place for decades and was generally ignored. Besides a large parklike portion on the western edge called the "North American Tract," which was constructed in the early 1900s, the only other element built from the plan was the Pagoda Garden, little evidence of which exists.[41]

Progress at the garden ramped up dramatically beginning in 1913, by which time St. Louis was the fourth largest city in the country with a population of nearly 700,000 residents. The landscape architect John Noyes was hired as an instructor at the Shaw School of Botany and, two years later, a new policy initiated by Moore established Noyes as the resident designer for the property and the first landscape architect hired to work there since the termination of the relationship with the Olmsted firm. Noyes was trained by well-known and celebrated landscape architects—Warren Manning, a protégé of Frederick Law Olmsted who designed numerous campuses, gardens, and parks, and the Danish-born Jens Jensen, who was a leader in the prairie-style design movement that celebrated the native flora and wide expanses found in the midwestern landscape. Noyes worked in this role for two busy decades, during which he drew up a revised general plan for the property and oversaw the addition of numerous features, including the Desert, Mediterranean, and Floral Display Houses and the Italian and Perennial Gardens.

Noyes designed the Rose Garden, the Shakespearean Garden, and the cylindrical Economic Garden. His vision also transformed a flat area originally designed by Gurney as an area to display formal plantings into a naturalistic rolling garden, bisected by a small stream, called the Knolls. Although his influence has been somewhat lost to history, he obviously played a significant role in the evolution of the garden and exemplifies the value of making in-house staff simultaneously responsible for stewarding the past, building the present, and planning for the future. When Noyes left in the late 1950s, he was replaced by Emmett Layton as the on-staff land-

Fig. 4. The Lily Pool and the naturalistic Knolls landscape surrounding it were designed by John Noyes and later removed to make way for the now iconic Climatron. (Missouri Botanical Garden)

scape architect. His influence is not as easily tracked, and he appears to have been the last in-house landscape architect in a design position. Following his departure, a series of chief horticulturalists filled the stewardship void.

Mounting Challenges

Despite significant improvements accomplished under his leadership, Moore's tenure suffered from a series of challenges beyond his control. In the 1920s, for instance, when St. Louis was known to have the filthiest air of any city in the country thanks to local industries and residences burning coal for energy and heat, the garden's plant material suffered the effects. In addition to the estimated 150,000 tons of sulfur emitted annually—much of which came down in the form of acid rain—nine hundred tons of solid matter fell in the city per square mile.[42] During colder months, the skies above the garden filled with dark clouds of airborne soot that would fall on the landscape and seep through the glass walls of display and commercial greenhouses. Poor air quality damaged or killed thousands of plants and trees, compelling the trustees to purchase thirteen hundred acres of land forty miles away to safely propagate new plants. Once the city's air was gradually cleaned thanks to regulation that emerged from the 1937 Smoke Ordinance, this new property was turned into an arboretum.[43] In 2000 it was renamed the Henry Shaw Nature Preserve.

A challenge shared by most institutions came at around the same time in the form of the Great Depression, which left the Missouri Botanical Garden financially strapped throughout the remainder of Moore's term. Some help appeared in 1939 with the creation of the Friends of the Garden, a volunteer group formed primarily of women with social connections and the ability to provide fundraising support. That same year, Moore sent a leaflet explaining the institution's financial crisis to the public, along with an attached pledge form and a plea for support in any amount.[44] Despite a healthy stream of donations, financial challenges persisted. In 1946, Moore stated with tempered optimism in his annual report to the trustees: "The Garden is neither in severe financial straits, nor is it overburdened with wealth."[45]

It was evident by this time that the resources provided in Shaw's will were insufficient and that supplementary support would be needed if the garden was to effectively compete with other leading botanical and horticultural institutions.[46] Per Shaw's will, portions of his property had been sold off from time to time to raise funds. Now there was essentially none left to sell. The Friends group had some success, raising around $4,000 or $5,000 a year, yet this was not enough to make a substantial difference. Age and insufficient maintenance were already taking their toll on the landscape, a situation made worse during the summer of 1946 when wind, rain, and a hailstorm left damage the *St. Louis Post-Dispatch* deemed "the worst in the history of the garden."[47] The storm destroyed cherished specimen trees, flooded streets and low-lying lands, smashed windows, and stripped blossoms from flowering plants. Money earmarked for maintenance and new gardens was diverted to cover emergency cleanup.

Moore's health declined, and he stepped down in 1953. He died three years later, his last days shadowed by the garden's still poor financial condition, rising operating costs, declining visitor attendance, and the visible effects of deferred maintenance. The trustees commissioned an appraisal of the garden that, possibly as expected, was dismal. In what could be interpreted as an affront to Shaw, the report recommended drastically cutting research funding and redirecting as many resources as possible toward desperately needed maintenance. The suggestion raised the ire of the scientific staff, who likely believed their work supported the garden's core mission. By this time, St. Louis was the eighth largest city in the country. It reached its peak population of around 850,000 at mid-century, then began a consistent decline.[48]

The Climatron and the Leadership of David Gates

In a somewhat ironic twist, it took a work of architecture to turn around the fortunes of the Missouri Botanical Garden. In 1960, one year after the institution's centennial celebration, the Climatron, a revolutionary geodesic dome greenhouse modeled after Buckminster Fuller's futuristic work of architecture, opened to the public with substantial fanfare. The structure's name was coined by the director at the time, Frits W. Went, a Dutch biologist and eminent botanist, who brought a new perspective to the garden. He believed the institution's primary goal should be to develop a heightened awareness and appreciation of plants through innovative displays.[49] The Climatron was surely one of these—a unique and innovative space age–looking open greenhouse with no interior spatial division. It was the garden's first major capital project in five decades and provided a needed promotional and financial boost. To build it, however, the Economic Garden, the Knolls, and half of the Italian Garden had to be removed. This seems not to have a been a problem for the public as intrigued crowds poured in specifically to see the Climatron, paying fifty cents per person and surging attendance well beyond recent totals.

The attraction helped bring in $2,000 a week in its first few months and a total of $250,000 by 1963.[50] This was the year Went resigned following a disagreement with the trustees over the garden's future. He wanted to build on the public's interest in the Climatron and reimagine the entire facility as a cultural center for the city, with the addition of a large pavilion that could host musical events, live theater performances, art exhibits, lectures, and restaurants. The trustees were more conservative than Went, whose proposal they considered neither appropriate nor affordable.[51]

Fig. 5. The futuristic Climatron opened to the public in 1960 and immediately became a successful feature credited with attracting never-before-seen crowds to the garden. (Missouri Botanic Garden)

As visitors poured in to see the Climatron and the horticultural display gardens, research at the institution continued to decline. This trend, which stemmed from the budget cuts of the 1950s, was eventually reversed under the leadership of David Gates, a scientist with a background in physics and ecology. Gates started as director in 1965 and within three years had boosted the number of professional staff and graduate students from around twenty to nearly forty. In addition, he instituted a charge for general admission for the first time, initiated a $3 million capital campaign, and pushed researchers to apply for grant money, all of which helped balance the budget and establish fiscal responsibility. As enthusiastically documented in the 1967 annual report, "Operating income for the fiscal year was $502,000, while expenses were $501,000!"[52]

Gates's most significant contribution, however, was arguably aligning the garden's focus with the burgeoning national environmental movement and directing research to address broader ecological issues.[53] As he noted in a 1968 report, "The environmental bomb has been lit. It is not a question of whether or not it will explode. It is already detonating and our beautiful green world of plants, animals, people, water, soil, and atmosphere is being ravaged and destroyed."[54] He was a vocal and influential environmentalist and gained national attention by raising awareness about humans' harmful impact on the natural environment. He spoke in front of congressional committees, lectured before the scientific community and the public, and wrote numerous journal and magazine articles on the impending environmental crisis. He also pressed the federal government for support for enhanced ecological research and touted the need for a formal discipline he called "environmental management."[55] Because of these efforts, the *New York Times Magazine* labeled Gates "a leader among scientific Cassandras who seek to alert the nation to impending doom."[56] Likely because of this greater calling, Gates left the garden in 1971 to return to academia.

The Influence of Peter Raven

In 1971, by which time the Missouri Botanical Garden employed eighty staff members and a volunteer force of three hundred, Peter Raven, an academic from Stanford University with no previous administrative experience, took over as director. Although there had been many changes to the institution in the decades preceding his arrival, the designed landscape had remained relatively unchanged. More than a century after its opening, much of the garden looked more like a park than a botanical garden. John Elsley, who started as the curator of hardy plants in 1970, described the landscape at that time as "just a rose garden, a little herb garden, lawns and paths."[57] This is a surprisingly negative statement, especially in light of all that had been accomplished under Moore and with Noyes on staff. It does convey how much of the garden remained relatively undeveloped. Raven, a scientist by training, clearly understood the landscape's potential and devoted much of his time and energy to planning its future. He hired the Pittsburgh-based landscape architecture firm Environmental Planning and Design (EPD) to collaborate on a new master plan, which was led by the noted garden designer Geoffrey Rausch. That plan, which was completed in 1974, fills the property to its edges with new gardens designed as an apparent representation of modernist-era norms rather than a period-based nod to the era of Henry Shaw.

Much of EPD's plan was implemented over the next several years, including the installation of fountains, a new formal entry, a Mediterranean House, and an English Woodland Garden designed by Elsley. None of this was envisioned by Shaw, yet it fit well within his goal of diverse horticultural displays. The same was true for Seiwa-en, the first Japanese garden built within hundreds of miles of St. Louis.[58] Bearing a name meaning "garden of pure, clear harmony and peace," the fourteen-acre Japanese-inspired landscape opened in 1977, by which time international gardens were popular additions to American botanical gardens.[59] Dr. Koichi Kawana, the landscape architect referred to by the *Los Angeles Times* as "the dean of traditional Japanese garden design in California," designed Seiwa-en, along with similar gardens in Chicago, Memphis, and Los Angeles.[60] The new garden replaced the Olmsted firm–designed North American Tract, the loss of which apparently caused little distress among historians or Olmsted lovers. One news service confirmed this lack of interest when it referred to the destroyed landscape as "an infrequently visited area of the grounds."[61] The aesthetic of Seiwa-en was not entirely new, for the landscape proferred such familiar features as a picturesque pond, open lawns, and gently rolling hills. Unlike the North American Tract, however, which, according to historical photographs, sought to replicate a natural condition, the new landscape was easy to understand as entirely confected. It also provided the garden the opportunity to share plants from a new part of the world while adding a beautiful new space for visitors.

Raven continued to move forward with EDP's master plan, remaining adamant about following it strictly. When he received an offer for a monumental art piece created by the famed sculptor Mark di Suvero, he rejected it because it was incompatible with the plan and, in his words, "too heroic for any space we have."[62] Such decisions highlighted a deliberate process that was lauded as being both systematic and orderly. One journalist described the Japanese garden as "the most visible manifestation of the implementation of a carefully-conceived, thoughtful long-range plan."[63]

In addition to the landscapes, a new visitor center opened in 1982. The 60,000-square-foot facility was designed to welcome 250,000 visitors per year, with an auditorium, classrooms, orientation theater, greenhouse, and revenue-generating

Fig. 6. The fourteen-acre Japanese Garden, named Seiwa-en, meaning "garden of pure, clear harmony and peace," was designed and constructed in the 1970s. It replaced the naturalistic North American Tract, an infrequently used space designed by the Olmsted Brothers firm. (Missouri Botanical Garden)

restaurant and gift shop. Much of the initial funding for these projects came from philanthropic donations, along with membership dues and admission fees, which were raised a couple of times in the 1970s. More reliable financial help arrived in 1983 when the citizens of St. Louis approved a four-cent property tax increase to incorporate the garden into the culturally focused Metropolitan Zoological Park and Museum District. The income boosted educational and scientific programs and enabled the garden to lower admission fees for adults and make admission free to children and senior citizens.

While working on transforming the physical landscape, Raven earned recognition for his contributions to and support of research to preserve endangered plants, as well as his advocacy of conservation and environmental sustainability. In 1999, *Time* magazine listed him as one of its "Heroes of the Planet."[64] At Raven's retirement in 2010, the chair of the board of trustees celebrated him, claiming Raven had helped to transform Shaw's garden "from a beautiful local attraction into one of the top three botanical institutions in the world."[65] Nearly one million people visited the garden annually, the full-time staff had grown sixfold since 1971, to nearly five hundred, and the institution's net assets had ballooned to nearly $170 million.

Contemporary Issues

A focus on environmental sustainability persisted beyond the efforts of Gates and Raven and has remained an explicit part of the Missouri Botanical Garden's mission since being codified in the institution's 2007 strategic plan as such: "To protect plants and ecosystems and thus enhance life for people and all living things, the Garden will elevate sustainability as a core principle in all its work—its programs of research and conservation, horticulture, and education and ongoing operations—and make the relevance of sustainability evident to all its visitors."[66] As was true for many institutions, the 2008 recession thwarted progress toward achieving these ambitious goals. A subsequent plan, which was completed in 2015, set a new path for the future. While it preserves a focus on research, sustainability, and conservation, the plan makes surprisingly little reference to the designed landscape. One subtle statement proclaims, "We also take responsibility for the preservation of the character and historic nature of the Garden and our sites with strategies and materials that model best practices."[67]

In 2020 the Missouri Botanical Garden broke ground for a new visitor center to replace the outdated 1982 building. With the ambitious goal of creating a "new treasure" for St. Louis and advancing the garden's role as a "global leader in plant science, and as an important cultural institution for the region and the nation," the $95 million facility was conceived to better accommodate the public.[68] Reflecting its status within the community and among garden lovers, the project was fully funded through philanthropy. Construction coincided with the peak of the coronavirus pandemic. For months, the landscape was closed to the public, and staff worked to maintain the gardens without help from volunteers. The garden's president, Peter Wyse Jackson, who began his tenure in 2010 with a focus on fundraising and an "if it's not broke, don't fix it" mindset, was determined to take care of both the living plant collections and his staff during the crisis.[69] With a loan from the federal Paycheck Protection Program, the organization cut expenses and shifted staff assignments to avoid extensive layoffs. Loyal donors reportedly "stepped up in dozens of ways" to increase their support of the garden and its ongoing mission.[70]

While the gardens were closed to the public, the 2,400-acre Shaw Nature Preserve experienced a boost in visitors. With ample space for visitors to practice social distancing, the property's managed prairie, glade, wetland, and woodland landscapes offered respite and access to nature for a community dealing with isolation and stress. From an accidental landscape born out of an environmental disaster, the preserve evolved into a perhaps unexpectedly critical resource for the citizens of St. Louis, who, through the taxes codified in 1972, contributed nearly $13 million to the Missouri Botanical Garden in 2020. This funding, along with another $13 million in support from donor contributions, memberships, and bequests, provided 60 percent of the garden's revenue for the year. With an endowment valued at around $160 million and total assets adding up to more than twice that amount as of September 2021, the garden remains well positioned to achieve its goals in science and conservation while providing a horticultural amenity for the region and the nation.[71]

The new visitor center opened with fanfare in August 2022, even as the landscape remained a work in progress. The building and the site were seemingly designed with permanence in mind—a new face for the over-160-year-old institution as it welcomes more than one million visitors per year. Massive blocks of natural stone anchor the site, and a towering glass atrium punctuates the stone building's façade with a glowing beacon that harks back to Shaw's Tower Grove house as it signals entry to the garden. The side opposite the main entry opens onto a beautiful plaza with water features that provides an appropriate transition to the historical and established garden areas. Created with a contemporary aesthetic, the new addition successfully builds on the continued layering of design eras throughout the garden. The passage of time, it seems, is celebrated as a positive force in the life of the garden and not something to regret or regard as a threat.

The visitor center, like the Climatron, introduces architecture as a means for fulfilling program needs while formally and semipermanently marking a place in time in the life of the garden. Both were the result of bold decisions approved by trustees and represent important points of inflection. They also exemplify the value of having consistent and committed leadership to lead fundraising and oversee the fulfillment of institutional goals and mission. Of course, top-level contributions would be less meaningful without the thousands of managers, supervisors, horticulturalists, laborers, administrators, and volunteers who have worked on behalf of the garden

Fig. 7. A new visitor center architecturally influenced by Henry Shaw's house introduced contemporary architecture to the garden in 2022. The entry landscape, along with a new garden behind the building, was designed by Michael Vergason landscape architects. (Photo by author)

since its founding. The aggregate impact of decisions made and actions taken at this level can be at least as meaningful and transformative as the impact of decisions and actions that filter down from above. More detailed stories of such instances and individuals are presented in later case studies.

Ultimately, the Missouri Botanical Garden is a prime example of a venerable, consistently well-run institutional landscape that both understands and values its legacy. Generations of leaders and staff have respected, supported, and implemented the vison Shaw defined in his early work and subsequently codified in his will, to be continued after his death. Without an approved original plan master guiding its evolution, the garden evolved from a nearly clean slate into a complex composite of individual spaces and buildings that successfully reads as a unified place. Visiting today makes its success obvious. The maintenance of its grounds and the variety and quality of its extensive horticultural displays rival those of any garden in the United States. With Shaw's house and mausoleum serving as the garden's historical touchstones, the landscape has expanded and adapted consistently over time, almost always in accordance with its founder's vision. In this sense, it serves as a model of successful stewardship for other gardens, affording knowledge and pleasure—and landscapes to follow.

Vizcaya Museum and Gardens, Miami, Florida

"What do you think of the name 'Vizcaya' for the Florida house?" asked James Deering, the millionaire industrialist, of his design adviser, Paul Chalfin, in a 1914 letter.[72] Construction had recently started on his thirty-four-room, now world-famous Italianate villa, which sits perched on the shore of Biscayne Bay, just south of Miami. It would emerge as a celebrated property in Deering's time and evolve into an enduring cultural jewel for South Florida. Although it still impresses crowds today, the villa succeeds more as an expression of designed excess and luxury than as an influential work of architecture. Likewise, the landscape—once a sprawling 180 acres of natural and manipulated native marshes and scrubby hammocks, now reduced primarily to ten acres of ornately designed formal gardens—reads as beautifully orchestrated garden whimsy more than important landscape design. Even so, the composite of architecture and landscape, each of which would suffer immeasurably without the other, is extraordinary for its unique photograph-worthy beauty. A pleasure garden like few others, it presents a history and an evolution distinct from the Missouri Botanical Garden's. With a design in place from the start, it is a story about preserving and restoring a discrete landscape more than a case of expansion. Nonetheless, striking similarities in experiences link the two landscapes as they have evolved.

Deering and Chalfin conceived the property in 1910, designed the villa between 1912 and 1914 with the architect Burrall Hoffman, and completed it in time for Deering to move in on Christmas day in 1916. The extensive landscape, designed in collaboration with the landscape architect Diego Suarez, would take another six years to complete. The property was a showplace, a hybrid of age-old European design influence and ostentatious new American wealth, embedded in a lush and exotic landscape. As the owner, financer, and design collaborator on what is without doubt among America's unique once-private homes, Deering is a significant figure in U.S. architectural and landscape history. What happened to Vizcaya after he died, when the private estate slowly transitioned into a municipal museum and public garden, makes its story more meaningful.

Establishing the Legacy

James Deering was born in the western Maine town of South Paris. His father, William Deering, was a natural entrepreneur who, like many other successful men of his time—Henry Shaw included—capitalized on the booming post–Civil War economy by converting vision into a fortune. In 1873, he moved his family to the Chicago area to take over an existing farm equipment manufacturing business that eventually became the Deering Harvester Company, where James and his half-brother, Charles, worked. By the turn of the century, the Deering family was among the wealthiest in the country. In 1902, Deering Harvester merged with McCormick Harvesting Machine Company to form International Harvester. A corporate restructuring marginalized James's role, and by 1908 he was essentially retired. At forty-nine, he was a wealthy bachelor, and commenced life as a member of America's idle rich, traveling extensively and regularly to Europe.

Deering suffered from pernicious anemia, meaning his body could not produce healthy red blood cells. As a remedy, his doctors prescribed spending winter months in a warm climate. Following the lead of his father and brother, he sought land for himself in South Florida.[73] By the summer of 1910, when New York City, Chicago, and Philadelphia had amassed populations of at least one million people, Miami had around 12,000 residents. Its swamps, coastal forests, and long, uninterrupted stretches of white sand beaches remained relatively isolated and beautifully unspoiled. Hundreds of square miles of flat and easily buildable land sat waiting for the hordes of developers that would soon come.[74] Shortly after their first meeting, Deering took Chalfin to Europe, where they toured numerous villas and bought art, furniture, and sculpture. Upon their return stateside, Deering purchased one hundred thirty acres of waterfront land between Coconut Grove and Miami on which to build his estate. The property contained a generous stretch of mangrove swamps and provided stunning views across Biscayne Bay. Chalfin hired the École des Beaux-Arts–trained Hoffman to design the ornate villa. The design process was fraught with drama and fueled by inflated egos and a commitment to excess—a story related in detail by Witold Rybczynski and Laurie Olin in their engaging book, *Vizcaya: An American Villa and Its Makers.*

As conceived, the property's formality and intricate detailing presented a surprising contrast to the wild and messy natural landscape, creating a juxtaposition that remains an integral component of Vizcaya's persistently unique character. The villa rests on the Miami Rock Ridge, a continuous outcropping of limestone running from Miami Beach south into the Everglades. Along the property's shoreline, the

Fig. 8. Southeastern Florida's native subtropical landscape established a wild and natural character that attracted James Deering and provided the backdrop to Vizcaya's ornate villa and gardens. (Vizcaya Museum and Gardens Archives)

unmistakable dendritic root system of mangroves emerges from the murky water and helps to establish a distinctive tropical tone. A few yards inland, the landscape changes into a rockland hammock, a tropical hardwood forest that, like the mangroves, is increasingly endangered. In its natural and healthy state, the biodiverse landscape can be home to over one hundred native trees and shrubs. Most of the large trees are found in the forest core, with the edges filled with nearly impenetrable evergreen thickets of intertwining vines, shrubs, palms, and small trees. Deering valued this native setting and once commented, "To me the charm of this part of the world is the jungle effect and anything done to interfere with this must be against my wishes and without my knowledge."[75]

He was adamant that trees not be taken down for the project and that there be no clear-cutting of vegetation for new plantings, only the removal of "dead trees, dead underbrush, and trees that might as well be dead."[76] Most of the initial layout and conceptualization of the site were complete when Suarez was hired as the landscape architect in 1914.[77] The intricately detailed gardens, a hybrid of classic French and Italian landscapes, with formal parterres, elaborate fountains, and individual spaces (including a Maze Garden, a Rose Garden, a Theater Garden, and a Secret Garden) delineated by Italianate architectural features, would seem more ordinary without the contrasting native backdrop. It is nearly impossible to assign singular credit for designing the landscape, as Deering, Chalfin, Suarez, Hoffman, and others all played a significant role. As Olin explains about the synergistic influence of the group: "Each individual contributed, in particular ways at various moments in the history of the project, to the creation of something none of them could possibly have done alone nor was ever to do again."[78]

When Deering first approached the villa on Christmas Day in 1916, the boom of cannon fire heralded his arrival from the bay on his yacht. The onset of World War I delayed the completion of the gardens, which were not completed until 1922. Estimates place the total cost to build Vizcaya between $10 million and $15 million in today's dollars. Deering typically lived in Vizcaya only between December and April but, because of the intensity of needed maintenance, he employed a large permanent staff of more than forty workers to keep things pristine even while he was away.[79] Employees lived across the road from the villa in a self-sustaining village. It comprised twenty-two buildings and a working farm that provided food for Deer-

Fig. 9. This view from the villa looking across the formal gardens at Vizcaya shows the juxtaposition of the designed landscape and the native environment. Over time, much of the surrounding natural landscape was lost to development. (Vizcaya Museum and Gardens Archives)

ing, his family and guests, and those who worked on the estate. He threw parties and luncheons at Vizcaya two or three times a week while he was in residence, and on Sundays and Wednesdays Deering generously opened the gardens to the public so visitors could "view his horticultural achievements," thus making the estate a semi-public landscape from the beginning.[80]

The Death of James Deering and the Great Hurricane

The first jolt to Vizcaya's maturation occurred in September 1925, when Deering died onboard an ocean liner returning to the United States from Europe. His health had been declining for years, so his death likely was not surprising. The value of his estate was not made public, but it was reported that he was the single largest payer of income tax in Florida at the time of his death, so it was surely considerable.[81]

Although there were rumors that Deering intended to bequeath Vizcaya to the city of Miami,[82] he instead gave it in equal shares to his brother Charles and his brother-in-law, Richard R. Howe. By keeping it in the family, he not only postponed a sustainable, long-term solution, he also created a burden for his heirs, neither of whom intended to live there. Deering's will failed to specifically address the future use of the estate, although there is evidence suggesting he wanted the villa converted into a public museum.

Staff stayed on to maintain the property even after Deering died. Because they were used to him being away for long periods while he was alive, it is likely that management remained relatively unchanged. Without his passion and deep pockets, however, the long-term future of the villa and its landscape was in question. Maintaining formal landscapes in South Florida's tropical heat and humidity is challenging, as well-adapted plants grow rapidly and in ways not always intended. The plantings in Vizcaya's lush gardens required substantial resources to keep them trimmed and neat per the design intent. It took as many as twenty-five workers to tend the grounds, which created an expenditure that had been acceptable to Deering because, as recalled by a friend, "His marvelously planted grounds were his great joy."[83] The new owners failed to enjoy the same sense of reward.

Tropical storms created another challenge for Vizcaya, as they do for the entire region. This was especially true before advance-warning technologies could detect impending storms. Such was the case on September 17, 1926, when South Florida was struck by the force of what was later called the Great Miami Hurricane. By this time the growing city had over 100,000 residents, most of whom lived in houses and neighborhoods structurally incapable of withstanding such a powerful storm. The Category 4 hurricane battered Miami, Coconut Grove, and other communities throughout the night, leveling many homes and other buildings. The American Red Cross later reported 372 people had died and over six thousand had been injured from the storm. The hurricane left in its wake damage totaling more than $100 million—the equivalent of around $1.7 billion in 2023.

Vizcaya's villa survived the storm intact since it was constructed primarily out of concrete. Much of the landscape was not so fortunate. Photographs taken shortly after the storm depict masses of debris, urns, columns, trees, and even several of Deering's boats strewn across the property. Most of the trees in the native rockland

Fig. 10. A devastating hurricane in 1926 stripped much of the foliage in the native landscape, flooded Vizcaya's gardens with saltwater, and left behind significant damage across the estate. (Vizcaya Museum and Gardens Archives)

hammock were either uprooted or stripped almost entirely of their leaves. The mangrove forest was also extensively damaged. As is typical in such situations, the native communities that had evolved in the presence of such storms were able to rebound relatively quickly. Although the structures in the formal gardens fared surprisingly well, it became apparent that some of the ornamental plantings selected by Suarez from his office in New York City for their exotic character could not survive exposure to saltwater. Roses, for instance, almost completely succumbed to the storm. Their removal from the property following clean-up from the storm was, at the time, the most significant design change in the formal gardens.

The Next Generation Steps In

Charles Deering's influence over the fortunes of Vizcaya was cut short by his death in February 1927. He left his share of the property to his daughters, Barbara Deering Danielson and Marion Deering McCormick, whose husband, Chauncey

McCormick, of the McCormick Harvesting Machine Company family, quickly became one of Vizcaya's most influential stewards. He was a wealthy businessman who, along with Marion, was active in the social and art worlds of Chicago, where they lived most of the year. Like many in their elite social class, the couple embodied the centuries-old connection among art, wealth, and philanthropy and seemed to take seriously their responsibilities as caretakers of the Deering estate. Howe, on the other hand, retained partial ownership, but neither he nor anyone on his side of the family seems to have taken an active interest in managing the property. Once Howe died in 1943, Charles Deering's daughters bought out their cousins for total ownership.

There is little documentation about Vizcaya between 1926 and 1933. Although no one lived in the villa full-time, one of James Deering's nieces would stay there during the winter months.[84] Neither family was severely affected by the Great Depression, although even the wealthiest Americans were forced to think differently about money. As was the case with other large estates built before the imposition of various federal taxes in the 1920s and 1930s, paying for the upkeep of an expensive second home like Vizcaya was a financial burden. To help defray the costs, Chauncey McCormick and the Deering estate's lawyer, Francis M. Miller, began discussions with the City of Miami in 1933 to periodically open the property to the public. Marion McCormick and Danielson had long wanted the estate to become a museum, believing that sharing it with the public was in keeping with their uncle's wishes and part of their understood role as stewards of the property.[85]

Opening to the Public and the Return of Paul Chalfin

By the early 1930s, when the Art Deco style was all the rage in nearby Miami Beach, the area had emerged as a popular tourist resort. Deering's family paid an exorbitant $32,000 per year in property taxes to the city, county, and state during this time.[86] Chauncey McCormick moved forward with a plan to open the villa and gardens to the public with an entry fee, as had been done at other grand residences in the Southeast, including George Vanderbilt's Biltmore Estate near Asheville, North Carolina, and Charleston's Middleton Place.[87] There was hope that gaining status as a cultural institution might reduce the property's taxes. Miami's mayor, Everest Sewell, said that although "opening [Vizcaya] to the public will be a great benefit to the City as a whole," it would not be practical to reduce the taxes on the property.[88] The Board of County Commissioners echoed the decision, stating that reducing taxes for the wealthy heirs of James Deering would not be fair to Miami's other citizens and would surely invite criticism.[89]

Admitting the public into Vizcaya was never exclusively about raising money. Deering's family understood the inherent educational, civic, and cultural value the estate offered and appeared ready to share it with the people of Miami. Even without the tax break, they decided to open on a trial basis, and McCormick temporarily relocated to Miami to supervise every detail of the preparations.[90] The villa was apparently in relatively good shape, but the landscape still suffered the effects of the hurricane eight years earlier and required restorative work before being presentable. Rather than contacting Suarez, who had completed most of the designs for the formal gardens, McCormick engaged Chalfin to evaluate both the landscape and the villa interiors. He made clear this was not to be an elaborate restoration of the original plan and instructed Chalfin "not to spend money except in those places and for such things as will in my opinion tend to draw ticket buyers at the gate."[91] Even with this caveat, his expectations remained high; as he conveyed to the designer, he believed that "family and visitors at Vizcaya are going to find it more beautiful this year than they have ever seen it."[92] McCormick was neither an artist nor a designer, yet he influenced and directed aesthetic changes at Vizcaya. During this time he was also a trustee of the Art Institute of Chicago and chair of the art committee for the 1933 World's Fair, so, although he deferred somewhat to Chalfin, he was not shy about expressing his opinion.

Chalfin was likely eager to return to Vizcaya. Since his earlier work with Deering, the designer had completed several commissions in New York and designed his own house in Connecticut, yet nothing equaled the magnitude or renown of Deering's estate. He was tasked with revisiting a landscape created for an era of opulence and adapting it to a new economic and practical reality. Following a trip to Miami in early 1935, he detailed a twenty-week restoration plan to McCormick, most of which would be spent on the "reconstruction and the erecting of salvageable property," with only around two weeks left for what he referred to as "aesthetic betterments."[93] Much of the work in the gardens involved grading, repairing walkways, and restoring urns, columns, gates, and other garden features damaged in the

Fig. 11. An expansive Lagoon Garden made up much of Vizcaya's original 180-acre estate. The villa appears at center and toward the top of the image. (State Archives of Florida)

hurricane. Although he eagerly claimed credit for adding considerable value to the objects he restored, Chalfin was unenthused by the venture, describing his satisfaction as being "of the modest order of the social worker who rescues a bath tub in a slum from use as a coal bin."[94] He took the work seriously, explaining the effort and resources required with handwritten notes, sketch concepts, and detailed plans that included simplified and modified plantings, specifying species better adapted to the climate and site conditions.

The family opened Vizcaya briefly before another significant storm—the infamous Great Labor Day Hurricane—hit the Florida Keys and the Miami area on September 2, 1935. The Category 5 storm made landfall with sustained winds reaching 185 miles per hour and a storm surge of nearly twenty feet, making it one of the most destructive hurricanes ever to hit Florida. As with the 1926 storm, the landscape at Vizcaya was devastated. Yet another hurricane hit in November that same year, and the tropical location and climate that made the gardens so special once again became unpredictable threats to its future. Chalfin returned to complete his restoration of the gardens and, in January 1937, wrote to McCormick that he was satisfied with the results, stating, "The place is far from ruinous."[95] This would be his last direct involvement with Vizcaya, as his fading vision ended his design career by the early 1940s.

Selling Vizcaya

Plans to open Vizcaya to the public were put on hold following the storms. Marion McCormick and Barbara Danielson continued to vacation at the villa into the 1940s even as maintaining the property, particularly its vast landscape, remained an economic burden. In 1945 the family followed a familiar path taken by the second-generation owners of other grand estates and sold off part of the land to generate income and reduce the expense of ongoing maintenance. The sale was partially altruistic as their buyer was the Catholic Diocese of St. Augustine, which at the time was the diocese for the entire state. They sold one hundred thirty of the estate's one hundred eighty acres, much of which was undeveloped slash pine rockland. The most tragic loss was the unique and expansive Lagoon Garden on the southern end of the property. The church subsequently filled in the landscape to build the Mercy Hospital, which opened in 1950 and provided needed hospital beds for the rapidly growing city. The remaining land was then sold to real estate developers, who turned the already disappearing native landscape into an unremarkable suburban neighborhood. Although unfortunate in many ways, the transaction enabled the family to hold on to the villa and main gardens, which otherwise might also have been lost.

In 1952, James Deering's nieces sold the remaining fifty acres of the Vizcaya property to Dade County for $1.4 million. The villa's furniture and art collections and garden sculptures—valued at another $1.4 million—were added to the deal at no extra cost. This was a coup for the growing region, which offered ideal weather and exquisite beaches but was essentially a cultural wasteland compared to New York City or Chicago. The art, architecture, and remaining ten acres of formal gardens

Fig. 12. Beginning in the 1940s, Deering's family sold off most of the estate's acreage. The top portion of this photograph shows the construction of a hospital on filled land that was once the Lagoon Garden. (Vizcaya Museum and Gardens Archives)

at Vizcaya would provide Miami with an instant resource and attraction for both its residents and visitors. The sisters accepted payment in bonds to be paid off over time through admission fees rather than an upfront payment. They did not need the money themselves, which was particularly true for Marion McCormick, who was among the ten wealthiest women in the country thanks to her stake in both the Deering and McCormick family fortunes.[96] Whether they considered other options is unknown, although it is obvious that doing something culturally beneficial with the property took precedence over making a lot of money. As McCormick noted in a 1953 letter to Hoffman, "Even up to the last minute, we could have sold the property of Vizcaya to be used as a subdivision at far greater financial benefit to us than having it preserved as a museum."[97]

Vizcaya, Dade County Art Museum

Vizcaya finally opened to the public in 1953 bearing a new name extension: Dade County Art Museum. It emerged as a self-sustaining operation of the Board of County Commissioners under the oversight of the county's parks department. A committee made up of the director of Dade County Parks, the director of the Art Institute of Chicago (when Chauncey McCormick was its president), and a local citizen referred to by McCormick as the family's "representative in Florida" was put in charge of running the museum.[98] The committee hired Robert Tyler Davis, an art professor at McGill University and former director of the Montreal Art Museum, as the director of Vizcaya. Although the family remained engaged, the stewardship of the property was now a civic enterprise.

What was left of the estate followed the example of the Biltmore Estate, which opened to the public in 1930 at the request of local city and county administrators, who thought doing so would help attract tourists and boost Asheville's struggling Depression-era economy. Unlike the Deering family, however, the Vanderbilts maintained ownership and lived in their mansion for decades to come, even as Biltmore lost $250,000 per year owing to the cost of upkeep.[99] Even with the over $1 million bond remaining unpaid, selling Vizcaya to Dade County enabled Deering's heirs to shed most of their responsibility for the property, thus preventing the ornate house and elaborate landscape from becoming their very own white elephant.

Miami was still relatively young and small at the time, and sustaining an expensive art museum proved a challenge. The county levied a $2 admission fee, which was almost on par with the Biltmore's fee of $2.50 around the same time.[100] In 1954 a group of unpaid women volunteers formed the Vizcaya Volunteer Guides to serve as the public face of the property by leading tours and relating history and information to visitors. This helped a great deal because over the next year, between two hundred fifty and three hundred people visited Vizcaya on a typical day. While attendance was promising, it did not provide the income required to meet county budgets.[101]

A caveat to the otherwise favorable deal exacerbated this shortfall by dictating that Vizcaya had to earn its way—there would be no county tax money spent to maintain the property. As Davis explained, "Never in the history of the country has anyone tried to operate something of this sort so as to be self-sustaining and pay off the cost, too."[102] If there was a bright side, it was the consistently fresh audience provided by the city's booming tourist industry. It also helped that Deering's nieces were lenient with their demands for payback.

As was reported about Vizcaya in 1955: "The physical property is in wonderful condition. During the thirty years since Mr. Deering's death his heirs have kept constant vigilance against the ravages of time."[103] It is unlikely everything was quite that rosy, however. The oppressive heat and humidity of South Florida continued to challenge plantings in the gardens. That year, ticket sales and limited fundraising brought in around $180,000, three quarters of which went directly to paying operating costs. Little money was left over to fund needed improvements. As the museum struggled financially, its future as a viable, self-sustaining museum lingered in uncertainty.

The Arrival of the Vizcayans

As has been the case with numerous cultural institutions, Vizcaya garnered needed support from a volunteer workforce eager to assist with public outreach and fundraising. In 1956, seventeen years after the formation of the Friends of the Garden at the Missouri Botanical Garden, a few dozen presumably well-to-do women concerned about the estate's deteriorating condition gathered with Davis on the villa's porch and committed to providing tangible support to the museum. Within a few months, county commissioners unanimously approved the group's charter and formally established the Vizcayans as a nonprofit corporation able to, among other things, "provide for the care, maintenance, support, continuance and success of Vizcaya as an art museum."[104] The Vizcayans soon became the estate's public face and established a direct connection to Vizcaya for people in the community who appreciated art and design and, importantly, had money to donate to what the group passionately believed was a worthy cause.

Within six months, the Vizcayans had recruited more than four hundred members and raised over $5,000 to help pay for needed facilities at the museum.[105] They began hosting concerts, galas, open houses, tea parties, and other fundraising events, including the Medici Ball, which instantly became one of the most celebrated events on Miami's active social calendar. Local newspapers, eager to fill their society pages with cultural goings-on, sent photographers to capture images of elite Miamians once again using Vizcaya in the glamorous ways intended by James Deering decades earlier. The group's initial focus was on interior restoration projects, while the landscape, which was also consistently filled with attendees for weddings and events, seems to have been of lower priority.

Davis, whose leadership led to the creation of the Vizcayans, left in 1957 and was replaced by Robert Hunter, another transplant from Canada with experience in fine arts, museums, and galleries and, like Davis, an apparent fixation on interiors over the landscape.[106] But frustration over Vizcaya's high costs persisted. Calls to sell the property became more frequent and more prominent, with one county commissioner decrying the villa as just "a collection of rooms furnished in bad taste."[107] Architects, designers, and art lovers reacted in anger and disgust, arguing loudly that the property was among the most culturally valuable works of design in Miami and certainly worthy of preservation. As was the case with City Hall Plaza in Boston, if landscape architects and garden enthusiasts expressed similarly strong reactions about the landscape, their efforts failed to attract the same attention.

The furor eventually died down, and the property was not sold. Tens of thousands of dollars were spent on improving the villa's interior and repairing several structures in the gardens. The Vizcayans contributed $2,000 per year to enhance the grounds with potted plants and annuals, while the county paid for basic maintenance.[108] A small work crew tended the gardens under the supervision of Sammy Sands, who started working at Vizcaya as a fifteen-year-old in 1914. He started as a "water boy" before moving into the greenhouses, where he learned about plants and was soon cutting and arranging flowers for Deering's friends and helping keep up with the constant demand for roses and floral annuals to be spread throughout the villa.[109] He worked as Vizcaya's head gardener well into his sixties before retiring in the mid-1960s. He was replaced by Roger Fosse, a Frenchman who had been a gardener at Versailles and had run a small landscaping business in Miami for several years before applying for a job with Miami's parks department in 1961 and, thanks to his garden-oriented background, was assigned to Vizcaya.

Fig. 13. By the 1970s, at which point the estate was the property of Dade County, landscape maintenance was reduced to minimal pruning and lawn mowing. Much of the ornate and horticulturally rich landscape created for Deering was lost. (State Archives of Florida)

Fosse worked for several years under Sands, who told him of Deering's desire for perfection and recalled how lavish the gardens looked during the days when there was no limit on money spent.[110] Once he took over as head gardener in 1965, Fosse managed a crew of ten undertrained landscape gardeners who, it appears, primarily mowed lawns, clipped hedges, and weeded beds. Aside from Fosse actively removing failing plants and replacing them with those better adapted to the region, the garden remained in relative limbo for several years. Things changed in the late 1970s, when several garden restorations were undertaken with fundraising and oversight led by the Vizcayans and design help from the Miami-based landscape architect John Riordan. Among the restored garden areas was the small Theater Garden, for which Riordan conducted extensive research and selected plant material better able to withstand saltwater impacts than what had originally been selected by Chalfin and Suarez. Restored gardens meant an increased need for maintenance, which was made more difficult by a reduction in staff by 1980, when Fosse's staff numbered only seven workers.

At his retirement in 1988, Fosse recalled that one of his best memories was designing the Garden for the Blind—the first new garden created since the days of Suarez. It was reportedly filled with fragrant plants and interpretive signs in Braille, but there is little documentation about its quality as a designed space or how long it lasted. The gardens were generally well kept during this time, although they displayed little of the colorful, flower-filled exuberance Deering loved. As Fosse once said, "It may not be exactly what Mr. Deering had in mind, but does it not look fine?"[111]

The Vizcaya Trust

Over the next few decades, Vizcaya remained under the direct management of the county. In addition to the Vizcayans and the volunteer guides, a third group, the Vizcaya Foundation, formed in 1979 specifically to create an endowment for the estate, which continued to struggle financially. During this time, the adjacent neighborhoods grew increasingly dense, engulfing the once private and isolated property within an expanding suburban fabric. The population of the Miami-Dade region reached two million by the early 1990s and, along with more locals and an exploding tourist population, Vizcaya's profile in the area continued to grow. The main property was listed on the National Register of Historic Places in 1970, with the adjacent village added in 1978. In 1994 the property achieved National Historic Landmark status, with its area of significance listed as Architecture, without reference to the gardens. While the villa was celebrated, the landscape continued to stagnate under the care of the underfunded parks department. Plantings were simplified and certain lawn areas were paved over to reduce maintenance and maximize event space. Following the pattern of previous storms, Hurricane Andrew inflicted nearly $2 million in damage on the property in 1992. The storm left behind fifty fallen trees and extensive damage to the Marine Garden, a popular space that was subsequently closed to the public for nearly twenty-five years.

In 1997, around 200,000 visitors paid the $10 admission fee to visit Vizcaya.[112] Although crowds were showing up, dissatisfaction with how the county was running the museum created more doubt over the estate's future. The landscape continued its decline, owing in part to a continued lack of dedicated staff. Deering had upward of fifty workers tending the landscape during his time; by the late 1990s, that number had been reduced to six. Along with the closed Marine Garden, the Maze Garden existed without its signature boxwoods because, like the roses in the Rose Garden, the plants had failed to survive the seawater inundation. For many years, the surrounding twenty acres of native rockland hammocks and mangrove swamps received essentially no care.

During this time, the museum's chief curator position remained vacant for nearly two years, deepening worries that a forthcoming review by the American Association of Museums would lead to a failed accreditation, especially because it had barely passed on its last attempt in 1987.[113] Recognizing that losing accreditation might mean the museum could lose access to millions of dollars in grant funding, the Vizcayans hired a consultant to advise on a course of action. The recommendation came back clear: Management of the museum and gardens must be taken from the county's parks department and placed under Vizcaya's volunteer board.[114] This advice proved prescient as, following its inspection, the American Association of Museums confirmed that Vizcaya's accreditation was indeed threatened if its governance system did not expand beyond the county department. The Miami-Dade Board of County Commissioners then appointed a municipal task force to study the governance structure. After considering options, the group recommended that Miami-Dade County retain ownership of the property and transfer museum-specific operations to a twenty-one-person trust within two years.

Although there was some opposition to the transfer of authority, the proposal eventually passed, and the Vizcaya Museum and Gardens Trust was created. The codified legislation asserts: "Miami-Dade County through the Board of County Commissioners will continue to own Vizcaya and oversee its operations" while the "Trust will establish and oversee Museum-specific policies to the extent required to operate, maintain, restore and preserve Vizcaya as an [American Association of Museums]–accredited institution and implement the Museum's mission statement."[115] Although focused on the museum, the change put the entire estate in a better position to function as a unique facility with specific needs rather than just one of many properties run by the parks department.

Among the first actions undertaken by the trust was producing a strategic plan and a master plan. While the strategic plan outlined broad strategies and goals, the master plan identified planning, design, and development opportunities for the entire estate, with a focused interest on the farming village across the street. The plan specifies around $100 million in needed restorations, renovations, improvements, and additions, nearly a quarter of them specific to the gardens. Restorations of sculptures, fountains, structures, lighting, irrigation, and drainage are specified, along with, apparently for the first time, an explicit goal to restore Suarez and Chalfin's historical planting scheme.

The Building Better Communities Bond and a Period of Restoration

Vizcaya's status as a municipal facility finally began to pay off financially in 2004, when at least partial funding for the initiatives outlined in the master plan became available. As Miami's metropolitan region grew toward 2.5 million people, the Board of County Commissioners proposed a $2.9 billion general obligation bond titled Building Better Communities to support needed improvements in the city and its neighborhoods. Utility and public realm infrastructure, parks, public safety measures, health care facilities, schools, libraries, and cultural institutions were all identified as priorities for the initiative, which passed by popular vote. Vizcaya was awarded $50 million to "restore and improve its historically landmarked main house and gardens to make them more accessible and enjoyable for future generations."[116]

Along with this new source of funding and a self-directed mandate to restore the gardens, the trustees retooled the maintenance staff to support the commitment to the landscape. In 2007, Ian Simpkins, who had most recently worked as the executive director at Shofuso Japanese Garden in Philadelphia, was hired as Vizcaya's first chief horticulturist.[117] Although there had previously been chief gardeners on staff with backgrounds in horticulture, none before Simpkins had held a formal degree. He was enticed by the opportunity to bring back such a significant historical garden and to start a horticulture division essentially from scratch.[118] When he arrived at the estate, the landscape looked to him like a poorly tended cemetery—sad and neglected, with basically no horticulture being practiced.[119] His initial challenge was to transform a group of park attendants trained for basic landscape maintenance tasks into a team capable of restoring the landscape's rich legacy.

That same year, the integrity of the estate's design faced a threat increasingly common to landscapes in dense urban areas. A developer proposed constructing three residential towers, each more than three hundred feet tall, less than two thousand feet from the formal gardens, generating fears of ruined views and dark shadows. The proposal drew the ire of many people, including Olin, whose recently published book about Vizcaya had sparked a renewed interest in the estate. In a letter to the *Miami Herald,* the landscape architect explained the importance of contained views within what was conceived as a "private tropical refuge," and argued it would be "a tragic mistake to destroy these iconic views with the addition of banal contemporary structures."[120]

The Vizcayans joined the fray and commissioned a viewshed analysis that included rendered perspectives showing the visual impact of the proposed towers from different locations within the property. The study provided compelling imagery and echoed Olin's opposition by concluding the imposing towers would destroy the gardens' early twentieth-century ambiance.[121] Although the argument struck many as convincing, the Miami City Commission initially approved the project for construction. Thanks to zoning technicalities, however, the project was rejected in 2008 and the towers were never built.[122] Not coincidentally, that same year the National Trust for Historic Preservation listed Vizcaya among its most endangered historic places because of this threat and the likelihood of future developments in the area. The organization also awarded its President's Award to the Vizcayans, celebrating the group as "an unstoppable group of dedicated volunteers who have made a positive mark on their community for over fifty years by protecting and preserving Vizcaya."[123]

While this drama played out, restoration of the structures and the plantings in the gardens continued. Simpkins referred to historical photographs and early drawings for inspiration. He wanted contemporary plantings to be as authentic as possible while also more effusive, filled with colorful plants able to thrive in the specific conditions of the site.[124] As was seen with the later work of Chalfin and then Fosse, stewardship of the formal gardens has never been explicitly about maintaining the original design with precision. More important has been finding ways to respond to the heat, humidity, salt spray, and seawater in ways that support Deering's vision of a formal European garden in the wilds of South Florida by increasing the site's horticultural diversity and complexity.

To boost the quality of the landscape, Simpkins instituted a policy requiring his entire staff to receive horticultural training, thus establishing a baseline of knowledge and expertise that had never previously existed. With an influx of financial resources from the bond issue, the staff embarked on a period of landscape improvement projects, an early one being the restoration of the Marine Garden, which reopened to the public in 2016. Simpkins also spurred a renewed focus on the remaining natural areas that Deering deemed integral to Vizcaya's landscape character. This meant diverting maintenance resources away from the ornamental landscapes to focus on eliminating and preventing harmful invasive plants that threatened the health and appearance of the entire landscape. In recognition of his positive impact on the gardens, Vizcaya's executive director promoted Simpkins to deputy director for horticulture and urban agriculture in 2016, a move that, according to the formal announcement, "reflects the importance of our formal gardens, natural areas, and living collection to the mission of Vizcaya."[125]

Public-Private Partnership and Hurricane Irma

Vizcaya presents an interesting take on the expanding role of public-private partnerships in which a municipal government rescues a private entity, rather than the more typical situation of a private organization forming to restore the condition and legacy of a public landscape. This arrangement changed somewhat in July 2017, when Miami-Dade County commissioners approved a governance transition at the request of Vizcaya. The deal codified a formal public-private partnership, similar to the one created between the City of New York and the Central Park Conservancy, in which the county retained ownership of the villa, grounds, and art collection, while management and operations were privatized in a not-for-profit organization called Vizcaya Museum and Gardens Trust, Inc. Under the new system, the county continued to support the museum with a $2.5 million annual appropriation. Since it was no longer a direct public entity, Vizcaya could now apply for more grants and attract and maximize the benefits of philanthropy, which soon played a larger role than ever before. The trust also collected the remaining $8 million owed from the

Building Better Communities bond, $30 million of which had been spent on capital improvements. At a practical level, all staff members now worked for the private organization, which meant that Simpkins gained even more control over his horticultural crews as they continued with restorations.

County commissioners approved Vizcaya's new master plan at the same 2017 meeting. The document emphasizes the restoration of the twelve-acre farming village on the west side of the property and recommends opening it to the public as an educational and cultural hub. For decades, various county agencies had used Vizcaya Village as office space. It was also home to a county-owned science museum that relocated to downtown Miami in 2016. Although the original buildings designed by Hoffman and Chalfin remained architecturally significant, they had deteriorated and required extensive restorations. Expanding green space, restoring the now rare native rockland hammock and pine groves, and preserving the property's massive and exotic banyan trees are identified as goals for the landscape.[126]

With ambitious plans in place, the 2017 summer storm season once again tested Vizcaya's fragile landscape. Hurricane Irma, one of the strongest storms ever observed in the Atlantic Ocean, pummeled Puerto Rico and the Caribbean before targeting Miami in early September. Even though the storm slid westward, avoiding a direct hit on the mainland, intense rain and a storm surge flooded the gardens, temporarily submerging them beneath four feet of saltwater. Irma left Vizcaya with millions of dollars' worth of damage and copious amounts of seaweed and debris piled across the landscape. Thanks to a previously established program of preventative pruning, not one tree was lost in the gardens. The native rockland hammock, however, once again suffered extensive damage. It took months of cleanup from staff and volunteers to get the landscape back in order.

On a positive note, the damage provided the impetus and opportunity to proceed with needed restorations in the gardens. Efforts included replanting the historical parterres at the base of the steps leading down from the villa's south terrace. Garden staff, with help from three professionals and an intern sent from Longwood Gardens near Philadelphia, removed the plantings that had, over decades, veered significantly from the original 1917 design. The parterres were originally planted with boxwoods, which failed horribly. William Broomfield, Vizcaya's first head gardener and the scion of a family of English gardeners, had trouble applying his knowledge of temperate zone horticulture to the semitropics, yet eventually learned a lot by trial and error.[127] He found success with a vigorous and native jasmine vine, which found a permanent home in the parterres. Simpkins and his staff used historical drawings as guides to accurately recreate the formal landscape, albeit with a contemporary brand of turf better able to withstand the conditions.

A similar process took place in the Rose Garden, which had been renamed the Fountain Garden, likely because the namesake plants loved by Deering for their vibrant flowers had not grown there since the original planting perished in the 1926 hurricane. After Irma, a new initiative was undertaken to reestablish the space's historical intent. With assistance and donations from Miami's Tropical Rose Society, numerous varieties of roses were tested in hopes of discovering species better able to adapt to the salt spray affecting the site's already alkaline soils. In September 2019 the garden reopened, filled with flowers in shades of pink, yellow, red, and fuchsia in a layout matching what existed in 1922.

Contemporary Issues

In 2020, Vizcaya suffered the impacts of the coronavirus pandemic along with the rest of the world. By that time the museum and gardens were thriving. Some 300,000 visitors and nearly one hundred lucrative corporate events and weddings helped bring in $4.5 million annually.[128] As the virus took hold, the museum and gardens closed for nearly two months in mid-March, at the height of the tourist season. After a partial reopening in May, visitation remained a fraction of what it had been the previous year. The museum cut salaries, drew money from its reserves, and reduced staff hours to make up for financial losses. Staff layoffs were avoided thanks to a greater than $1 million loan from the federal Paycheck Protection Program.[129] Vizcaya's annual operating budget at the time was over $12 million, $1 million of which went to maintaining the extensive gardens. By the fall of 2020, the institution had a $3 million deficit and faced an uncertain future. A plea for support from public donations produced another $1 million, improving the museum's financial outlook and confirming the depth and loyalty of Vizcaya's public support.

While COVID-19 delivered a short-term challenge, climate change looms as a constant and existential threat to Vizcaya and its region. As oceans warm and sea levels rise, more frequent and increasingly wet and powerful tropical systems are predicted to bring catastrophe to South Florida. King tides, the uncommonly high tides that typically arrive with full and new moons, are already causing more flooding across Miami's low-lying neighborhoods as freshwater pushes up from the subterranean aquifer through the sewer system. These tides increased 400 percent between 2010 and 2020, and if they continue to escalate at a similar rate, it could mean parts of the city will experience around 150 tidal floods per year by 2045.[130] Even with updated building codes and standards anticipating a future of changing conditions, there is a legitimate question as to what, if anything, Miami can do to effectively stem the impending damage. In Coral Gables, Vizcaya's community and home to the area's most expensive real estate, tropical storms pose peril because most of the mangrove swamps that act as nature's protection against storm surges were destroyed to build highly coveted neighborhoods. Additionally, the sea level could rise an estimated two feet as soon as 2060, which would submerge some of Miami's coastal neighborhoods.

Vizcaya's willingness to adopt innovative methods to protect the villa and the gardens from the impacts of climate change presents a healthy example to other institutions facing a similarly uncertain future.[131] Even so, it is likely the gardens will continue to be assaulted by tropical storms and rising sea levels until it no longer resembles the landscape created by Deering, Chalfin, and Suarez or, worse, is gone for good. Despite the economic, managerial, and environmental resilience it has shown over time, there will surely come a time when decisions extend beyond how to steward the property's legacy and become more dire. At some point, questions of how best to move forward or adapt will be replaced with decisions about whether to sacrifice the property to the inevitable forces of nature.

As South Florida bides its time in the face of pending environmental doom, Vizcaya exists as a sparkling attraction, a snapshot depiction capturing a unique era in American history. As is the case with many similar historic properties, the goal is to capture the essence of a specific place in time with a certain degree of authenticity. In this case, it was a period when excessive wealth became translated into landscape and architectural expression. While many things have changed since Deering's age, visitors from all backgrounds can get a glimpse of what life was like for one of Florida's wealthiest residents on grounds designed by a talented group of men and subsequently managed and stewarded by countless staff members, elected officials,

Fig. 14. After a series of restorations partially funded by a public bond, the formal gardens were restored to their original quality beginning in the early 2000s. (Photo by author)

and volunteers. As is often a primary goal with such places, the coiffed grounds and polished structures experienced today disguise the litany of challenges faced by individuals struggling for more than a century to manage and preserve, to some extent, a landscape created at the whim of a single man.

As is true for the Missouri Botanical Garden, Vizcaya's continued relevancy will depend on how well it balances a desire to respect and celebrate its design legacy with the need to adapt to changing climatic and cultural realities. As of today, both gardens are showcases of design, horticulture, and history, with neither exhibiting overt scars of past struggles. Certainly, not all historic gardens are fortunate enough to have such a solid institutional backbone or to have been rescued by a flourishing municipality. The gardens that manage to survive and thrive beyond their original design or patronage are likely outnumbered by those that have been lost forever. As seen in the ongoing advocacy and work of the Garden Conservancy and TCLF, the challenges that threaten many designed landscapes persist. Fortunately, gardens are uniquely loved as American landscapes and, whether they are intended for knowledge or pleasure (or both), their display aspect and the recognized preciousness associated with their history often promote a stewardship ethic and a level of respect not always enjoyed by parks, campuses, or community spaces.

2

CULTURAL SITES

A wave of nostalgia, like a convulsive shudder, is passing across the United States, leaving in its wake hundreds and possibly thousands, before it's over, of reconstructed, resurrected, restored, or preserved buildings and landscapes.

—Grady Clay, editor of *Landscape Architecture Magazine,* 1976

Broadly drawn, *cultural sites* can refer to many different landscapes that serve or reflect a specific community or culture. The U.S. Department of the Interior, home to the National Park Service (NPS), defines a *cultural landscape* as a "geographic area (including both cultural and natural resources and the wildlife or domestic animals therein), associated with a historic event, activity, or person or exhibiting other cultural or aesthetic values."[1] Accordingly, the term may apply to landscapes with cultural significance, such as the birthplace of an important historical figure, a religious site, agricultural land, or the site of a historic event, such as a battlefield. More narrowly, the term "cultural site" refers to *historic designed landscapes,* which are more relevant to this book and are what the NPS defines as places "consciously designed or laid out by a landscape architect, master gardener, architect, engineer, or horticulturist according to design principles, or an amateur gardener working in a recognized style or tradition."[2] Because such categorization typically applies to landscapes addressed in other parts of this book—gardens, parks, and campuses in particular—this chapter focuses on places that do not easily fit within other type categories.

It would be nearly impossible to dissociate the subject of America's cultural landscapes from the work of the NPS, which is the most vital and influential steward of the United States' combined cultural, historical, and natural landscapes. As of 2022, four hundred parks, historical sites, battlefields, recreation areas, monuments, and other natural and cultural resources managed by the NPS covered more than 84 million acres across the country. Although it is best known for managing the iconic scenic parks, monuments, and parkways that attract tourists from around the world, since the later decades of the twentieth century the NPS has also been a leading force in the stewardship of designed and cultural landscapes. During the 1970s an emerging interest in historical parks, gardens, and other sites made evident that many of these landscapes, both famous and obscure, had been lost, destroyed, or, in relatively fortunate situations, merely left to suffer in a spiral of degradation. Before plans to manage these places could be considered, the NPS embarked on a lengthy process that produced troves of methods for and policies bearing on their rescue and restoration. Intertwined with these efforts was the concurrent birth of the contemporary practice of landscape preservation and, ultimately, a deeper understanding of the need for responsible stewardship.

Without attempting to dissect the labyrinthine composite of acts, programs, executive orders, and reorganizations and the reams of documents and policies that make up the complex history of the NPS, it is important to recognize certain events that explain the agency's pivot toward designed landscapes. Aside from Congress's establishment of Yellowstone National Park in 1872 as the first national park, the most significant early action was the 1906 passage of the American Antiquities Act, which was the first law written to explicitly protect historical features on federally owned land. As enacted, the law authorizes the president of the United States to "declare by public proclamation historic landmarks, historic and prehistoric structures, and other objects of historic or scientific interest that are situated upon the lands owned or controlled by the Government of the United States to be national monuments." This language effectively acknowledged that even though the country was still young, at least in its cultural makeup, there were historical places that deserved protection.

A decade later, Congress passed the Organic Act of 1916, which created the NPS expressly to promote and manage national parks, monuments, and reservations in such a manner as to "conserve the scenery and the natural and historic objects and the wild life therein and to . . . leave them unimpaired for the enjoyment of future generations." This was a time of international war, and American citizens were encouraged to "See America First" and avoid traveling abroad. National parks, though as yet ill-prepared to handle large crowds, were touted as attractive destinations. A *New York Times* article supporting the act declared, "In future years it is to be hoped that the parks may become . . . the playgrounds and meeting places of the country" where people "may meet in a common admiration of the greatness of their country . . . and draw a new appreciation and a deeper understanding of their common heritage."[3] Although this "common heritage" surely did not include Indigenous peoples or those brought to the United States or raised as slaves, the act represented a positive step toward using public lands as unifying cultural features. Prior to this, management was assigned to an assistant in the office of the secretary of the interior with a small, primarily part-time staff. Now, as an official entity within the federal government, the NPS was a formidable bureaucratic agency committed to its evolving mission.

In 1933, then president Franklin D. Roosevelt issued an executive order that expanded the purview of the NPS by consolidating more than sixty existing federal parks, monuments, and historical sites under its control. This move meant that in addition to stewarding the famous natural parks, the service was now a curator of the nation's history, with increasing oversight of such cultural sites as cemeteries, battlefields, memorials, and historical parks. That same year the Historic American Buildings Survey (HABS) began as a Depression-era NPS program to employ hundreds of out-of-work architects, historians, and photographers to document the country's historic architecture. In 1934 the NPS entered into a partnership with the American Institute of Architects (AIA) and the Library of Congress to formalize HABS as a tool for researching, photographing, documenting, and cataloguing these historical resources in a national archive. Although a version for historic landscapes remained decades away, an ethic was developing that would one day apply to them as well.

According to R. Newton Mayall, a landscape architect and civil engineer who started working on HABS at its inception, more than one thousand historic structures built before 1840 had been measured and recorded by 1935.[4] During this same period, only one landscape received the same treatment.[5] This began to change beginning in May of that year, when funds were made available to document gardens in Massachusetts. Mayall notes the complexity involved in tracking the history of residential gardens that may have changed over time. "Where the present layout is somewhat changed from the original, we endeavor to trace back and gather sufficient material to make possible a truthful restoration if anyone wishes to do so," he wrote in a 1935 issue of *Landscape Architecture Magazine*.[6]

Momentum continued during the summer of 1935, when the Preservation of Historic Sites Act made it official federal policy to "restore, reconstruct, rehabilitate, preserve, and maintain historic or prehistoric sites, buildings, objects, and properties of national historical or archaeological significance."[7] The charge was not only to protect these resources but to research and document them for the sake of making them available for future generations. Over time, assumptions about historical resources shifted. Previously viewed as objects of static preservation, sites and features managed by the NPS became better understood as parts of evolving systems.

The first NPS Historic Grounds Report defining treatment plans specific to landscapes was adopted in the 1960s, by which time the service was overseeing three categories of places: natural, historic, and recreational. In 1964, then secretary of the interior Stewart Udall signed a policy memorandum stating, "Each of these categories requires a separate management concept and a separate set of management principles coordinated to form one organic management plan for the entire [Parks] System."[8] Natural landscapes would be managed to be as close to their natural state as possible, allowing for the preservation of any historical artifacts or features where possible. The reverse was to be true of historic sites, where cultural heritage was prioritized over nature. Preservation was growing in stature within the NPS at the time, as can be seen in the 1966 passage of the National Historic Preservation Act, which established a national register of historic districts, sites, and structures—a program to be owned and maintained by the NPS. The definition of historic preservation was expanded to include "the protection, rehabilitation, restoration, and reconstruction of districts, sites, buildings, structures, and objects significant in American history, architecture, archeology, or culture."[9] Reaching an acceptable balance between preserving nature and history while promoting landscapes for scenic tourism and recreation proved challenging, as seen in a 1969 memorandum written by Udall's successor, Walter J. Hickel, which states, "There are serious gaps and inadequacies which must be remedied while opportunities still exist if the System is to fulfill the people's need always to see and understand their heritage of history and the natural world."[10]

During his presidency, Richard Nixon was an ally of national parks and the NPS. In 1971, in reference to his "Legacy of Parks" program, which greatly increased the federal budget for parks and open space, he proclaimed: "It was in 1872 that Congress established our first national park at Yellowstone and laid the foundation for the creation of other parks in later years. That was a time when Americans were more interested in taming the wilderness than preserving it; yet farsighted and sensitive men and women were able to begin the great work of preservation even then."[11] By this time, the population of the United States had grown to 205 million people and the number of recorded visits to national park areas had soared to 170 million per year. As expanding cities and new suburban developments quickly ate into land that had once seemed limitless, Nixon's program sought $200 million to purchase land in both urban and wilderness areas for conservation and recreation. He also signed an executive order that expanded the National Register of Historic Places to strengthen protection of threatened historic resources. "The Federal Government shall provide leadership in preserving, restoring, and maintaining the historic and cultural environment of the Nation," it stated.[12]

Sixty-one historic sites were added to the parks system between 1952 and 1972. These were generally landscapes tied to a historical event or to the life of a notable person rather than recognized works of design. They included twelve sites specifically related to presidents, along with an assortment of forts, battlefields, birthplaces, and other locations deemed historically significant enough to be placed on the National Register. This growing interest in history was boosted by the patriotic fervor that accompanied the nation's forthcoming Bicentennial in 1976. This was the same year the Secretary of the Interior's Standards for Historic Preservation Projects were created to establish consistency in the acquisition, protection, stabilization, preservation, rehabilitation, restoration, and reconstruction of historic buildings. Although historical and cultural landscapes would not be given the same standards for decades to come, interest in them was growing.

Even as the government's prioritization of natural parks and open spaces raised a significant hurdle for those promoting the value of historical, cultural, and designed landscapes, momentum proved to be on the side of history. As expressed by the art and architecture critic Wolf Von Eckardt in the *Washington Post,* "The federal government is not the major force in preservation. What has made the historic preservation movement strong enough to change the face of our cities, is precisely that it is a movement, a mighty popular movement."[13] An early example is a conference on the preservation and restoration of historical gardens and landscapes that convened in 1975 at Dumbarton Oaks in Washington, D.C. Jointly sponsored by the National Trust for Historic Preservation and the American Horticultural Society, the event sought to professionalize the emerging interest in historical landscapes and emphasized the need to approach restoration using scientific methods.

In June 1978, twenty landscape-related professionals from the United States and Canada gathered for a small symposium on the future of landscape preservation. Held in New Harmony, Indiana, and supported by a grant from the National Endowment for the Humanities, the symposium proved influential enough to produce a new organization, the Alliance for Historic Landscape Preservation. From its beginning, the alliance made interdisciplinary membership a core part of its mission and included landscape architects, historians, archaeologists, geographers, architects, and lawyers. Its focus on education through conferences and publications helped raise awareness and build support for landscape preservation.[14]

This interest in historical places also expanded the discipline of landscape architecture. The purview of trained landscape architects, once known primarily for designing new places, increasingly involved the care of established landscapes. As expressed by the renowned preservationist and architect James Marston Fitch in 1976:

> It is apparent that the preservation of historic landscapes . . . confronts the profession [of landscape architecture] with new and exciting challenges. At the same time, however, it demands conceptual changes . . . subtle but critically important shifts in perspectives, from emphasis on the art of creating to that of

> care-taking. It involves intervening in the life of extant organisms, each with its own pre-history, as a medical doctor works with a patient; and not creation *de novo,* like the sculptor or painter.[15]

As the broader profession evolved, so did the NPS. In 1980, the year the Central Park Conservancy was established in New York City, an amendment to the National Historic Preservation Act dictated a more fluid policy for administering historical resources. Whereas language in the 1966 version refers to the "preservation, acquisition, and development" of properties, the revised act speaks in broader temporal terms by dictating a "spirit of stewardship for the inspiration and benefit of present and future generations."[16]

Momentum continued throughout the 1980s. In 1981 the NPS's first codified reference to "cultural landscapes" appeared in the second release of the *Cultural Resource Management Guideline.* Jerry Rogers, who worked at the NPS for thirty-four years and served under three presidents, was appointed associate director of cultural resources in 1983. Among his first moves, he gathered around twenty academics and private and public practitioners to consider the service's role in the preservation of cultural landscapes. Out of this meeting came a proposal to document two prototypical Washington, DC, landscapes. First came Meridian Hill Park, the formal, neoclassical landscape initially designed by George Burnap and later modified by Horace Peaslee between 1912 and 1940. The start of this study was followed one year later by the study of Dumbarton Oaks Park, the twenty-seven-acre naturalistic landscape in Georgetown designed in the 1920s by Beatrix Farrand (see chapter 5). These two seminal projects energized landscape architects and helped instigate nearly two decades of lobbying and advocacy for more stringent processes for preserving and managing cultural landscapes. The NPS responded by introducing and codifying the Cultural Landscape Report (commonly known as a CLR) in the 1985 release of the *Cultural Resource Management Guideline.* CLRs soon became the official means for documenting and proposing treatment options for historical and cultural landscapes.[17]

Designed landscapes finally came to the forefront in 1987 with the publication of a technical article titled "How to Evaluate and Nominate Designed Historic Landscapes" and published in the NPS's *National Register Bulletin.* It lists seventeen types of designed landscapes, including various forms of gardens, parks, campuses, zoos, planned communities, cemeteries, estate grounds, and recreational fields. Because the landscapes in this list differed from the expected NPS landscapes made special by their natural scenic beauty or relation to a specific historical event, such as parks or monuments, they required new criteria. They were typically decades- or even centuries-old landscapes whose value was concealed beneath the residue of natural degradation or human neglect. Determining their relative merit meant identifying and describing what they were when supposedly at their best—their period of significance—and not their current condition. Beyond historical significance alone, these landscapes would require a certain degree of authenticity in relation to their original state to be considered. Preservation, therefore, required first performing some version of rescue to establish a historically responsive baseline to which new interventions would relate.

By the end of the 1980s, the NPS's commitment to designed landscapes was set, and management policies had been crafted to direct work specifically for this new type.[18] An important change during this time was the increased recognition that a site's natural features could be as influential as its built or architectural ones. In 1992 the service created the Historic Landscape Initiative under its Heritage Preservation Services Program and hired Charles Birnbaum as its first coordinator. Historical and cultural landscapes were increasingly given the same attention as other properties managed by the NPS, and Birnbaum spent the next few years leading the creation and documentation of processes dictating the preferred treatment of designed landscapes. That same year the NPS established the Olmsted Center for Landscape Preservation as a resource for protecting and preserving cultural landscapes under the service's management. It began as an emergency measure to prevent the deterioration and loss of parks, gardens, and other landscapes by providing technical support and advice. In time, it became a leading expert in the documentation of CLRs and similar reports.

As the NPS honed this aspect of its mission, landscape professionals more broadly were intensifying their interest in historical and cultural sites. Public and private practitioners, historians, and educators in the American Society of Landscape Architects (ASLA) formed the Historic Preservation Open Committee, which helped guide the implementation of work being defined by the NPS. Critical to the development of an effective interdisciplinary field of landscape preservation was the inception of a common language and ethic that could unify disparate efforts. Historic landscapes, once considered as set in one place in time, like a building, were increasingly understood as inherently dynamic. As the noted preservation landscape architect Patricia O'Donnell said in 1994, "The whole issue of freezing a landscape in time is passé."[19] Figuring out the practical implications of this change fell to the NPS, which released its "Guidelines for the Treatment of Cultural Landscapes" in 1996 to "provide guidance to cultural landscape owners, stewards and managers, landscape architects, preservation planners, architects, engineers, contractors, and project reviewers prior to and during the planning and implementation of treatment projects."[20]

Both internal staff and outside consultants produced CLRs for numerous landscapes across the country. These documents became proof that a more rigorous process was needed to determine when and how to replace missing features, remove those deemed incompatible, and, importantly, make compatible alterations to accommodate changes that occurred over time. To formalize this documentation, the NPS created the Historic American Landscapes Survey (HALS) in 2000 to document historic landscapes in the same manner as HABS did for buildings. A year later the NPS signed a memorandum of understanding with the ASLA and the Library of Congress to formalize HALS as a joint effort to "preserve, protect and interpret America's significant and threatened historic landscapes."[21] While the ASLA provides technical guidance and the Library of Congress preserves documentation and makes it available to the public, the NPS is tasked with selecting and approving sites for documentation, standardizing processes, and administering the program's planning and operation.

Perhaps the most significant aspect of HALS and the work the NPS performs related to designed landscapes is that the sites selected and documented are not limited to places designed by noted designers or those embodying an expected aesthetic. In this sense, the process represents forgotten or threatened landscapes that are important, on some level, to the history of the country, regardless of provenance or design quality. The case studies in this chapter reflect this as well as those landscapes with significant cultural histories that do not fit easily within the other chapters of this book. The first, Alcatraz Island, is a prime example of a landscape with a near mythical cultural history yet without a notable design pedigree. Famous as the site of the notorious former federal prison in San Francisco Bay, it is one of the most intriguing places that have benefited from the NPS's rehabilitation and preservation efforts. By using historical photographs and correspondence as guides, NPS staff recovered and meticulously reestablished gardens that once thrived and later suffered in accordance with the island's changing population. Now, through emerging challenges wrought by climate change, the Gardens of Alcatraz are cared for by committed professionals and hordes of dedicated volunteers determined to maintain the landscape for the sake of historical integrity and public enjoyment.

Of course, the NPS is not solely responsible for America's cultural sites. There are many other places of significance maintained and managed in other ways. One of these is the second case study, Mount Auburn Cemetery, the beautiful, 175-acre Boston-area landscape that serves as this chapter's second case study. While it shares obvious visual traits with botanical gardens and urban parks, its interment function and mission to inspire visitors, provide solace for the bereaved, and commemorate the dead in, as it self-describes, a "landscape of exceptional beauty" makes it a unique place.[22] It was created in response to a critical community need and has evolved into one of the most beloved landscapes in the country. Like the Missouri Botanical Garden, it is a venerable institution that has thrived for nearly two centuries with a clearly articulated purpose and a stability maintained by effective leadership and management. This consistent governance, along with its traditional gardenesque character and location in a densely populated urban region, makes Mount Auburn distinctly different from Alcatraz, which sits isolated in San Francisco Bay and is reachable only by ferry. Considering the two landscapes in tandem makes evident that despite the differences between them, there are shared issues and experiences relevant to all cultural landscapes.

Alcatraz Island, San Francisco, California

Alcatraz Island is among the most intriguing cases in the NPS's ever-expanding archives. Best known as "The Rock," the inescapable Prohibition-era prison site, Alcatraz is unique among the landscapes and stewardship stories in the United

States and holds an iconic place in the annals of American history. It is one of several islands that emerge from San Francisco Bay, each a hill or a ridge left unflooded by the rising waters that came with the end of the last ice age. The 22.5-acre landform was an exposed and barren ledge of greywacke sandstone, home to nesting cormorants and pelicans, when Spanish explorers encountered it in the 1770s. The Mexican government laid claim to the island after it acquired California and gained its freedom from Spain in 1822. The United States took control in 1846, a couple of years before the dramatic influx of wealth from the Gold Rush. Recognizing its strategic location at the mouth of the bay, President Millard Fillmore signed an executive order in 1850 that reserved Alcatraz for public use. Over the next decade, massive earthworks transformed the island with roads and infrastructure and the erection of the first lighthouse on the West Coast. It served as a fortress, protecting the growing region in an increasingly globalized world.

In addition to the famous cellblock building perched on the island's high point, Alcatraz possesses surprisingly lush gardens spreading within the shadow of prison cells and scaling the steep slopes of the windswept landscape. The gardens' design legacy is not particularly noteworthy, having attracted no signature or even professional designer. Instead, its significance is rooted in the island's cultural history, exemplifying the civilizing power of landscape and the importance of advocacy and collaboration in the stewardship process. Today the island is a tourist magnet that attracts more than one and a half million visitors per year. While the gardens remain relatively subtle within the magnitude of the famous site—a surprise more than a draw—they represent a landscape anomaly with a singular history. Without the intervention of the NPS and the efforts of hundreds of volunteers, however, it is unlikely the gardens would still exist.

Establishing the Legacy

Beginning in 1861, Alcatraz was converted into a military and civil prison that housed soldiers convicted of crimes and infractions and people identified as disloyal secessionists during the Civil War.[23] During this time, the island faced decades of construction, and space became constricted with the addition of barracks, officers' quarters, prison buildings, shops, utility plants, and other utilitarian structures. When the iconic cell house, perched prominently on the island and seen clearly

Fig. 15. This photograph from the early 1900s shows Alcatraz Island's development as a military prison and the difficult conditions for establishing gardens. (National Park Service)

from San Francisco, was built in 1912, it was the largest steel-reinforced concrete structure in the world. The land around the buildings gradually evolved into a series of gardens that provided officers and their families, inmates, and prison officials access to immediate beauty in what was otherwise a difficult environment, albeit one with great views across the bay.

The original planting medium consisted of soil transported across the bay from the mainland and nearby Angel Island, not to cultivate plants but to secure cannons in place. Embedded in the soils was an abundance of seeds from various indigenous and adaptable plants, including coyote bush, blackberry, and the vivid California poppy.[24] Over the years, a variety of ornamental flowers, shrubs, and trees were selected and planted based on availability and in response to individual tastes. With no source of freshwater aside from moisture from the frequent fog banks and winter rains, water for drinking, bathing, laundry, cooking, and irrigation was barged in to fill a variety of cisterns.[25]

There were different periods of influence in the gardens. During the island's military years, the wives of army officers brought with them highly ornamental flowers such as roses and irises to plant in ornate Victorian-style gardens next to the houses in which they raised their children. Residents planted other gardens to brighten spaces across the island, and the prison inmates who had gained the status of (prisoner) trustee tended compact plant beds filled primarily with annual and perennial flowers grown from seed. A letter published in 1918 in the Alcatraz newsletter, *The Rock,* describes the gardens as such: "The visitor who comes here expects to find a barren rock, but as he strolls over it he is surprised to find roses in bloom, sweet peas, lilies, and a large variety of other flowers in all their beauty and fragrance. . . . In this way barren wastes are converted into garden spots, and ugliness is transformed into beauty."[26] This beautification effort was enhanced in 1924 when, as part of an initiative undertaken to improve views to the island from across the bay, prisoners planted hundreds of trees and shrubs and spread many pounds of nasturtium and poppy seeds donated by the California Spring Blossom and Wildlife Association.

Life as a Federal Prison

When crime rates in America soared during the Great Depression, and the culture of gangster violence emerged out of the Prohibition era, the Bureau of Prisons took over Alcatraz and turned it into its most famous iteration as home to convicted criminals deemed too dangerous to incarcerate elsewhere. Federal prison employees arrived on the island in 1934 and encountered an environment shrouded in fog and described as "drab and threatening."[27] Remnants of the earlier gardens could be seen, especially on the large slope facing San Francisco, which had been covered with a dense mat of pink Persian Carpet ice plant as part of the earlier beautification work. As it had been to previous island inhabitants, gardening became a form of therapy. "Planting flowers in rocks and tending them like a baby was helpful to the general morale of people working on the island," a former worker recalled.[28] In the late 1930s, Fred Reichel, the warden's secretary, convinced his boss to allow prisoners to garden on the property. He had no training, yet still became an expert gardener dedicated to improving the appearance of the island.[29] He sought advice from respected

California horticulturalists, from whom he gained knowledge and experience about which plants could work in the region and, with the help of inmates, planted many succulents, ice plants, and bulbs that proved adaptable to the bay's unique climate and the island's difficult conditions.

Reichel, who left Alcatraz in 1941, later recalled: "I kept no records of my failures, for I had many—the main thing was to assure some success by trying many things and holding on to the plants which had learned that life is worth holding on to even at its bitterest."[30] From his informed perspective, these plants were an apt metaphor for the lives of those prisoners sent to Alcatraz. Several inmates then began their own legacy-building work in the gardens. These were still prisoners, many considered dangerous, so only a few were allowed out on the site, and never in groups. Their primary garden was on the west side slopes, within view of the guard tower above. No plants could be higher than waist height, lest prisoners hide behind the shrubs to avoid notice. Among the notable inmate gardeners who continued the work started by Reichel was Elliott Michener, a convicted counterfeiter serving an

Fig. 16. A federal prisoner stands looking over the flourishing gardens in the 1950s. (National Park Service)

almost ten-year sentence on the island. He was sent to Alcatraz following a failed attempt to escape from the penitentiary in Leavenworth, Kansas.

Michener worked as a full-time gardener for six years, during which he transformed a barren hillside into a lush, terraced garden bursting with daisies, poppies, irises, red-hot pokers, snapdragons, and other flowers.[31] Gardening became his passion, an effective and fulfilling way to pass his time in incarceration. Looking back on his time there, he wrote: "The hillside provided a refuge from the disturbances of the prison, the work a release; and it became an obsession."[32] He left in 1950 to complete his sentence in Leavenworth, and when he was finally released from that prison several years later, he turned his passion for plants and gardening into a career as a landscaper. He wrote to the warden at Alcatraz asking to return to the island to see his gardens, but prison rules prohibited him from doing so.[33] Other inmates carried on his legacy of stewardship of the island's gardens until 1963, when Robert F. Kennedy, the nation's attorney general, shut down Alcatraz. This was not because conditions inside were inhumane but because it was too expensive to service and maintain and would cost upward of $5 million to fully repair.

The National Park Service Takes Over

For several subsequent years, Alcatraz remained empty except for security personnel. Debates over what to do with the island persisted, while the gardens suffered a fate familiar to that of other neglected landscapes. Many plants died without maintenance or regular watering, while others were buried beneath untended vines and aggressive plants better able to survive the harsh environment. The federal government, burdened with an $1,800 per month maintenance fee for the deserted island, put the property up for sale. The City of San Francisco reportedly wanted to purchase it for $1 million and lease it to a private developer. The Texas businessman Lamar Hunt came up with a proposal to buy the island for $2 million and turn it into a shopping destination and tourist resort containing, among other attractions, a space museum with a 364-foot tower topped by a replica of the *Apollo 8* moon rocket.

In 1969 the city's board of supervisors passed Hunt's plan by one vote, angering many in the community. Ten days later, around eighty Native Americans from thirty tribes mounted Alcatraz and peacefully occupied the former penitentiary, claiming the island as home to an Indian cultural center, school, and museum. They lived there for nineteen months, during which the federal government announced that Alcatraz would be transferred to the Department of the Interior and become incorporated into the planned Golden Gate National Recreation Area—a 34,000-acre continuous shoreline park running north and south from San Francisco's Golden Gate Bridge.[34] Tribal members stayed on Alcatraz until June 11, 1971, when U.S. marshals escorted those remaining from the island. Although the protest was unsuccessful in its primary goal, consistent news coverage of the occupation galvanized support for the Native American community. Afterward, the U.S. government returned millions of acres of ancestral land and passed more than fifty legislative proposals supporting tribal self-rule. The incident also added a new layer to the island's already dense mystique as a cultural icon. Throughout the drama, the gardens remained forgotten as evidence from the previous periods of horticulture was lost.[35] This marked the end of the landscape's period of significance as designated by the NPS.

Fig. 17. When the National Park Service took over Alcatraz in the 1970s, the island's landscape was in a state of deterioration. This slope, shown covered in debris, once boasted lush gardens planted by prisoners. (Library of Congress)

In October 1972, President Nixon signed legislation formally establishing the Golden Gate National Recreation Area and signifying the island was no longer merely a prison site or a vestige of the past; it was now part of an extended system of open spaces. Efforts emerged to repurpose the island and maximize its value as a tourist attraction and a cultural resource. When it opened to the public under NPS management in 1973, evidence of the gardens' existence or history was still not readily apparent. Few of the 50,000 visitors who clamored to see inside the infamous cellblocks during its first year likely paid much attention to the landscape as they walked from the ferry stop through the once-blooming slopes that now carried the look of abandonment and neglect.

Five years later, the property was listed on the National Register of Historic Places with, not surprisingly for the time, barely a reference to the landscape as part of the justification for its significance, which was limited to its "military" and "socio/cultural" significance. Early studies conducted by the NPS include the 1979 Historic Resource Study and, a year later, the Alcatraz Development Concept Plan. The documents for both studies concentrate on buildings and structures while barely acknowledging the richness of the island's landscape history.

Forming Partnerships

In line with broader national trends, the support of nonprofit organizations buoyed the NPS's efforts at Alcatraz. The first of these organizations was the Golden Gate National Parks Conservancy (frequently shortened to the Parks Conservancy), which was founded in 1981 to support the Golden Gate National Recreation Area. The organization offered tours, recruited volunteers, and raised money for the vast expanse of parks. Its initial work at Alcatraz was the creation and implementation of a self-guided audio tour of the prison. In 1988, it produced a planning document titled "Alcatraz the Future—Concept Plan and Guidelines," to build on previous studies by the NPS and implement a comprehensive management plan. In conjunction with this effort, the Parks Conservancy and the NPS hired a team of experts led by the California-based landscape architect Lawrence Halprin to create a master plan for transforming the island into a public park. Almost nothing from the plan was realized.

The gardens moved to the forefront once the NPS began studying the design features of the landscape beginning in 1991, nearly two decades after the agency took over management of Alcatraz. Over the next year, during which the agency created the Historic Landscape Initiative in Washington, DC, cultural research staff in California completed a botanical study titled "Historic Gardens of Alcatraz" and a draft version of the "Alcatraz Cultural Landscape Report," which was undertaken to determine what features needed to be preserved and interpreted in concert with emerging national guidelines. Combined, these documents signified a dramatic shift in focus toward the gardens and their cultural heritage.

In 1996, the gardens gained recognition with a book sponsored by the Parks Conservancy titled *Gardens of Alcatraz*, which contains essays by the environmental journalist John Hart, the historian Russell Beatty, who was teaching landscape architecture at the University of California, Berkeley, at the time, and the landscape architect Michael Boland, who played an important stewardship role at the nearby Presidio in Golden Gate Park from 1989 until his retirement in 2023. The three men pored through archives for letters and photographs that would provide insight into the gardens' previously unpublished heritage. In addition to raising public awareness of the gardens, the book provided the basis for yet another document, "Landscape Stabilization and Maintenance Guidelines," produced by the Olmsted Center for Landscape Preservation. This document identified the urgency of existing threats to the structure and integrity of the gardens and recommended immediate action to prevent further deterioration of the gardens' remaining features. An official at the Parks Conservancy estimated it would take an investment of $100,000 per year for three to five years to restore the gardens to some semblance of accuracy.

The Garden Conservancy entered as the third impact player in 2003 when it joined forces with the NPS and the Golden Gate Parks Conservancy in a collaborative effort that became known as the Alcatraz Historic Gardens Project. The Garden Conservancy's early portfolio was filled with primarily East Coast residential gardens, and to expand its reach, it searched for opportunities specifically on the West Coast and sought to include landscape types beyond residential gardens. In Alcatraz, the group was presented with a landscape both uniquely challenging and culturally significant. A partnership with the NPS and the Parks Conservancy made sense because, although both groups offered specific expertise, neither had the necessary practical experience relevant to garden preservation. Because of its unique set of skills in this regard, the Garden Conservancy took the lead on the restoration efforts. Its staff also led the fundraising effort, tapping into the conservancy's own donor base and those of foundations across the country. Important to this endeavor was a 2006 grant totaling $250,000 awarded by Saving America's Treasures, the innovative program administered by the NPS that funds the preservation and conservation of the country's "irreplaceable and endangered historic properties, sites, documents, artistic works and artifacts."[36] Out of 340 total applications that year, forty-two grants were awarded, with the Alcatraz project the only garden restoration. The Garden Conservancy raised matching funds, boosting the total to $500,000, which provided much of the needed funds for three years' worth of work. Speaking on behalf of the collaborative team, Antonia Adezio called it "a gratifying vote of confidence in the teamwork that is making it happen."[37]

The Alcatraz Historic Gardens Project

The stated mission of the project to restore Alcatraz Gardens is "to preserve and maintain the gardens created by those who lived on the island during its military and prison eras, and to interpret their history, horticulture, and cultural significance for visitors."[38] Realizing that it would take too long to research and produce a CLR to address immediate needs in the garden, the NPS authorized a Level 2 Cultural Landscape Inventory (CLI) to guide the early efforts. This document essentially identifies short-term fixes that can be tackled while continuing to work on longer-term solutions. Although not as thorough as a full CLR, the CLI establishes an overall treatment philosophy and strategy for restoration while providing "information to establish historic significance by determining whether there are sufficient extant features to convey the property's historic appearance and function."[39]

Because there are no historical planting plans or drawings to consider, deciphering the landscape's heritage required poring over correspondence and photographs to determine general character and, in some fortunate instances, uncover specific plant species and varieties. As plants succumbed over time, various gardeners acquired and planted new species based on availability and arranged them according to personal taste. Roses proved particularly able to survive the rough conditions, as fifteen varieties were discovered on the site following the decades of neglect. The CLI identified specific projects, and treatment plans were produced for five named gardens. The plan documented site history and existing conditions and included planting and structural repair plans, maintenance considerations, timelines of proposed work, and interpretive plans for each of the five spaces. A preservation approach outlined for the area once tended by Michener reflects a standard approach for the whole landscape: "Stabilization and rehabilitation of the West Side Gardens is necessary to preserve these unique historic gardens, to let visitors experience a landscape similar to that of the prison years, and to demonstrate sustainable gardening principles to be applied today."[40] Because only the portion of the landscape created and tended by residents and prisoners as gardens was deemed culturally significant according to the provisions of the CLI, most of the island would not be part of the restoration work.

Work began on the first five gardens in 2005 after a period spent clearing masses of brush and brambles. There was no full-time garden staff in the beginning, although both the Parks Conservancy and the Garden Conservancy had horticulturalists on site. Carola Ashford, who received a fellowship from the Garden Conservancy, became the site's first project manager, working part-time. Colleagues lauded Ashford for her dedication, meticulous research, and her ability to "see a garden through all of the overgrowth."[41] Undoing the damage inflicted by forty years of neglect was a daunting task. Teams of predominantly retiree-aged volunteers took the ferry to the island from the mainland on Wednesdays and Fridays each week to remove debris and vines. A crew of ten inmates from nearby Alameda County's Santa Rita Jail also rode out to the island every week, ready to work.[42] They uncovered historical features such as railings, paths, terraces, and foundations hidden by time and in need of extensive repair. English ivy, initially planted to stabilize slopes, had overtaken the gardens, which also boasted blackberry, sweet peas, nasturtium, honeysuckle, and other plants. Remarkably, remnants of around two hundred species of ornamental plants had survived.

After the first year, the prominent garden beds beside the main road leading from the ferry stop to the cellblock had been restored. Work was not complete, however, because once an area was restored, it transitioned into the cycle of routine maintenance as clearing of the next area got under way. Anticipating the growing workload, the Garden Conservancy hired Shelagh Fritz in 2006 as Alcatraz's first full-time, on-the-ground gardener. She brought experience gained from an internship at Longwood Gardens and two years working in England on the landscape of Syon

Fig. 18. Following restorations begun in the early 2000s, the slopes were once again covered in ornamental plants and flowers. Historically accurate plantings were used where possible according to availability and conditions. (Photo by author)

House, which proclaims itself to be "one of the last great houses in London."[43] Once at Alcatraz, she worked five days a week restoring the gardens and leading teams of volunteers, which included groups of corporate professionals from San Francisco who used the hands-on work for community engagement and team-building exercises. These teams proved particularly effective, often accomplishing more in one day with their labor alone than was possible with staff and regular volunteers.[44]

The restoration of the first five gardens was completed in 2009. More than four acres of tangled brambles and bare dirt were transformed into lushly planted beds filled with perennials and shrubs. Because potable water for human use was shipped daily to supply the entire island—water for drinking, hand washing, and cleaning—a rainwater catchment system was installed as a sustainable source of landscape irrigation. Repurposing the concrete cisterns that once recycled water from the former federal prison showers provided on-site storage for up to 12,000 gallons of collected rainwater. This resource was critical to staff and volunteers as they worked to reestablish and maintain the tapestries of plants and flowers that had been created during the site's colorful past. Returning visitors to the island reported being amazed to find acres of well-tended gardens where none had been before, thus fulfilling the project's goal of interpreting this previously little-known aspect of Alcatraz's history for the public.[45]

In 2010, Alcatraz attracted around 1.3 million visitors annually. Joining them were more than seven hundred volunteers, who provided nearly eight thousand hours of free labor maintaining and restoring the gardens. These volunteers included some who had been working on Alcatraz since the project started in 2003, as well as groups from more than forty corporations and organizations. The CLR was completed that same year. The document applies lessons learned from the earlier projects and lays out a twenty-year plan for maintaining and managing the entire 22.5-acre landscape. Although the entire island is managed by the NPS as a cultural landscape, only the areas originally created as residential and prisoner gardens are maintained by Parks Conservancy staff and volunteers.

Sustainable resource management became a focus as restoration and maintenance work proceeded steadily over several years. Evolving strategies for composting, water harvesting, and planting drought-tolerant species supported an NPS mandate for ecologically driven practices. Fritz continued her weekday commute from just north of San Francisco, taking the last stretch through the city on her bicycle and then riding the fifteen-minute ferry back and forth across the bay. She became the project manager in 2012, at which time she was more involved in planning the work and purchasing the plants than just managing the physical work. Two years later, her situation abruptly changed when the Garden Conservancy downsized its role at Alcatraz, and she transitioned into being an employee of the Parks Conservancy.

Garden Stewardship

The NPS and the two conservancies publicly reported in 2014 that "the restored gardens are now moving into their next stage of evolution—one of ongoing stewardship and interpretation under local management—and will continue bringing history to life at Alcatraz."[46] The Parks Conservancy assumed management of the

Fig. 19. Today, staff from the Golden Gate National Parks Conservancy lead tours of the restored gardens. Pictured in the photograph is Shelagh Fritz, who began working at Alcatraz in 2006 and became one of the landscape's most important stewards. (Photo by author)

restored gardens, including all staffing, volunteer organization, and educational programs. The Garden Conservancy continued in an advisory capacity for a few more years and continued to help maintain existing donor relationships. Fundraising was challenged by a familiar reality: donors are more eager to give financial support to restorations and beautification initiatives than to maintenance. This unwelcome realization was exacerbated by a continued depletion of the usual donor base, which had yet to rematerialize following the 2008 economic crisis.[47] In response to these financial adversities, the NPS agreed to shift some of the revenue from visitor ticket sales to the ongoing care of the landscape.

At this point, what began as the Alcatraz Historic Gardens Project in 2003 evolved into the Golden Gate National Parks Conservancy's Alcatraz Gardens

Stewardship Program, which was established to manage, maintain, and preserve the gardens while adding "life, color, and vibrancy to the Rock."[48] The actual work remained heavily dependent on volunteer labor, with forty regular volunteers and a "hard-core" ten consistently trekking out to the island twice a week.[49] Fritz's role continued to change as she took on a broader version of stewardship. In addition to looking after the physical needs of the gardens, she became more involved in community engagement and outreach, educating the public and getting people excited about gardens and gardening.

Contemporary Issues

"For me to leave [Alcatraz] something would have to be a really good opportunity and challenging," Fritz remarked in 2019.[50] Challenges soon came. In 2020, the apparent stability in the gardens was shaken by the onset of the coronavirus pandemic. San Francisco initiated a shelter-in-place order, forcing the island to close to tourists and staff. For several months the landscape sat untended as a phase of unintended neglect triggered a renewed assault by weeds and aggressive species able to outcompete their neighbors. Because the lack of visitors also meant a loss of revenue from ticket sales, the Parks Conservancy was forced to permanently lay off more than one hundred employees, or 27 percent of its total workforce. When the outdoor areas of Alcatraz reopened briefly in August of that year, it was on a limited basis, with ferries operating at 20 percent capacity.[51] Whereas a typical summer day before the pandemic might see more than five thousand visitors, there were now only around two hundred. Following another shutdown in December 2021, Alcatraz reopened to an eager public willing to shell out $41 per person for tickets. Volunteers once again arrived to tend the gardens, pulling weeds, watering plants, removing invasive species, maintaining the path system, tilling compost, picking up garbage, and assisting in the ongoing restoration of structures and features.

This period of uncertainty coincided with California's driest year since 1924. A 2021 report from the California Department of Water Resources struck an ominous tone when it reported that extreme drought conditions were occurring with increased frequency. Even with periodic relief brought by occasional wet years, such patterns are not likely to change anytime soon. "California's climate is transitioning to a warmer setting in which historical relationships among temperature, precipitation, and runoff are changing," the report states.[52] Even though Alcatraz Island is essentially a tall rock in the middle of a bay and therefore is less susceptible to climate change than many of the sprawling, water-guzzling landscapes throughout mainland California, the gardens still face threats. In 1973, the year the NPS took over Alcatraz, around twenty-eight inches of rain fell on San Francisco. Over the next three decades leading up to the start of the restoration project, the annual rainfall averaged more than twenty-three inches, consistently filling the cisterns and providing the water needed for irrigation. That figure remained steady until 2011, when extreme drought hit the city, the state, and the region. Although dry spells were not unprecedented, the next decade produced a regional megadrought, which scientists attributed principally to the warming climate. This exceptionally dry period meant there were times the cisterns were never replenished, forcing the irrigation system serving the historic gardens to be shut off and leaving the plantings to suffer in the dry heat.

The lack of rain is not the only environmental threat. The famous bank of fog that regularly rolls across San Francisco Bay and enshrouds the Golden Gate Bridge has decreased almost 35 percent over the past century. Although the cause of this decrease is not entirely understood by scientists, it has been attributed in part to climate change.[53] While a lack of fog affects natural ecosystems and increases the risk of wildfires on the mainland, it amplifies the impacts of drought on Alcatraz. Less fog and rain, combined with the negative impacts of higher temperatures, may create a revised reality that propels a move away from historical authenticity and toward the use of a palette of plants better able to adapt to xeric conditions. While this would likely have minimal effect on the landscape's beauty, it would compromise its historical integrity, which is a paramount focus of the NPS.

Alcatraz is a micro-scale example of the climate change–related issues faced by the NPS and the massive land holdings it manages across the country. In 2010, its director created the National Park Service Climate Change Response Strategy to guide agency-wide actions on four climate-related components: science, mitigation, adaptation, and communication.[54] Although much of this strategy involves vast natural landscapes such as Yellowstone, it does address designed and cultural landscapes. "As leaders in cultural resource preservation, the NPS must contribute to a larger national effort to aggressively seek short- and long-term solutions to ensure the preservation of cultural resources most at risk to the adverse effects of climate change," it states.[55]

In the environmental context, NPS tour guides continue to highlight Alcatraz's gardens to the hordes of mostly unsuspecting visitors keen to explore the famous cellblock and glimpse what life was like on The Rock. Almost everything on the island displays an understandable patina, with weathering concrete, stone, and metal all exhibiting a visible degree of disintegration or rust. Fortunately, the site elements were originally constructed for solidity and durability rather than appearance and seem prepared to maintain their integrity, even if only as a relic. Depending on the season, the landscape is accentuated by an abundance of flowers that cover the site, adding color and vibrancy to the otherwise drab buildings and muted terrain. Because the entire island remains strictly managed as a cultural landscape, only the areas originally created as residential and prisoner gardens are maintained by Parks Conservancy staff and volunteers as such. A commitment to historical accuracy dictates that those portions of the island not formally identified as historic gardens be managed as prison grounds, which have only a minimal design legacy and therefore often appear unkempt and undermaintained. In contrast, the restored gardens shine like oases within the broader landscape.

While the gardens offer little by way of interesting design or horticultural excellence, their unique, human-centric story set within a brutal cultural and physical context makes Alcatraz one of the most unusual landscapes in the country. With a history rooted in a specific period, the island will likely not face significant change in the future apart from environmental impacts associated with climate change. Maintenance is more about preserving history than modifying it, yet cannot escape the need to adjust over time. Ongoing stewardship will continue to involve managing a volunteer workforce, securing ongoing funding, and finding innovative means for supplying the gardens with water. Where the gardens might have been now had the NPS not stepped in will never be known. It is likely they would have been lost forever.

Mount Auburn Cemetery, Cambridge, Massachusetts

Mount Auburn Cemetery is easily among the most beautiful and impeccably tended designed landscapes in the United States and about as far from Alcatraz as one could imagine. Looking in from the passing street, gazing through the sturdy iron pickets of a tall perimeter fence, a passer-by might mistake it for a park, a garden, or an arboretum, though its specific function distinguishes it from each. Situated on the border separating Cambridge and Watertown, a suburb of Boston, the landscape unfolds as a lushly planted oasis spreading beneath the canopy of thousands of trees. Wandering the circuitous route of meandering paths, up and down hills, and around a series of picturesque-inspired ponds and meadows, one might think that losing oneself in the landscape is part of Mount Auburn's program. Much of its architecture carries the veneer of age and tradition, with an unsurprising penchant for the Gothic Revival style. Although this recall of past styles is translated into the landscape and its legions of traditional granite markers patinated by long exposure to the elements, a closer look at the cemetery's landscape reveals a more progressive institution than might be expected. While history is respected, even cherished, the cemetery's focus seems to be as much on innovation and adaptation as on preservation.

Like the Missouri Botanical Garden, Mount Auburn Cemetery has thrived for many decades under a consistent governance structure that was established at its founding. Led by an engaged board of trustees and a series of potent and effective presidents, the institution's managers have successfully facilitated change, restoration, and, where appropriate, preservation. While its mission has remained fundamentally unchanged, how it is reflected in the designed landscape has evolved in significant ways. It is a beloved place, cherished alike by Bostonians, New Englanders, and landscape lovers from further afield. Although taking care of it may seem easy thanks to nearly two hundred years of practice, the cemetery has faced numerous significant challenges.

Establishing the Legacy

Before Europeans settled in and around what is now urban and suburban Boston, the natural landscape consisted of wooded hills and scattered islands adjacent to a sinuous shoreline of wandering tidal streams, salt marshes, and shallow mudflats. The Indigenous Massachusett tribe inhabited the region before the Puritans, the first Europeans to colonize the area, arrived in 1630 and settled on Shawmut Peninsula—an eight-hundred-acre hilly landform connected to the mainland by a narrow isthmus. After several years spent establishing new homes and businesses, the settlers named their emerging community after the English city of Boston. The community grew faster than its neighboring communities, including Watertown and Cambridge, both of which sat across the Charles River. By the middle of the eighteenth century, the region had grown into a vital player in the exportation of fish, lumber, and produce to Europe.

As was common in industrializing cities during the time, Boston simultaneously became more crowded and less sanitary. Without an effective sewage system, communicable illnesses spread quickly. The death rate increased as diseases such as yellow fever took hold, and how to dispose of the mass of bodies became an increasingly critical issue. Although never as serious a problem in Boston as in the larger cities of New York and Philadelphia, it was a problem that needed to be solved.[56] In 1823, Josiah Quincy, a former congressman and future president of Harvard College, was elected Boston's second mayor. He seemed to recognize the challenges that came with urbanization and focused his leadership on making the city safer and healthier. He instituted criminal reform, organized a fire department, established municipal water and sewage systems, and attempted to address urban burials, for which he formed an advisory committee to investigate solutions.

The early history of Mount Auburn is conveyed in the author and historian Blanche Linden's exhaustively researched and beautifully written and illustrated book, *Silent City on a Hill.* Jacob Bigelow, a Harvard-trained physician, writer, and botany enthusiast, recognized the problem with burials and offered as a solution the creation of a rural cemetery on the outskirts of Boston. In 1825, he invited several colleagues to his house to present his proposal for a cemetery "composed of family burial lots, separated and interspersed with trees, shrubs, and flowers, in a wood or landscape garden."[57] The health benefits of moving burials out of the crowded city were apparent, and he asserted that instilling beauty and nature into the processes of death and grieving would help console surviving friends and family members. The group agreed with the scheme and eventually selected a hilly and wooded seventy-two-acre property on the border between Cambridge and Watertown that had long been used as a place of refuge for Harvard students, who fondly referred to the landscape as "Sweet Auburn." George Brimmer, a friend of Bigelow's from college and a committed nature lover, had purchased the land to preserve its large mature trees and natural beauty from the encroaching development absorbing the region.

Formally establishing Mount Auburn fell to the Massachusetts Horticulture Society, which was incorporated in 1829 to improve the diversity and quality of horticulture grown in the New England region. Henry Dearborn, a lawyer and former soldier interested in creating an experimental garden, cofounded the organization and became its first president. In June 1831, the Massachusetts legislature enabled the society to purchase land for "a rural cemetery or burial ground."[58] With the Horticulture Society acting as a fiduciary agent, Brimmer sold his property for $6,000, or over $200,000 in 2022 dollars.

Like the Gardens of Alcatraz, Mount Auburn had no definitive designer. Edward Everett, the famous American statesman who was the secretary at the 1831 meeting that approved the purchase of Sweet Auburn, wrote at its founding, "Little will be required from the hand of art to fit it for that purpose. Nature has already done almost all that is required."[59] Bigelow's own documented account emphasizes his role in the process, yet Dearborn is most frequently credited.[60] The first formal plan, drawn up in 1831 by the surveyor and civil engineer Alexander Wadsworth, was a collaboration among him, Bigelow, and Dearborn, who was more interested in developing a horticulturally rich garden than a naturalistic cemetery. Wadsworth's plan depicts around two hundred gravesites strung out along a series of roads and paths that meander through lawns and woodlands across and around the seven hills that span the property, the highest of which climbs 125 feet above the adjacent Charles River. Ponds and dells are shown in the low areas, creating an overall picturesque-style vision that seems to anticipate later park work by Olmsted and Vaux. Construction began quickly and took several years to complete.[61] As initially conceived, Mount Auburn's landscape was split into three entities: the existing woodland, the cemetery, and the experimental garden.

Mount Auburn Opens

On September 24, 1831, a beautiful fall day described as having an "unclouded sun and an atmosphere purified by the showers of the preceding night,"[62] a crowd of more than two thousand people participated in the ceremony to consecrate the new cemetery. In his speech, Joseph Story, a justice of the Supreme Court and soon to be the first president of Mount Auburn, emphasized for the crowd the inherent connection between the cemetery's physical landscape and its sacred purpose. "Mount

Auburn, in the noblest sense, belongs no longer to the living, but to the dead," he said.[63] The landscape would be a place of beauty and serenity as well as a setting for mourning and remembrance. Because this was before cities began regularly setting aside land for designed public parks, the Boston area community eagerly used the landscape as a place of enjoyment.

Bigelow was confident about the success of Mount Auburn and stated in an 1831 address, "No place in the environs of our city will possess stronger attractions to the visitor."[64] An account published in the *Boston Courier* around the same time proclaims similar confidence:

> The natural features of Mount Auburn are incomparable for the purpose to which it is now sacred. There is not in all the untrodden valleys of the west, a more secluded, more natural or appropriate spot for the religious exercises of the living; we may be allowed to add our doubts whether the most opulent neighborhood of Europe furnishes a spot so singularly appropriate for a "Garden of Graves."[65]

Both were proven correct almost instantaneously, as locals showed up in droves. Word spread about the cemetery, and it became a popular attraction that soon drew the attention of the entire country. Writers, including Henry Wadsworth Longfellow (Alexander Wadsworth's first cousin), Henry James, and Nathaniel Hawthorne, found inspiration in the landscape and spread its reputation as a place of beauty. Other cities took notice of the benefits that rural cemeteries could provide, and similar landscapes soon appeared in Philadelphia (Laurel Hill), Baltimore (Green Mount), and New York City (Green-Wood). The movement spread, prompting the noted Andrew Jackson Downing to write in 1849 that, as a result of Mount Auburn's example, "there is scarcely a city of note in the whole country that has not its rural cemetery."[66]

While Bigelow understood that Mount Auburn would be popular with the public, he and other leaders were apparently unprepared for the complications that accompanied the crowds. For one, they had wrongly believed that creating and maintaining a naturalistic landscape such as Mount Auburn Cemetery would not be a challenge because, as in nature, the landscape would take care of itself. In addition to failing to anticipate ongoing maintenance needs, the early leaders apparently failed to foresee the variety of ways people in the rapidly densifying region would use the landscape. Alongside the visitors who came, as intended, for quiet reflection and repose, hordes of people arrived in carriages and on horseback for picnics and recreation, kicking up dust, damaging roads, and creating erosion problems. As early as 1832, regulations were adopted, and operations changed to better control the number of visitors and dictate what they were allowed to do once there. These regulations were amended over time to include a long list of rules taken for granted in public places today, such as an interdiction against discharging firearms, a reduced speed limit for vehicles ("no faster than a walk"), expulsion for unsavory behavior, and prohibitions against refreshments, the removal of plants, and vandalism.[67] For a short time in 1833, access was limited to plot owners holding nontransferable tickets for entry. This was not a popular move, however, and the cemetery once again opened to the public, this time for good.

Early Governance and Evolution

By 1834, debates were flaring over the viability of having an experimental garden and a naturalistic rural cemetery share the same property. It became increasingly clear to Bigelow that the cemetery would take precedence and that the Massachusetts Horticultural Society was not the best entity to be in charge, in light of the burial functions involved.[68] Following a series of negotiations, the society conveyed Mount Auburn and its property to the newly incorporated Proprietors of the Cemetery of Mount Auburn. A board of trustees was established, with Joseph Story as its first president.

From its inception, Mount Auburn's interment function set it apart from other designed landscapes. Selling burial plots to the public has always been fundamental to its mission, as well as a primary source of its funding. The purchase of parcels generates money for, among other things, tending the landscape. At the same time, the more acreage sold to individual proprietors, the less public space is available in which to maintain a cohesive aesthetic character. During its early years, plot holders of Mount Auburn Cemetery could select any location they wanted within the general plan laid out by Dearborn, "without regard to the economy of the land, or the size and shape of the intermediate spaces."[69] Following trends found in other rural cemeteries, owners defined the limits of their property with iron fences. There was no formal mandate or plan for maintenance at the time, so if proprietors wanted to maintain their parcels, they had to pay for it themselves. This lack of management led to a haphazard appearance that ran counter to the original vision of Bigelow and Dearborn.

Recognizing the visual impact of these changes, the trustees rightly insisted on more control over how the cemetery developed. In an 1843 vote, it was decided that money paid by proprietors for the upkeep of lots would be used for a centrally managed Fund for Repairs, which was later renamed the Trust Fund. Trustees also committed to a more unified approach to design, laying out lots and paths as features in concert with logical and efficient patterns for expansion. The space between the plots, which was once ignored, became organized and maintained as an important contribution to the visual quality of the landscape. Sales of plots continued at a high rate, providing needed revenue to build more roads, plant more trees, and prepare new land. As had occurred during the early development of Boston, the hilly landscape changed as soil from high points was shaved off and used to fill low areas and expand the available and sellable property.

The cemetery was not the only open space in the city at the time—the Boston Common, for instance, was nearby—and the popularity of both landscapes proved how much people craved access to designed green space. "Judging from the crowds of people in carriages, and on foot," Downing wrote, "I think it is plain enough how much our citizens, of all classes, would enjoy public parks on a similar scale."[70] Boston was far from the only place facing crowded, filthy, and unsanitary conditions. Civic leaders and citizen advocates in numerous cities joined Downing in arguing for parks. New York City, probably the least healthy of all urban areas in the United States, led the way when, in 1853, city commissioners received authorization to acquire land to create Central Park. This was twenty-five years before Olmsted began designing Boston's "Emerald Necklace."

In the meantime, Mount Auburn continued to draw many citizens in search of fresh air and a beautiful, albeit fabricated, natural environment. The cemetery became accessible to a broader population in 1856 with the introduction of the Cambridge Horse Railroad, which connected Mount Auburn to downtown Boston. Although it was still relatively new as a destination, the landscape continued to evolve, creating change that sparked concern over how best to maintain it moving forward. Bigelow noted that "the only safety against the ultimate impoverish-

Fig. 20. Many of the early gravesites in Mount Auburn were family plots corralled by stone curbing and fences. Individual plot owners were responsible for providing their own maintenance. (Mount Auburn Cemetery)

ment and decay of the Cemetery must consist in a rigid determination on the part of future Boards who may be entrusted with the management of this property, to reserve annually, such a sum from the receipts of the current year . . . to keep the Cemetery in perpetual repair after the sale of the land shall have been completed."[71]

By mid-century, three thousand burial plots and more than seventeen hundred iron fences filled the landscape. In a relatively short period of time, the grounds appeared as a collection of randomly strewn headstones and markers corralled together as a series of cages. During this time, Mount Auburn's aesthetic began to shift. No longer was nature in fashion, as a preference for the formality and ornamentality of the Victorian style took hold in America. Taking cues from the class-conscious British, American cemeteries adopted the notion that the burial condition somehow translated to social status in heaven, and gravesites were increasingly embellished with elaborate monuments, obelisks, and sculptures and surrounded by curbs or low walls made of granite.

Horticulturally, the landscape borrowed from the Victorian preference for collecting and displaying ornamental plants. Burial plots became individual gardens as arrays of formal borders and expansive beds appeared in the cemetery. The conservatory was packed with nonhardy exotic plants in the winter months that expanded the options for annual bedding plants in the spring. The creation of more complex plantings prompted the hiring of a cemetery gardener, who, according to Bigelow, was "charged with the care and repairing of lots for those proprietors who may desire that service; and who shall keep for sale shrubs, trees and flowers, at the Cemetery."[72] This was 1857, the same year the trustees successfully responded to Bigelow's concerns about finances by creating the Permanent Fund, which would produce income "adequate to cover the cost of maintaining in perpetuity the grounds and buildings after receipts from the sale of lots shall have ceased."[73]

A year later, the trustees embarked on a plan to spend $1,000 each year for the grading of new paths, roads, and burial sites, as well as the enhancement of portions of the cemetery yet to be sold. By the end of the decade, the cemetery had accrued annual expenses exceeding $10,000. The long-term viability of the institution remained far from guaranteed, and proprietors were encouraged to hire only maintenance staff employed by the cemetery and to help recruit other people to purchase plots.[74] Maintenance was becoming more of a problem as the second generation of proprietors inherited plots that they may not have had the means or interest to tend to properly. In some cases, current owners of deteriorating plots could not be located, forcing the trustees to condemn certain parcels, remove fences or curbs, and take over maintenance.

As policies related to operations, personnel, and finances continued to evolve, so did the overall landscape. The woodland areas that initially made "Sweet Auburn" such a desirable property densified and became less penetrable. Without active maintenance, a preponderance of seedlings and thickets closed in what were once open woods. Although the imagery of the forest was still important to the character of the cemetery, Bigelow considered the growth of the woodlands a "serious evil" that was being compounded by the planting of new trees. He promoted removing half the trees to create "broad vistas and open spaces, through which the works of art could be seen; and the light of the sun might be admitted to the grass and cultivated flowers."[75] In the 1860s, more than half the trees were felled. As reported by Bigelow, "Many of the superfluous trees have been removed, and the branches of

Fig. 21. The cemetery's character changed over time, as a more naturalistic aesthetic took over in the early twentieth century. Trees were removed to open up extended views, and open lawns became a defining feature of the landscape. (Library of Congress)

others trimmed away, so that the whole ground has a more finished and ornamental appearance than it before possessed."[76] The Victorian aspect of the landscape was highlighted, with parcels of turf interrupted by bedding plants and occasional trees.

The Civil War and the Birth of the Lawn-Park Cemetery

The Civil War triggered a dramatic change in how people viewed death and burial, and, in turn, altered general perceptions about cemeteries as designed landscapes. During the war, expansive burial grounds were often created near general army hospitals to accommodate concentrations of the estimated 620,000 people from both sides who died in the war. Such makeshift sites filled to or beyond capacity, leaving thousands of other soldiers to be buried in small cemeteries near battlefields scattered across the country. In 1862—just one year into the war—Congress passed legislation creating a national cemetery system. Within ten years, more than seventy national cemeteries buried nearly 300,000 Union soldiers, about 90 percent of all

Northern soldiers who died in the conflict.[77] The magnitude of violence on native soil and the dramatic death rates within a population of otherwise healthy young men altered the concept of what a cemetery should be. Perceiving burial grounds as quaint romantic settings no longer seemed appropriate, and a novel concept of the cemetery emerged. The established imagery of rolling bucolic hills, lawns, and natural woodlands found in places like Mount Auburn was accompanied by a far more dramatic vision—stark and barren landscapes set aside for commemorating the mass gravesites holding those lost to the brutality of war.

More than nine hundred people who served in the Civil War are either buried or commemorated in Mount Auburn. The changing perceptions on graves and cemeteries that emerged from the war became reflected in the landscape. As Linden observes, "Reverence for old interpretations of the past gave way to progress," while "Americans put behind them cataclysmic and cyclical concepts of history."[78] During this period, Bigelow designed, funded, and commissioned the sculptor Martin Milmore to create a monument to honor the Union soldiers who had sacrificed "to achieve the greatest moral and social results of modern times."[79] The sculpture, which measures fifteen feet long and eight feet tall, depicts a Sphinx sitting atop a granite base on which is inscribed "American Union Preserved / African Slavery Destroyed / By the Uprising of a Great People / By the Blood of Fallen Heroes." Although it was in keeping with the Egyptian Revival, a style found widespread in cemeteries even before the Civil War, it was an oddity compared to what more frequently took the form of obelisks, temples, tombs, and pyramids.[80] By the time the monument made its way to the cemetery grounds in 1872, Bigelow's worsening eyesight from cataracts left him unable to see what had been created.

Changing Paradigms

Along with changing attitudes about death and the burial process, the rapid industrialization that emerged following the Civil War introduced paradigm-changing inventions and innovative tools that streamlined and mechanized the maintenance and management of designed landscapes. Around 1870, an Indiana inventor named Elwood McGuire patented a hand-operated lawnmower that improved on a much heavier British version developed forty years earlier.[81] No longer did people need horses to mow large lawns; homeowners and maintenance crews alike could easily use McGuire's model for routine care. Mount Auburn was one among many American landscapes that took to the invention, which helped to instigate a widespread conversion across the grounds to manicured turf more in keeping with the lawn-centric and parklike character that was emerging as a national trend.[82]

This new, grassy look for cemeteries was propelled by Adolph Strauch, a Prussian-born landscape architect whose work at Spring Grove Cemetery in Cincinnati broke from long-standing picturesque-era norms by using expanses of turf to establish wide, uncluttered vistas.[83] Strauch rejected the Victorian predilection for ornamentation and individual expression and prohibited the use of fences or curbs to delineate plots at Spring Grove. He developed an open park setting in which a comprehensive landscape character took precedent over the objects placed in it, favoring the creation of an atmosphere of, in his words, "cheerfulness, luxuriance of growth, shade, solitude, and repose amid scenery designed to *imitate* rural nature."[84] Such thinking could easily be applied to Olmsted and Vaux's vision for Central Park, which was being developed around the same time. Out of this vision emerged the influential lawn-park cemetery style, which could be seen as a hybrid of an urban park and a gardenesque cemetery.

Mount Auburn eagerly adopted this same aesthetic. In the 1860s and 1870s, it expanded its borders by purchasing adjacent property, gaining land on which it could experiment with this new style. Its 1875 annual report describes spaces added to the grounds as "lawns, ornamented with flowers and shade trees; where the monuments are not obtrusive, the boundaries of the lots only marked by sunken posts; and where from the absence of stonework and iron fences, a general aspect of rural beauty, and quiet is the characteristic feature"[85] In addition to this work, fences and curbs were removed on older parts of the landscape and trees, shrubs, and other plantings were added across the grounds. In 1882 alone, 650 trees and almost 1,900 shrubs were planted on the property—a substantial increase over a typical year's planting.[86] The fund established for the care and preservation of the lots totaled nearly $350,000 that year, a considerable amount for the time.[87] By the end of the century, a staff of around two hundred workers maintained the grounds. The average hourly labor rate at Mount Auburn was nineteen cents and the typical laborer earned $1.60, working a ten-hour day.[88]

The lawn-park style held firm in Mount Auburn well into the twentieth century,

Fig. 22. Willow Pond was created in the 1920s in the parklike Lawn Cemetery style, which emphasized uninterrupted views of the landscape by minimizing the appearance of gravestones and other markers. This 2019 photograph shows a recent restoration that replaced lawn area with a vegetated rain garden designed to improve water quality. (Photo by author)

a period during which seemingly little concern was expressed over how the appearance of the historical grounds had changed. In the late 1920s, the development of Willow Pond in the southwestern corner added to the landscape a lake surrounded by an extensive lawn punctuated by trees and shrubs. The space was designed by the landscape architect Laurence Caldwell, who influenced the look of Mount Auburn while consulting there between 1929 and 1937. Following Strauch's precedent, he established a directive for the landscape surrounding Willow Pond that all markers must be laid flush with the ground, with no individual or family monuments allowed to protrude above grade. The space, unblemished by clutter, fostered an appropriate serenity that was "favorably received by present owners and by those buying lots," according to Charles Almy, the cemetery's president between 1918 and 1934.[89]

The evolving character was not just about aesthetics, however. Twentieth-century societal changes, including increased mobility, more economic uncertainty, and smaller families, meant that people often looked for one or two burial sites rather than the large family plots that defined earlier eras at the cemetery. This distinction worked well for Caldwell, who believed that traditional cemeteries were "uninteresting and repelling because of the hodgepodge of objects and the monotonous view of tombstones."[90] To him, the modern cemetery should function more as a beautiful park, "a location of venerable trees and excellent landscapes . . . a safe haven for birds, and a spot where visitors would have seclusion and quiet for a full appreciation of the beauty of nature."[91]

Oakes Ames and the Great Hurricane of 1938

The planting of the landscape around Willow Pond continued into the 1930s, by which time Mount Auburn Cemetery was more than a century removed from its consecration. Through its early phases and changes, it evolved into one of the most cherished and significant designed landscapes in the country, at once venerable and dynamic. Following the death of Almy in 1934, Oakes Ames became the institution's eighth president and the first to serve in a full-time capacity. While the trustees retained ultimate control, Ames was deeply involved in the management of the institution and its grounds. In a move that formalized the more hands-on approach to leadership, he relocated the executive offices from downtown Boston to an administration building on cemetery grounds. Ames led the institution for three decades, making him Mount Auburn's longest-serving leader since Bigelow. He was a horticultural enthusiast who, like Dearborn, saw value in the grounds as an intentional garden, with its plant collections at least as important as its scenic landscape. "It is hoped that eventually all of the more desirable species of trees that thrive in this climate will be represented in Mount Auburn, thereby carrying out the original plans of the founders for a botanical collection of real interest," he once said.[92] It was not just about collecting for the sake of numbers, however. Ames wanted only those trees and shrubs that "will blend most harmoniously with these beautiful surroundings."[93] Under his direction, flowering and evergreen plantings were planted in large numbers both to provide visual interest throughout the seasons and to establish a rich and diverse habitat for local and migrating birds, his other passion.

On September 21, 1938, not long into Ames's tenure, Mount Auburn, along with much of New England, was caught unprepared for what historians call the Great Hurricane of 1938, which was the most powerful storm recorded in the region. Early predictions showed the storm coming ashore much further south before sliding harmlessly out to sea. However, it ultimately followed a straight path northward before slamming into Long Island at seventy miles per hour with sustained winds of over 120 miles per hour, then moving through southern Connecticut and western Massachusetts. The storm left almost seven hundred New Englanders dead and catastrophic damage to property. Estimates of felled trees reached as high as two billion, representing one-tenth of all trees in the region at the time.[94] The Boston area was spared a direct hit yet suffered significant damage. Mount Auburn lost more than eight hundred trees, approximately 16 percent of its entire collection.[95] Nearly as many suffered "substantial damage."[96] Ironically, as devastating as it was, the storm brought with it a unique opportunity to remake the landscape with a deliberate and strategic planting of diverse species that supported Ames's vision. The visual impact of tree loss was mitigated by the opportunity to deliberately establish the next-generation canopy. In a single year, between 1939 and 1940, the cemetery planted more than fifteen hundred trees and shrubs, many of which were species or cultivars not previously found on the grounds.

During that same time, World War II triggered a period of financial uncertainty. Pent-up demand for once-rationed goods after the war led to a nationwide spike in inflation, and the cost of labor soon steadily outpaced the generation of income. Workers were hard to find during the war and difficult to afford afterward. Between the late 1930s and 1951, labor hours spent annually on cemetery operations dropped from 200,000 to around 131,500.[97] To increase efficiency, the cemetery charged proprietors more for maintenance and prioritized tasks across the broader landscape. Although this initially translated into a decline in the quality of the grounds, a smaller workforce did not mean longlasting doom for the standard of care. According to Ames, the difference was made up over time by "the introduction of labor-saving machinery and equipment such as power mowers, electric trimmers, load luggers, leaf collector, power digger, etc."[98] As had been the case with the introduction of the push mower in 1870, technological advances streamlined maintenance practices.

Unlike the nation's somber response to the end of the Civil War, the end of World

War II in 1945 sparked national pride and unbridled optimism. Bold and dramatic military cemeteries, primarily in Europe, conveyed the imagery of thousands of white crosses aligned in rows on acres of green lawns. In America, landscape architects and designers reflected more intimately on the past. They frequently engaged precedent from the colonial era, creating small, ordered, rectilinear gardens defined more by their horticulture than by their built features. At Mount Auburn Cemetery in the 1940s, the landscape architect and Olmsted protégé Arthur Shurcliff helped to create several such spaces that expressed the individuality of specific plots rather than have them fade into their context as with previous styles. Shurcliff was an expert in the then popular Colonial Revival style and collaborated on the recreation of Colonial Williamsburg in the late 1920s. Following trends at the time, he designed single and companion gravesites arranged in garden rooms delineated by formal hedges. Because this new direction reflected a "very different design philosophy" that distinguished it from the philosophy of previous eras, the 2001 National Historic Landmark designation for Mount Auburn places areas created after 1945 as outside the cemetery's period of significance.[99]

Although numerous spaces were redesigned under his watch, the tenure of Ames is generally regarded as a time of quiet adaptation rather than of major change. New technologies boosted maintenance, but the quality of the landscape still fell below desired levels. In the thirty years between 1921 and 1951, operating costs continued to outpace income, even as the trust fund for grounds maintenance rose during the same period from over $2.2 million to around $3.5 million.[100] Mount Auburn was not destitute, yet neither was it in a secure place financially. Frustrated by what he considered governmental insanity regarding the post–World War II economy and its detrimental effect on the cemetery, Ames quoted William Gladstone, a former prime minister of Great Britain: "Show me the manner in which a nation or a community cares for its dead and I will measure with mathematical exactness the tender sympathies of its people, their respect for the laws of the land, and their loyalty to high ideals."[101]

Alan Chesney and the Friends of Mount Auburn Cemetery

In 1968, Alan Chesney, a geology graduate and former sales manager at the Head Ski Company, succeeded Ames as president of Mount Auburn. By this time the cemetery had grown to 164 acres, with nearly 70,000 people interred on its grounds. Chesney's previous professional expertise suggests the trustees sought a business leader with management savvy who could effectively manage Mount Auburn as a small business, which it was. With an annual operating budget of nearly $800,000, cemetery staff planted more than seven hundred trees and shrubs and erected seventy monuments, 328 headstones, and two mausoleums just in Chesney's first year.[102] During that same time, more than five hundred people were buried, prompting worries that less available land could mean a future loss of income.

In response to these fears, the cemetery expanded once again to its current size of 175 acres. Fifteen acres of the property were transitioned into a landscape designed for single graves and family plots. This area, called the Meadow, was developed to mimic the Colonial Revival style, with companion graves identified by upright markers aligned against formal hedges and next to single gravestones laid flush in the lawn.

In 1975, about halfway into Chesney's tenure, the National Register of Historic Places added Mount Auburn Cemetery to its roster. In addition to citing essential works of architecture and the "exceptional and diverse collection of nineteenth- and twentieth-century monuments and sculpture," the nomination relies heavily on the quality of the cultural landscape as justification for the recognition. The beauty of the environment, the presence of legacy trees, the preservation of roads and paths, and the layering of features such as iron gates and granite curbing are all referenced as contributors to the quality of the historic place.[103] As was later continued in the landmark nomination, it limited the cemetery's period of significance to the years before World War II, ignoring the contribution of newer spaces to the significance of the landscape.

Even with mounting honors and accolades and robust lot sales, Mount Auburn Cemetery failed to regain the popularity it once enjoyed, thanks in part to the addition of parks and other open spaces in Boston and surrounding areas. To better engage public interest as well as to assist in the stewardship of the property, the trustees capitalized on the energy and passion of volunteers and established the Friends of Mount Auburn Cemetery in 1986. Chesney announced that the group would support the conservation of the cemetery's natural and cultural resources while also promoting outreach and educational opportunities.[104] Within a few months, more

Fig. 23. Mount Auburn expanded its property with the purchase of land that became known as the Meadow. Using the colonial style that was popular in the 1930s and surged in the 1970s in anticipation of the nation's bicentennial observance, companion graves with upright markers were placed against a hedge or shrub bed to create the look of a small private garden. (Mount Auburn Cemetery)

than three hundred people had paid their dues as charter members; one year later the Friends of Mount Auburn had seven hundred members. Rather than setting steep fees affordable only by the wealthy, trustees made them, according to Chesney, "purposely low so that everyone interested in history, horticulture, birds, and art can enjoy the benefits of membership."[105] The group immediately began leading and facilitating tours, hosting lectures and other educational programs, producing a newsletter, and generally elevating public interest in the landscape.

By 1986, Mount Auburn had a full-time staff of forty-three, with an additional thirty seasonal grounds and maintenance workers.[106] While Chesney officially led the institution as president, much of the operational leadership came from Duncan Munro, who held the position of superintendent from 1967 to 1989. Upon Munro's retirement, Chesney reflected, "By any statistical measurement, Duncan's tenure saw the greatest growth and improvement in the operation of this venerable and honorable institution."[107] In addition to ongoing facility maintenance and the conversion of land into burial lots, Munro experimented with innovations in machinery and best practices to streamline hand labor where possible. He also oversaw the transition to more sustainable landscape management practices, including a move to integrated pest management, an ecosystem-based strategy that focuses on the long-term prevention of pests rather than on the use of toxic pesticides. By 1986, after nine years of employing the program, Munro and his staff had reduced the use of chemical spraying on the grounds by 80 percent.[108] Innovation was a key component of the cemetery's management since its inception, and this move to more sustainable maintenance was a precursor to future practices.

When Chesney retired from Mount Auburn in 1988, he left behind a legacy that included a commitment to preserving significant aspects of the historical landscape. This emphasis resulted in part from a 1980 controversy over the removal of a piece of an iron fence that had been installed in 1844. After one portion was taken down by order of the trustees, members of the Cambridge Historical Commission and the Society for the Preservation of New England Antiquities protested the proposed removal of the remainder of the fence, which Chesney claimed posed a danger because of its corroded joints, missing segments, and loose pickets.[109] Mount Auburn was a private institution and had the right to manage its property, but this situation made evident its importance as a publicly accessible landscape. People in the community cared about it and had developed a sense of responsibility for its legacy. The trustees eventually relented on removing the fence. During the process, the Friends of Mount Auburn became the conduit through which the institution communicated with the public. The controversy also contributed to the eventual selection of Chesney's replacement, as the trustees recognized the need for a president who could effectively lead the institution as a cultural and historical resource.

William Clendaniel and the Mount Auburn Master Plan

William Clendaniel took over as the institution's twelfth president in 1988. In time he would become Mount Auburn's arguably most influential leader since Bigelow. He grew up in Vermont and maintained deep roots in the New England landscape. He attended the elite Choate School in Connecticut, Williams College in the Berkshires as an undergraduate, and then Harvard Law School. Before starting at Mount Auburn, he was the deputy director of Trustees of Reservations, the nonprofit conservation organization and landowner created to protect the scenic, historical, and ecological value of one hundred parks, gardens, farms, and other "special places" in Massachusetts.[110] Clendaniel was also a member of the advisory committee for the state's Olmsted Historic Landscape Preservation Program, making him well versed in the emerging practice of historic landscape preservation and stewardship. This expertise was desirable because by the time he started at Mount Auburn, the landscape was showing its age. It was still a beautiful place with a rich history and an incredible collection of mature trees, but, as stated by one staff member in hindsight, it was mired in a "sleepy phase, with low public interest, cruising at low gear."[111]

During the tenures of Ames and Chesney, maintenance seems to have been task-oriented and reactive. Per standard practice and with limited resources, there was no discernible strategy for responding to the unique history of the landscape and monuments.[112] Clendaniel came in with the benefit of the growing base of shared knowledge from the preservation community and recognized the need for a new approach to restoration. He also understood that the evolving landscape, unlike the static sculptures, structures, and monuments, required an approach other than strict preservation. "At a site like Mount Auburn," he once said, "you're never going to go back and recreate the landscape that was here at any particular time. We're not a landscape that lends itself to a single period of significance, so you keep adapting it for current uses, while trying to still retain a sense of its past."[113]

The continual loss of land for traditional burials remained a persistent issue. Success in the sale of lots was beneficial from a financial perspective, but it also produced ominous implications for the future of the property. The trustees were concerned about this and, in 1986, they formed a committee focused on long-range planning. In weighing options, the group even considered the potential for one day transitioning from a working cemetery into a more traditional and passive public garden, arboretum, and museum.[114] In early 1990, Clendaniel, in probably his most influential action, instigated Mount Auburn's first comprehensive master planning process since the days of Dearborn. Cemetery staff and trustees engaged a consultant team that included the Halvorson Company, a Boston landscape architecture firm, and the historian Shary Page Berg, to carefully document the institution's past and create an implementable plan for its future. Although such planning was typical for constantly expanding university campuses at the time, it was less common practice in the management of gardens and other landscapes. In that sense, the effort at Mount Auburn Cemetery was innovative and became an example for others to follow.

Producing a master plan can be an expensive endeavor and is typically undertaken with the realization that exponentially more money will be needed for the plan's implementation. The cemetery had built up a relatively substantial permanent fund for future operations by this time but was unable to dramatically increase its spending on new initiatives. Thanks to the success of the Central Park Conservancy and other groups, philanthropy was increasingly understood to be a critical aspect in the stewardship of cultural and historical landscapes. Clendaniel and the trustees surely recognized this and, in 1990, formally established the Friends of Mount Auburn as a nonprofit charitable trust. At the time, the group had 850 members and generated $18,000 per year in membership dues alone.[115] This money, along with many volunteer hours, had enabled the group to successfully achieve its goal of establishing programs that further engaged the cemetery with the public. With the Friends group formally separated from the fiscal and administrative umbrella as a stand-alone organization, members could raise funds and seek grants on behalf of the institution. Generations of New Englanders and others had visited Mount Auburn, and this move facilitated the ability of those who loved it to support the institution and its landscape.

Compiling the master plan took several years and involved numerous meetings, extensive research, surveys of visitors, and inventories of all landscape features and structures. The plan set as a primary goal improving the "integration of cemetery, landscape design, and the horticultural collection" while acknowledging that "the challenge of planning for the future is to respect the spirit of the past while continuing to be responsive to changing needs."[116] It detailed how incorporating innovative and nontraditional interment methods could expand the cemetery's burial capacity without compromising the integrity of its landscape. Importantly, the plan mandated that preservation and enhancement of the landscape should take precedence when leadership pursued new development. In other words, expanding burial plots should not threaten the visual quality of the landscape.

One fundamental recommendation of the master plan was to rebrand Mount

Auburn as a "horticultural landscape" rather than an arboretum, as had been the case since the Oakes Ames era when an extensive planting program boosted both the number and variety of trees. By the 1990s, more than four thousand trees representing three hundred species and nearly six hundred varieties covered the grounds. As had been the case in Bigelow's time, the success of the trees came at the expense of the rest of the landscape as they blocked views and created a lack of space and sun to support plantings beneath. The master plan recommended a strategic reduction in the canopy by a quarter and to "develop and maintain . . . a documented and selective collection of woody and herbaceous plants from around the world that are hardy in the Boston area. . . ." This shift in focus underscored the difficulty of assigning a traditional typology to Mount Auburn. No longer calling it an arboretum was logical since this more comprehensive view of plantings incorporated nonwoody shrubs, perennials, groundcovers, and annuals. Using the less specific term "horticultural landscape" enabled the institution to emphasize the role of plants for their beauty rather than their botanical qualities.

Beginning in 1993, a series of new landscapes was created, and others were restored or enhanced. Per the directive of the master plan, design decisions responded specifically to the needs of the landscape rather than to the desires of customers, as was the norm in the cemetery's early days. Seeking innovation while respecting historical precedent became imperative. A multidisciplinary team of staff members, the Cemetery Development team, began considering proposals for new burial sites based on historical images and records from the archives and the goals articulated in the master plan. Respecting history while reflecting contemporary practices around memorialization and commemoration became the goal.[117]

An example of meeting this goal is Halcyon Garden, a landscape designed in the late 1990s in collaboration with the Cambridge-based landscape architecture firm Reed Hilderbrand. The project transformed an existing service road along the property's perimeter into a beautiful and functional landscape. A relatively flat, three-hundred-foot-long roadbed became the site for two hundred concrete crypts stacked two and three deep in neat rows under a carpet of lawn. Over time, however, the lawn became difficult to maintain owing to high use and densifying shade beneath the overhead tree canopy, and was replaced with mulch. Although perhaps not as attractive as the lawn, the adaptation proved more practical from an operations perspective and exemplifies the types of decisions owners and managers must make as landscapes evolve in unpredictable ways.

As the landscape at Mount Auburn Cemetery continued to improve and evolve, recognition of its cultural significance grew. Four years after the cemetery was designated a National Historic Landmark in 2003, the National Trust for Historic Preservation honored Mount Auburn with its Trustees' Emeritus Award for Excellence in the Stewardship of Historic Sites. This recognition coincided with a yearlong celebration of the cemetery's initial consecration. By that time, nearly 100,000 people had been interred at Mount Auburn, their lives commemorated by tens of thousands of monuments spread across the grounds. The cemetery had also reestablished its popularity as a destination, with more than 200,000 annual visitors by 2007. A year later, with his legacy firmly established, Clendaniel retired.

David Barnett and Mount Auburn's Sustainable Landscape

In 1993, toward the end of the master planning process, Clendaniel and the trustees created the position of director of horticulture to assist with the realization of the horticulturally focused initiatives laid out in the plan and to instill more curatorial practices across the historical landscape. The successful candidate for the job was David Barnett, who at the time was working at Planting Fields Arboretum, the contemporary iteration of a luxurious 1920s era estate designed by the Olmsted Brothers firm on Long Island, New York. Even though Barnett loved the job and had every intention of remaining there, Mount Auburn successfully lured him away with the opportunity to steward what he regarded as one of America's premier landscapes.[118] Over the next fifteen years he worked closely with Clendaniel, who became an influential mentor. During this period, Barnett was promoted to vice president of operations and horticulture, and then executive vice president. His rise in the institution continued, and in September 2008, following a national search, and with the apparent enthusiasm of the trustees, he was appointed the thirteenth president of Mount Auburn.

Barnett grew up in Connecticut in a horticulturally oriented family—his father was a landscape architect and his grandfather owned a garden center and nursery. He earned an undergraduate degree in environmental horticulture at the University of Connecticut and a master of horticulture and a PhD in ecology at the University of California, Davis. His motivation to work specifically in the realm of public gardens emerged out of a summer internship at Longwood Gardens, in Chester County, Pennsylvania.[119] In addition to this experience and several years at Planting Fields, he had also worked at the Morton Arboretum in the suburbs of Chicago, building expertise directly applicable to his work at Mount Auburn.

Once in charge, Barnett maintained hands-on oversight of horticultural practices and the design and implementation of new projects. He was also committed to maintaining the positive momentum established by Clendaniel. Enhancing the visitor experience, expanding options for cemetery services, preserving buildings and structures, and being "worthy stewards of the natural and ornamental landscape" all remained priorities.[120] Because his tenure started amid the 2008 recession, however, the implementation of the master plan would not happen as quickly as hoped. One project already underway was the restoration of Consecration Dell—the four-acre sylvan valley in which Joseph Story gave his famous 1831 speech and which was now the last remnant of the property's original rural landscape. Even though it was revered as the ceremonial heart of the cemetery, the Dell had been minimally maintained for a century and was overrun with aggressive and non-native Norway maples. These trees, along with an abundance of native red oaks, created the type of mature, densely forested space that Bigelow had long ago asserted was detrimental to the character of the landscape. As is typical in such situations, the canopy was full while the shaded ground plane remained relatively sparse aside from tree trunks. The steep, once lushly planted slopes sliding into the vernal pool were left essentially barren, which led to erosion problems, a lack of biodiversity, and an aesthetic character significantly worse than desired.

The ensuing restoration involved several phases over two decades. It was sensitively managed, with a focus on reestablishing a specific character rather than replicating history. Removing the Norway maples and other non-native species was a priority, as was stabilizing the slopes with compost-filled burlap tubes. An extensive planting added more than one hundred twenty native species selected by Mount Auburn's horticultural curator, Dennis Collins. In 2004 alone, nearly three hundred shrubs, more than fifty trees, and over 3,400 herbaceous plants were planted as part of the restoration. By 2019 the plantings had filled in and the urban woodland appeared similar to the day Story said to the gathered crowd: "All around us there breathes a solemn calm, as if we were in the bosom of a wilderness, broken only by the breeze as it murmurs through the tops of the forest, or by the notes of the of the warbler pouring forth his matin or his evening song."[121]

Much of the financial support for Consecration Dell came from the Anthony J. & Mildred D. Ruggiero Memorial Trust, which was established in 1994 to support Mount Auburn's wildlife habitat enhancement and related educational programs. The site is a rich parcel of natural habitat within a densely urban region and functions as a vital haven for local and migratory birds and a mating spot for resident spotted salamanders. As with several other projects funded by the trust, diverse plantings foster nesting space, protection, and food sources for a variety of animals. This habitat work bolsters the implementation of the institution's strategic plan, which seeks to "create a vision . . . that reflects both goals for growth and ecological imperatives, while staying true to our founders' vision" and lists as an initiative to be "a model of environmental stewardship."[122]

Even though Mount Auburn's landscape always responded to the environment from a conservation perspective, it was not until Barnett's presidency that it became a core focus of the institution. The timing was right, as responsible, "green" design had become common by the 2010s.[123] From the start of his tenure, Barnett committed to employing sustainable practices where possible to save money, improve the habitat for wildlife, and be a positive steward of the planet.[124] In everyday practice, the staff switched to more efficient irrigation, mulched fallen leaves on site rather than removing them, essentially ceased using toxic fertilizers, and transitioned to planting turf species that required less water and mowing. In addition, broad stretches of manicured lawn were systematically converted into native meadows.

Moving away from the water, energy, and labor burden inflicted by pristine turf not only created more habitat and biodiversity, it also signified a partial rejection of the once celebrated lawn-focused imagery of the twentieth-century American cemetery. Once again, allowing the landscape to evolve responsibly took priority over striving for historical accuracy. Although certain formal garden areas and ornamental Victorian-style landscapes were deemed culturally important enough to save, removing vast lawns and reducing the cemetery's carbon footprint was an overarching goal for Barnett and his staff. The naturalistic meadows convey a version of the past's rural landscape aesthetic and present an innovative solution for the future. In

Fig. 24. In conjunction with broader institutional goals for environmental sustainability and efforts to reach carbon neutrality by 2050, low-maintenance turf practices such as replacing mown lawn with native grasses were increasingly employed. (Photo by author)

some areas in deep shade or requiring a more formal look, mass monoculture plantings of low-growing sedges provide a lower maintenance option to traditional turf.

Simultaneous with this amplified focus on environmental initiatives was a continued search for innovative approaches to land management and creative options for burials. With fewer parcels available for burial plots, contemporary stewardship involved strategically and sensitively tucking pocket gardens into existing spaces. An example of this solution is Hazel Path, a meandering pedestrian route up a steep slope that leads to the iconic, Bigelow-designed Washington Tower structure perched atop the property's highest point. The project, a collaboration with the Cambridge-based landscape architecture firm Michael Van Valkenburgh Associates, envisioned the sloping site as a woodland sanctuary densely packed with native trees and shrubs and the cremated remains of approximately four hundred people. The naturalistic aesthetic is a departure from the fussy ornamental plantings of previous eras and presents a new type of space created in response to contemporary criteria without the apparent need to pay homage to the ghosts of Dearborn and Bigelow.

Contemporary Issues

By the time Hazel Path was completed in 2019, Barnett had been at Mount Auburn for more than a quarter century. He oversaw fifty-five permanent staff and almost the same number of seasonal helpers, with an operating budget of around $10 million.[125] The landscape, surely more beautiful than at any other time in its history, attracted more than 250,000 visitors per year. More than 100,000 people were interred in the cemetery, and sales of new burial space generated over $4 million per year.[126] Unlike with many public landscapes, lack of funding was not an issue. The permanent fund had grown to more than $160 million and the institution's assets totaled more than $200 million. Although space remained limited, careful planning ensured there would be room to accommodate new burials for decades longer.

At the meeting of the Mount Auburn Cemetery Board of Trustees in the fall of 2020, Barnett formally announced his upcoming retirement as president. This was during the coronavirus pandemic, when the venerable institution proved its value to the community as a place to enjoy nature and tranquility. As the government-mandated guidelines for social distancing went into effect and many Boston-area businesses closed their doors, people flooded through Mount Auburn's gates and crowded its paths and landscapes, forcing the cemetery to temporarily close to the public. While most employees worked remotely, rotating crews handled an increase in casket burials and tended the landscape in preparation for reopening. Barnett reported via newsletter: "After experiencing the last 27 spring seasons working at Mount Auburn and walking the spectacular grounds every day, it has been extremely difficult for me to stay home."[127] As conditions improved, the cemetery returned to a semblance of normalcy.

Among the last and most significant initiatives undertaken during Barnett's tenure was the formal adoption of the Climate Action and Sustainability Plan in the summer of 2021. This was a continuation of Barnett's leadership over twenty-five years, during which a focus on sustainable maintenance practices, greenhouse gas emissions reduction, and biodiversity guided stewardship efforts. With the immediacy and rising stakes of climate change, Mount Auburn Cemetery committed to becoming 100 percent carbon neutral by 2050. This ambitious goal is articulated in the Climate Plan, which details numerous initiatives. Among these are increasing biodiversity in the plant collections, using Mount Auburn's five thousand trees to store and sequester carbon storage, modifying turf management, improving wildlife habitat, responsibly managing water quality and consumption, and expanding public engagement, outreach, and education. The heavily researched plan is packed with detailed science. To some, it might seem a surprisingly forward-looking and data-driven document for an institution as established and traditional-looking as Mount Auburn. Understanding the institution's history, however, makes evident just how expected the initiative should be.

Pursued concurrently with the production of the Climate Plan was an eighteen-month process to complete the Strategic Bridge Plan to prepare for Barnett's retirement and provide the framework for a smooth and effective transition to new leadership. Because much of Mount Auburn's ongoing success can be attributed to the contributions of strong leaders, finding the right person to steward the landscape into its next phase was critical. In expressing what was surely a common opinion, the chair of the board of trustees commented about Barnett: "Dave doesn't just work at Mount Auburn, he lives Mount Auburn. He epitomizes the values that we came up with during the Vision and Values process. The Statement of Vision and Values talks about welcome, compassion, dignity, stewardship, sustainability, and integrity: all of which Dave stands for as a person."[128] Like Bigelow, Ames, Chesney, and Clendaniel before him, Barnett left behind a legacy strong enough to endure and flexible enough to accept change.

In September 2021, Mount Auburn introduced Matthew Stephens as its fourteenth president. Stephens had earned horticulture degrees from the University of Delaware and Illinois State University and worked as the garden director for the San Francisco Botanical Garden prior to taking the new position. In a letter to members and friends of the cemetery, Stephens emphasized the evolutionary nature of the cemetery and the institution's rich legacy, writing:

> Our founders set literal and figurative roots for what this organization could look like and in that moment redefined what a cemetery can be. It seems fit-

ting that we should all play a part in evolving the tradition of redefining what cemeteries can be. I don't know exactly how we are going to do that right now, but this incredible tradition of redefinition is in our DNA as an organization and that is our story. Wise people came before us and it seems only fitting to take the inspiration for them onward. It would be wonderful to walk around the Cemetery today and have a conversation with our founders, to ask them questions about what they see today and what they had envisioned—and how they would think the Cemetery might keep evolving.[129]

In the same manner in which Barnett carried forth Clendaniel's focus on master planning, Stephens appeared eager to follow his predecessor's board-sanctioned agenda on sustainability. Such consistency can be critical for ensuring that long-term challenges and initiatives are carried through to completion rather than being altered at the whim of a new leader. In the summer of 2022, Stephens penned an opinion piece for the nonprofit journal *CommonWealth* in which he prodded the Massachusetts legislature to change regulations to permit natural and innovative burial methods that are more environmentally friendly than lined coffins or cremations.[130] In addition to aiding Mount Auburn's goal of achieving carbon neutrality, this flexibility would allow the cemetery function to evolve along with science and society—an important factor for cultural landscapes.

Visitors to Mount Auburn may not recognize this continued emphasis on sustainability, which is not necessarily a bad thing. Either way, they experience a landscape that fulfills, from any angle, Jacob Bigelow's plans for a pastoral landscape meant to provide beauty and a sense of nature to comfort the bereaved. It is a spectacular place. Dearborn's fingerprints remain evident in the older parts of the cemetery as the brilliant canopy of mature trees spreads over an evolving landscape balancing the preservation of historic character with the addition of contemporary spaces. The public seems to recognize the cemetery's sacred aspect and for the most part respects it as a place for quiet reflection more than for boisterous ribaldry, which can often be found in large urban parks. Had it not enjoyed consistently deft oversight that kept things on track, the cemetery may have devolved into disjointed mess. Fortunately, the spirit of stewardship runs deep, and commitment to the landscape seems never to have wavered.

Like the Gardens of Alcatraz, Mount Auburn is a unique and significant American landscape that defies convention and flourishes thanks to successful stewardship. Both places leverage the passion and expertise of generations of leaders, managers, staff, and volunteers to create richly drawn histories while conveying a willingness to innovate in the face of change. Even with wildly different stories, aesthetics, contexts, and organizational structures, the two places similarly exemplify the inherent power of landscapes to inspire, heal, and comfort while maintaining the character and historical details that make them special.

3

CAMPUSES

Because the American university campus is a homegrown product, its physical form has grown from its own needs and setting. It is part of the larger community as well as a community unto itself, and it has evolved along with its changing landscape and environs.

—Laurie Olin, landscape architect, 2013

There are perhaps no designed landscapes as simultaneously venerable and dynamic as higher education campuses. Often steeped in tradition to the point that they gain unfair reputations for unimaginative staidness, these significant places exist in all fifty states, some enmeshed in the urban fabric of dense downtowns, others sprawled across sites encompassing up to thousands of acres. Often known for tree-lined quadrangles and historic campus cores, these landscapes can vary dramatically in appearance as their growth extends for decades or, in many cases, centuries. Scale, form, materiality, and character differ across the grounds of massive public universities, small liberal arts colleges, community colleges, and other institutional types. While famed and prestigious eastern schools with hallowed lawns and predictable historicist architecture embody the traditional ideal of what these places are *supposed* to look like, most American campuses respond uniquely to their own physical and cultural context.

Despite their differences, many colleges and universities similarly rely on their campus environs to attract prospective students, maintain cherished legacies, and reinforce social, environmental, and pedagogical missions. In many cases the landscape simultaneously becomes the postcard image used to market the school, the front yard for hundreds to thousands of young adults, and a living laboratory in which such matters as stormwater resiliency, climate change readiness, and spatial equity are tested and solutions sought. Whether the campus is beautifully manicured, barely maintained, or somewhere in between, the devotion students maintain for their alma mater's landscape seems fairly universal. This is especially true for residential schools, where campuses often become the first places away from students' families that they consider home. The subsequent bond that develops between people and place creates a proprietary relationship that separates campuses from other landscape types.

Unlike gardens and parks, where the landscape is frequently the focus, campus grounds are often forced to compete for funding against other institutional needs. These needs can include building maintenance, expanding facilities, and a myriad of other academic, operational, and administrative priorities. While some schools reap the rewards of hefty endowments and generous donors, others struggle to cover basic landscape maintenance costs. Regardless of the institution's funding capacity, the extent to which university boards and administrators value campus landscapes and prioritize their ongoing care can determine their relative success as contemporary places. Whether the campus landscape is managed by in-house staff or through outside consultants, any resultant stewardship is often complicated by institutional bureaucracies weighed down by multitiered decision-making processes.

Campuses are typically complex places that often function as small cities, with layers of physical development, infrastructure, transit facilities, and a composite of buildings and landscapes resulting from both good and bad planning and design decisions. Because of this complexity, it can be difficult to assign to them a singular design legacy. Some campus landscapes, such as that at the University of Virginia (UVA), have a definitive space—in this case the Lawn, the heart of the Academical Village famously designed by Thomas Jefferson—to which all subsequent architecture and landscape architecture pay homage. There is no escaping Jefferson's shadow while wandering UVA's grounds. More common, however, are the colleges and universities that either grew incrementally from one or two initial buildings—Princeton University and William & Mary College, for instance—or were built from a master plan generated by someone without the gravitas of Jefferson—the Olmsted Brothers at Duke, for example. Even though landscape architects and scholars value the work of the firm, others may not recognize the significance of an "Olmsted Brothers landscape."

Whatever their evolutionary journey, successful stewardship of these places typically relies on the ability of people embedded in the institutions to recognize and articulate which aspects of history matter, determine how best to respond to that tradition moving forward, and then garner the necessary support to make it happen. As with stewarding any landscape, this should not be a one-time decision-making effort but rather a process that reengages repeatedly as infrastructure ages, expansion takes place, and the ways in which students and faculty use campus spaces changes over time.

The following two case studies represent the distinctions and commonalities that can exist between campuses. At first glance, the two private universities—Stanford University in Palo Alto, California, and Tuskegee University in Tuskegee, Alabama—appear to have very little in common. Stanford's endowment surpassed $36 billion in 2022, making it one of the wealthiest educational and research institutions in the world. Located in the prestigious Silicon Valley, it regularly invests millions of dollars in maintaining and improving its landscape, which has become a model of planning, design, and sustainability envied by many other schools. Tuskegee, on the other hand, is much smaller and lists a 2022 endowment of around $110 million. Sitting in rural Alabama, the historically Black university (HBCU) feels introverted and somewhat isolated, even though the energy experienced on campus is no less dynamic than at Stanford. Its well-maintained landscape may lack the impressive design pedigree found at Stanford but is obviously valued as a contributor to the character of the school.

Other than UVA, Stanford has perhaps the most celebrated campus design of any college or university in the country. Monumentally formal, with a series of rectilinear quadrangles spreading axially across the flat valley floor, the plan was a joint effort between Frederick Law Olmsted and the railroad magnate and patron, Leland Stanford. The "Olmsted Plan," as many people call it today, provided an unmistakable road map for development, although it was not always followed as such. Tuskegee, which looks nothing like Stanford, developed in a less organized manner. Buildings, many of which were designed by faculty and constructed by students, were sited incrementally and somewhat organically along the tops of several conjoining ridgelines. Whereas Stanford's legacy is tied to the physical form of the campus, Tuskegee's is closely linked to the spirit of its founding principal and president, Booker T. Washington. As with UVA, this connection to its founder fuels pride and precedent, absent the controversy inherent in Jefferson's slave-owning past.

Although the differences between them are significant, Stanford and Tuskegee share a passion for planning and a commitment to shaping the grounds to meet the needs of their students. They also have common goals in striving for excellence and employ similar tactics for preserving cherished legacies while accommodating and facilitating change. As with many mature campuses, both have struggled with how to accommodate continued growth. Stanford has sought optimal ways to spend its ample money to create a beautiful contemporary campus, while Tuskegee has, for much of its history, been forced to maximize limited resources.

Perhaps the most significant similarity between Stanford and Tuskegee, at least in relation to the focus of this book, is the impact the in-house planning and design staff have had on each campus. Dating back to the campus design's origins with Olmsted, Stanford has consistently engaged the country's most notable and celebrated architects and landscape architects to create new buildings and spaces. At the same time, the university is credited with creating one of the first and most robust planning offices of any academic institution in this country. Generations of professionals have provided critical oversight and, importantly, a consistent approach to addressing both expansion and preservation. At Tuskegee, founder Washington understood the benefit of hiring people who could simultaneously teach and practice, thus saving money by having staff fill roles while increasing the commitment to the institution and its students. Many of the buildings and landscapes on campus today were designed by professionals who both taught and practiced. This embedded expertise has benefited the development of both institutions.

Stanford University, Palo Alto, California

Stanford University is among the elite educational institutions in the world. It has one of the best reputations, largest endowments, and historically significant campuses of any college or university in the United States. The school was founded on over eight thousand acres of ranch land on the San Francisco Peninsula as the Leland Stanford Junior University in 1885 by then California senator Leland Stanford and his wife, Jane Stanford, as a memorial to their beloved son, Leland Jr. Its legacy as a landscape is tied to both its patrons and its credited designer, Olmsted, who modified his established campus planning model to accommodate the demands of his wealthy and opinionated clients. Leland Stanford was rich, powerful, and stubborn, and proved more than a match for the venerated Olmsted, who was in the waning years of his career and full of confidence and ego. The two men agreed on little at the beginning but managed to successfully compromise on an elegant Beaux-Arts–inspired master plan from which to build and expand the campus. What makes the university's evolving story particularly compelling is how generations of administrators spent nearly a century and hundreds of millions of dollars moving away from Olmsted's original plan, only to spend billions more trying to re-create it in the twenty-first century.

Stanford sits in California's richly planted Santa Clara Valley, nestled between the Santa Cruz Mountains and the San Francisco Bay in an area better known today as Silicon Valley. The notoriously precarious San Andreas Fault and the lesser-known Hayward Fault run in parallel through the valley on either side of the twelve-mile-wide bay. Before Europeans settled the area, the native landscape was made up primarily of loosely spattered groves of oak woodlands interspersed with swaths of grassy savannas, with extensive marshes and wetlands at the water's edge. The first people to inhabit this part of California were members of around fifty Indigenous tribes collectively referred to as the Ohlone. Their lives were forever changed in 1769, when the Spanish explorer Gaspar de Portolá led the first land-based expedition up the California coast, ending near present-day Palo Alto. Spanish missionaries soon permeated the land and forced the Ohlone to adopt the culture and religion of the colonial empire.

By the mid-nineteenth century, an estimated 90 percent of the Indigenous population was gone. Missionaries systematically transposed the valley floor into an agricultural landscape, planting rows of beans, corn, and peas along with large fields given over to wheat and gridded orchards filled with apple, pear, apricot, and peach trees. With its fertile soil and serene climate, the region later earned the nickname "Valley of Heart's Delight." The area's population grew with the construction of the railroad line connecting the rapidly growing cities of San Jose and San Francisco. Tracks reached the Palo Alto area in 1863, spurring wealthy San Franciscans, many made rich by the 1849 Gold Rush, to build large estates along the route. Among them were Leland and Jane Stanford, who purchased 650 acres of existing ranch land to create the Palo Alto Stock Farm, on which Leland fulfilled his passion for raising and training some of the fastest racehorses in the world. Over time, the couple purchased several other farms and parcels that eventually totaled more than eight thousand acres.

Establishing the Legacy

Leland Stanford was an archetypal Gilded Age entrepreneur who, like Henry Shaw, made his millions from the nineteenth-century westward expansion across the continent, first by selling goods to gold-seeking miners and then as one of the builders of the transcontinental railway. After establishing his reputation as a businessman, Stanford entered politics. An early organizer of the Republican Party in the city and state, in 1861 he was elected governor of California and president of the Central Pacific Railroad Company. He was ambitious and sometimes unscrupulous, known to exploit the power of his office to benefit the railroad.[1] On May 10, 1869, in front of workers, onlookers, and photographers, Stanford pounded in the famed ceremonial golden spike in Promontory Summit, Utah, symbolizing the completion of the railroad that connected the opposite coasts.

The Stanfords became extremely wealthy in a relatively short period. Their son, Leland Jr., was born in 1868, and they lavished on him privileges unknown to most children. The family traveled widely, including two Grand Tours across Europe. On the second trip, in 1884, the fifteen-year-old boy fell ill and died of typhoid in Florence, Italy. Many accounts attest the couple never got over their son's death. No longer able to leave their vast fortune to him, they committed themselves to creating a memorial to him instead.[2]

In 1885, by which time Leland Stanford was a U.S. senator and the seventh-wealthiest man in the country, he and Jane created the founding grant that formally and legally established the Leland Stanford Junior University. The couple sought to create a progressive nondenominational and coeducational institution at a time when most colleges and universities were all-male and tied to a religious organization.[3] They followed the advice of their friend Francis Walker, president of the Massachusetts Institute of Technology, by hiring Olmsted and the architect Charles Coolidge to design the university.[4] Leland Stanford had preconceived ideas about the campus and firmly established a process in which control of the design of the landscape and its buildings belonged to him and Jane.[5] The ensuing quarrel between Stanford and Olmsted over the layout of the campus is well documented, notably by the architect and art historian Paul Venable Turner in his book, *The Founders and the Architects: The Design of Stanford University.*

In the summer of 1886, Olmsted and Walker visited Palo Alto to "select a site and form a building and landscape scheme for that good man Governor Stanford's University."[6] Planning sessions in the fall led to a direction for the campus plan with which Olmsted did not agree. In September, he wrote, "The site is settled at last—not as I had hoped."[7] He favored an organic arrangement of buildings tucked into the foothills, "near the edge of an elevated table-land, commanding a fine characteristic California distance."[8] Leland, on the other hand, believed the campus belonged on the valley floor, with topography better suited to the quadrangle-based scheme he believed would create a more impressive monument to his son.[9] He asserted his authority as the client and made known that even though Olmsted might be disappointed, Leland was going to arrange the buildings the way he wanted them.[10]

The final plan is organized around a formal geometry, with a series of quadrangles stretched out along an east-west axis and intercepted by a dramatic perpendicular access road connecting to a central open space. One-story, mission-style buildings with clay tile roofs are connected by a series of arcades that establish a pattern logically expandable in the future. The concept of extended and unbroken views span-

Fig. 25. The original master plan for the Stanford campus was a collaboration between Frederick Law Olmsted and Leland Stanford. The monumental, quadrangle-based formality of the plan was Stanford's choice and went against the more organic campus layout typically favored by Olmsted. Fulfilling the plan proved to be a challenge during the university's first century. (Stanford University Libraries, Department of Special Collections and University Archives)

ning the length of the developed campus in both orthogonal directions is a critical element of the plan. The layout is distinctly monumental and rigidly architectonic, both characteristics unfamiliar to most of Olmsted's body of work. The campus edges, on the other hand, are buffers of naturalistic landscapes more in keeping with his style. An arboretum flanks the entry drive to the south, and the spreading native foothills he described as *forest plantations* grow to the north. The opposing aesthetics of formal versus informal or refined versus rustic remain at the heart of the campus's dynamic character.

Fig. 26. The only landscape designed in detail by Olmsted's firm was the Inner Quad. Although Leland Stanford wanted a traditional lawn in the space, Olmsted favored paving the surface and filling the circular planting beds with plants adapted to the Mediterranean climate. (Stanford University Libraries, Department of Special Collections and University Archives)

While Olmsted's plan determined the layout of the campus, the only detailed landscape he designed was the Inner Quad—a paved open space framed by arcaded buildings. Leland Stanford, who, like his wife, admired the heritage and aesthetics of traditional campus landscapes, initially favored covering the ground in turf. Olmsted, on the other hand, preferred laying paving blocks in concrete to create the impression of "one impervious stone and at a little distance its appearance is that of granite."[11] During construction, Leland switched to a less expensive bituminous product as a means of cost saving. Upon learning of this, Olmsted blasted the choice, describing the product's "sad color and soft, slimy texture" and complaining it would turn what should be a celebrated space into something "unpleasing."[12] His objection, however, was apparently ignored, as the quadrangle was covered with a heavy coat of asphalt along with the eight beds filled with semitropical trees and plants. Had Olmsted intended to design other spaces, Leland had different ideas.

He reached a point when meeting a deadline to open the university took precedence over design, and he was not shy about making decisions to meet the schedule. In the fall of 1889, by which time Olmsted was beginning work at the Biltmore estate, Leland was confident he could execute the landscape architect's plan with his own people, and essentially ended the relationship with the firm. In a response to a personal note of congratulations from Olmsted following the opening of the university, Leland congenially reported, "We are gradually improving the grounds in accordance with your plans."[13]

The Death of the Stanfords

On October 1, 1891, under characteristically clear blue skies, thousands of people showed up for the opening ceremonies of Leland Stanford Junior University. More than four hundred students, ninety of whom were women, stood in line to register for the first class. This was the culmination of Leland and Jane Stanford's dream, for which they had endowed "the munificent sum of twenty million dollars."[14] At the helm was David Starr Jordan, who had moved west from Indiana to become Stanford's first president and, in his words, "develop in California a university of the highest order."[15] Jordan was a scientist who shared Leland's progressive views that education needs to be accessible to all students and that adherence to religion should not be required. He was an influential leader who spent his time doing the university's business of hiring faculty and establishing operations while Leland continued to build the physical campus until he died unexpectedly of heart failure on June 21, 1893.

Per the founding grant, authority was transferred directly to Jane Stanford as the institution's sole trustee. She forged ahead erecting buildings with, as one writer characterized it, "the commanding meddlesome love which an unbridled maternal instinct thrusts upon an only child."[16] In addition to constructing a second quadrangle per the master plan, she authorized several buildings that contradicted the intended pattern of lateral development, thus signaling the end of any meaningful adherence to Olmsted's scheme.[17] In 1903, following continued legal proceedings triggered by Leland's death, she formally relinquished the powers reserved to her as surviving founder.[18] Per the grant, a board of twenty-four trustees would manage and control the university and "maintain on the Palo Alto Estate a farm for instruction."[19] Two years later, Jane died mysteriously from apparent strychnine poisoning, ending the familial connection that had created the university.

The First Quake and Post-Stanford Development

Months after Jane's death, the infamous 7.8-magnitude Great Earthquake of 1906 struck along nearly three hundred miles of California's San Andreas Fault, leveling an unprepared San Francisco and killing thousands of people. One student and a firefighter at Stanford died, while numerous campus buildings either were destroyed or suffered significant damage. Aside from a toppled chimney, however, the buildings in the original quadrangle remained intact.[20] The architecture firm that designed those early structures credited Leland, saying, "He told us his theory for withstanding earthquakes was to have broad footings under the walls."[21] The university set about rebuilding the campus, though now without the aid of the founders' considerable fortune. With funds now directed toward reconstruction, ambitious plans to expand with new quadrangles were put on hold.

In 1914 the board of trustees engaged James Frederick Dawson, a principal in the Olmsted Brothers firm, to evaluate the early progress of Stanford's landscape. The resulting twenty-two-page typed report expresses unequivocal support for Olmsted Sr.'s original plan and design intent. According to Dawson, new work on campus "ought to be loyal to the important ideas of design of Mr. Frederick Law Olmsted, Senior, because they were almost always sound and admirable."[22] He recommends selectively thinning trees and increasing species diversity in the arboretum and planting the grounds "in harmony with the Mission style rather than in the English landscape gardening style, by using small growing shrubs, vines and perennial 'ground cover' rather than turf."[23] He complains repeatedly about the presence of "bare ground," especially in the quadrangles, and remarks how distressing it must be for students to "have constantly forced upon their attention these uneven, unkempt, ugly areas of dirt" that "have apparently been in this unimproved condition ever since the erection of the buildings."[24] The report adds explicit planning recommendations, such as where essential buildings such as the library and museum should sit in the master plan. Although the firm offered proposals for numerous additional services, it does not appear it was engaged for much more work on campus aside from two small projects.

The university grew over the following decades. Once it became evident that the endowment was insufficient to meet the financial needs of the growing campus, administrators and trustees broke with Leland and Jane Stanford's wishes in the 1920s by charging tuition fees. They also embarked on a fundraising campaign to raise $3 million to endow faculty positions, support the medical school, and build new dormitories, academic buildings, and athletic and dining facilities on campus.

In 1920 the Scottish-born and -trained horticulturalist John McLaren, who helped to transform barren dunes into San Francisco's lush Golden Gate Park, was brought to Stanford as a consulting landscape architect. In addition to specific site designs, McLaren led planting initiatives across the campus. An unsigned 1920 letter to "C.S. Sargent of the Arnold Arboretum" reads, "You will be interested to know that Mr. McLaren is helping us prepare plans covering our entire 9000 acres."[25] How much of this work was accomplished is difficult to discern, yet it is evident there was some broad-based thinking about the campus landscape beyond the historical core. A 1927 newspaper article states: "The consent of the Board of Trustees of Stanford University having been obtained to appropriate 1000 acres of campus territory for the development of the greatest botanical gardens in the world whenever sufficient funds are within reach for assuring the success of the project."[26]

The Great Depression seems to have permanently thwarted these ambitions while also deflating both the university's endowment and its plans for campus expansion. One initiative allowed to proceed was the controversial redesign of the entry onto campus at the iconic Palm Drive. The project, which was announced in the fall of 1929 and was designed by the San Francisco architects Arthur Brown and John Bakewell, cost $30,000 and included the construction of two large sandstone arches with red-tiled roofs and low stone walls. The features were erected in 1930, and a year later the Olmsted Brothers firm was hired to enhance the surrounding landscape. The editorial board of the student newspaper questioned the relative need for the "much-heralded and widely discussed" beautification project, writing, "It should not take a great deal of thought to suggest a better way to spend such an amount with so many of the University's departments held back by a lamentable lack of money."[27] Despite the downturn, the university continued to build, slowly, including the less controversial construction of a gymnasium for women in 1931, for which the Olmsted Brothers firm was once again hired, apparently for the final time.

Internal Planning to Manage Post–World War II Campus Growth

A lack of real estate has never restricted Stanford's growth. Possessing land was important to Leland and Jane Stanford and a fundamental component of their founding grant. The couple's land holdings were so extensive that possessing space on which to expand would likely never be a concern, making "the Farm"—as the campus became known—a vast blank slate on which the university could plan and build seemingly indefinitely. There was, however, a caveat: trustees were prohibited from selling Stanford property for any reason, which meant that monetizing the vast landholdings through a traditional sale was not an option.

The need to develop the campus increased exponentially following World War II when, thanks to the signing of the G.I. Bill in 1944, more than 2.2 million veterans flooded colleges and universities across the country. At Stanford, nearly one-third of the 4,500 undergraduate students enrolled in the winter of 1946 were funded by the bill.[28] The university had yet to gain a significant national reputation and continued to struggle financially. A boost came during the early years of the Cold War when faculty in science and engineering programs were called on to aid the federal government in addressing evolving national defense priorities. Research grants provided a new revenue stream and accelerated the need to add faculty and update facilities. From a design perspective, modernism was taking hold, and innovations in material technologies drove aesthetic trends that dramatically altered the look and scale of Stanford's architecture. The low-slung, intricately detailed, and craftsman-built stone mission-style buildings and their neoclassical counterparts were not practical to meet changing needs, and with an immediate need for large structures for laboratories and classrooms, poured concrete became a more expedient and economical material.

In anticipation of imminent campus growth and with the understanding that "it is costly to plan, but it is more costly not to,"[29] Stanford's then president, Donald Tresidder, established an in-house planning office in 1945. Such an office was a relatively new concept for universities at the time. Tressider hired Eldridge T. Spencer, a Beaux-Arts–trained architect from San Francisco, as the first director of planning and supervising architect on a part-time basis. Spencer believed that Stanford needed a flexible, principles-based document that could direct the expansion of the campus for decades, rather than the typical graphic master plan, which tends to remain static.[30] He was a modernist and favored architecture that broke with Stan-

Fig. 27. This 1950 photograph shows the primary approach drive through the Arboretum, the oval turn-around, and, aligned with that zone, the first phase of buildings in the quadrangle based on Olmsted's master plan. As shown toward the upper left of the photograph, the lateral, uninterrupted axes critical to the plan were beginning to be blocked with new buildings. (Stanford University Libraries, Department of Special Collections and University Archives)

ford's historicist predilection for red-tiled roofs, arcades, and sand-colored façades. One of his guiding principles was that "architectural unity . . . achieved by an observance of the fundamentals of architectural composition is more important than the pseudo architectural unity which results from a simulation of the University's original architectural style."[31] In other words, emulating the form, scale, and proportion of existing buildings took precedence over mimicking their appearance.

Once Spencer developed planning studies and principles, Tresidder hired the urbanist, architecture critic, and former Stanford humanities professor Lewis Mumford to review the documents. In his 1947 "Memorandum on Planning," Mumford notes the incoherence of the university's recent expansion and suggests administrators look back to the "compactness, concentration, [and] unity" of Olmsted and Leland Stanford's original plan and focus on creating "a concentrated urban group in a permanent rural setting."[32] Even though his respect for Olmsted is evident from a reference to the historical campus plan as the "work of the most mature and effective mind America has so far produced in city planning and landscape design,"[33] Mumford, like Spencer, followed modernist norms and favored siting buildings as individual objects rather than as parts of quadrangles.

Of particular concern to Mumford was the negative impact cars had on the rapidly densifying campus. This was not a new concern for him as a year earlier he had complained about the broader downside of America's booming car culture, commenting that an automobile enjoyed a "sacred right to go anywhere, halt anywhere, and remain anywhere as long as its owner chooses."[34] This certainly appeared to be the case at Stanford by the 1940s, when roads were widened and paved and ad hoc parking areas appeared across the growing campus, even adjacent to the historical quadrangles.[35] He discusses the need for a thoughtfully designed circulation system that pushes automobiles to the perimeter to "create a unified university quarter, free from the invasion of traffic."[36] His proposal for a loop road around the campus to alleviate traffic in the interior was one of his main proposals that stuck, though it would take decades to accomplish.

A Modernist Layer in the Landscape

By the late 1940s, the campus had devolved into a relative mess as surface parking lots and poorly sited buildings fought against the intention for an ordered landscape. Following Mumford, planners and consultants produced volumes of plans, proposals, maps, and initiatives charting the future of Stanford's extensive landholdings. The university built an office park and a shopping mall on its property and entered into ongoing negotiations with local municipalities to determine what it could and could not do with its land. On campus, hurried design and construction processes bypassed careful planning to build residence halls and classrooms to accommodate enrollment, which had doubled to around eight thousand students since passage of the G.I. Bill. Not unexpectedly, the strong framework inherent in the historical master plan gradually disappeared. Following the burgeoning modernist movement's push toward innovation and away from tradition, Stanford began constructing an array of buildings designed as stand-alone objects and with little architectural cohesion. They were frequently sited fronting streets rather than framing quadrangles, blocking carefully planned axes and vistas, and ignoring the original plan's goal of creating meaningful open spaces. Parking lots, paved and not, continued to appear in scattered locations as more students, faculty, and staff brought their cars to campus.

Such was the state of Stanford's campus when the California landscape architect Thomas Church began his lengthy involvement with the university. He was quickly earning his status as an icon of modernist landscape design, having completed his most famous landscape, the Donnell Garden in Sonoma County, in 1948, the same year he worked on his first project at Stanford. Whether explicitly stated as such or not, it fell to Church, aided by assistance and oversight from Stanford's planning staff, to tie together the fractured campus through a series of landscape interventions.

An early challenge presented to Church was the ongoing deterioration of the only landscape designed in detail by Olmsted—the historic Inner Quad, which, thanks to its significant heritage, remained the campus's most historically significant designed landscape. Even with this distinction, however, the space was not immune to the detrimental impacts of neglect. In 1947, a student described the landscape as "drab, grey gravel that becomes one huge puddle in the winter."[37] Poured concrete paths ran diagonally through broken asphalt and gravel to accommodate popular pedestrian routes. The plantings in the circles were untended, with one writer for the *Stanford Daily* reporting in 1948: "The innocent vegetation has grown and . . . has become a devouring monster." [38] To address the deteriorating conditions, Church proposed repaving with asphalt, widening the planters by six feet, pruning the trees, and replacing the shrubs and lower plantings with ivy. He also recommended encircling the beds with low seat walls to control pedestrian circulation and make the space more functional.

Although there was some support for the plan, no money was allocated for the improvements aside from "taking out trip roots" that had crept into the path of travel.[39] A year later, the planning office committed to a new standard for landscapes. From then on, the "Inner Quad style of landscaping," with prominent trees and minimal shrubs planted in open paved areas, would be the norm for new buildings.[40] In addition to preserving views of the architecture, the style would be more cost-effective with respect to long-term maintenance and would respond to the arid climate in the manner recommended by Olmsted. It also adhered to Mumford's suggestion that "urban" spaces be intermingled with more "rural" ones.

In addition to the fifty-six-acre grounds of the new hospital and medical center built in the mid-1950s, Church planned and designed numerous campus landscapes. Between the time he started at Stanford in 1948 and his death thirty years later, he was credited with completing "about two-thirds of the executive landscape work" on campus.[41] His workload at the university increased substantially in the 1960s when a significant flush of money drove an extended period of expansion and notably reversed Stanford's early financial struggles.

Two Decades of Growth

Stanford has not always been among the country's wealthiest universities. Traces of its fiscal turnaround date back to 1939, when noted alums William Hewlett and David Packard founded their eponymous company in Packard's Palo Alto garage. The men were at the forefront of the computer age and instigated the technological innovation that would become the university's trademark, as well as the birth of Silicon Valley. Their eventual generosity toward their alma mater in the form of millions of dollars' worth of donations foretold the rise of super-wealthy donors who would fuel the university's emergence as an elite institution and help fund its seemingly endless growth.

Among Stanford's early fundraising heroes was its fifth president, Wallace Sterling, who took over after the sudden death of Tresidder in 1949. A decade later, under Sterling's leadership, the university received $25 million from the Ford Foundation, which at the time was the largest unrestricted grant ever given to an American university.[42] In order to receive the gift, however, the university needed to raise an additional $75 million. In 1961, Sterling formally announced a development campaign called "The Stanford PACE Program—A $100 Million Plan of Action for a Challenging Era," which he trumpeted as "our boldest venture since the university was founded."[43] More than $40 million of the total was earmarked for new buildings and facilities, making the physical campus the campaign's greatest beneficiary.

Sterling hired Harry Sanders as planning director in 1959. For the first time, the position was a full-time job.

Sanders was committed to Stanford and was known to work eighteen-hour days during the busy 1960s. According to his son, his father's "greatest personal belief was that there should be a unifying model for the planning and design of the university."[44] While acknowledging the importance of major building projects, Sanders understood the significance of creating and preserving human-scale open spaces because "these are the things which visually most affect our everyday lives."[45] Under the fifteen years of his leadership, the planning office, which he bolstered with more in-house architects, engineers, landscape architects, and project managers, oversaw more than five hundred building projects. Likely because Sanders was a planner and not a design architect, Sterling formed an advisory council to help oversee the design of new buildings. Its first appointed members were three well-known modernist architects, Gardner Dailey, Milton Pflueger, and John Carl Warnecke, each of whom had designed buildings for the campus. Joining them was Church, who, even as a landscape architect, was considered by Sanders and others "one of the finest architectural critics in the United States."[46]

Over time, Church became embedded in Stanford's culture. People were accustomed to seeing him on campus, wearing one of his trademark fedoras, looking at trees, and sketching designs on the back of an envelope.[47] He carried pruning shears with him, prepared to snip and trim plants when he saw fit. Sanders considered Church his confidante and a person he "knew well enough so that I could call him at any time to get his opinion on architects or other landscape architects, get his opinion on designs."[48] The landscape architect communicated regularly with President Sterling and even presented his plans directly to the trustees, the first consulting designer to do so. Perhaps as important as these high-level interactions was his regular collaboration with the planning office, which at the time had two landscape architects on staff, Dan Rolfs and Oscar Nelson.

Rolfs completed numerous small landscape designs of his own and would walk the campus with Church at least weekly, discussing plans for projects and developing solutions to specific landscape issues.[49] Nelson, on the other hand, worked on a broader, campuswide scale as a master planner. He worked at Stanford from 1955 to 1982 and is credited with establishing a sense of order and continuity across the campus by unifying standards for signage, lighting, and site furnishings.[50] As the owner's representative, Nelson would sit in on project design meetings and inform design teams and project managers what they could and could not do on Stanford's campus.[51] This oversight helped prevent the "ragged edges" that can occur at seams between projects designed by different design teams and create a sense of unity in new landscapes.[52] As the planner responsible for roads, he also worked with Sanders and Church to transform the core of the campus into a "pedestrian preserve" by pushing out cars, as Mumford had advocated.[53]

Despite the presence of Church and committed and focused landscape architects on staff, the campus landscape remained an underfunded work in progress in a condition that drew criticism from the university community. As one student put it in a 1966 letter to the editor of the *Stanford Daily,* "Let's face it, Stanford is as ugly as a dirt road border town, with none of the picturesque quaintness. Walking down the magnificent asphalt paths stunningly bordered in dirt and brownstone, one can't help but feel something might be salvaged if there were a better utilization and landscaping of the space between the barren buildings."[54]

Church produced a master plan for the campus core in the 1960s that reads today as a blatant affront to Olmsted's scheme and suggests he was not swayed by the gravitas of his predecessor's legacy. In contrast to the historical plan's monumental gestures and simple geometries, Church's version takes a more intimate approach by carving up the landscape into an extensive collection of intricately detailed plazas, gardens, and courtyards. Sinuous planting beds deviated from the original geometry. Reflecting the modernist ideal that he had helped create, the campus as envisioned by Church is a collection of spaces unique unto themselves that seemingly promote individualized ornamentation more than an attempt at unity. While his plan arranges most new buildings with orthogonal precision, he perpetuates existing problems by ignoring Olmsted's long axes and blocking vistas with groups of trees and circuitous walks. In perhaps his most blasphemous move, he sites buildings and surface parking lots within the native and naturalistic arboretum flanking Palm Drive that was considered a no-build zone by Olmsted and Leland Stanford.

It is difficult in hindsight to discern how much of Church's plan was realized, but the dissolution of the Olmsted plan clearly persisted. One the most egre-

Fig. 28. This 1970 photograph shows how the axes intended in Olmsted's plan to be left open to create visual corridors were systematically blocked by development. Here, Escondido Mall is shown blocked by Meyer Library, which was razed in 2007 and replaced with an open landscape. (Stanford University Libraries, Department of Special Collections and University Archives)

gious offenses against the plan was the J. Henry Meyer Memorial Library, a nearly 150,000-square-foot concrete behemoth that, once opened in 1966, squarely blocked Escondido Mall, one of the primary east-west lateral axes intended to be left unobstructed. In addition to this questionable planning move, the library's architecture became a target of ridicule, quickly earning the nickname "UGLY" (a contraction of "undergraduate library") and inspiring one writer for the *Stanford Magazine* to call it a bastard child of fascist architecture and California mission style.[55]

The campus construction boom extended into the 1970s with the announcement in 1973 of an $83 million, ten-year building program intent on reversing the pattern of outward expansion and making the historical core "the focal point of campus academic activity."[56] This included a commitment to move cars from the center of cam-

Fig. 29. White Memorial Plaza was designed by the landscape architect Thomas Church, who worked on the Stanford campus for many years. The space conveys his modernist aesthetic, with curved benches and a central fountain that contradict the geometry of the original plan. This is the best preserved of all the spaces Church designed for the campus. (Photo by author)

pus by completing a preliminary, loosely realized iteration of the long-planned loop road and erecting parking structures along the perimeter. Church worked extensively during this era, creating designs for individual building sites and new open spaces, most notably the fifteen-acre student activities center anchored by White Memorial Plaza, his signature space on the campus and the best remaining example of his influence. Sanders believed at the time that Church had more influence in creating and shaping the Stanford campus than any other single person apart from Olmsted.[57] Since much of his work was destroyed over time to make way for new development, the remaining spaces he designed have added a layer to the campus and are considered somewhat sacred today. It is difficult to tell how long they will last, however. As Oscar Nelson presciently stated in 1973 and remains true today:

"I don't think the campus will ever be in a finished state so long as our demands are constantly changing."[58]

Restoring Olmsted's Inner Quad

By 1980, Stanford was finishing up two decades of successful fundraising and extensive construction across campus. Lip service had been paid to Olmsted's legacy, while in practicality, it seemed to mean little. As the national economy slowed with a nearly 15 percent inflation rate and an unemployment rate stagnating at over 6 percent, Stanford, like its peers, had less money to spend and anticipated a quiet decade ahead.[59] In a cost-saving mode, administrators cut back on grounds maintenance, a common institutional practice that solves short-term economic problems yet inevitably leads to a systematic deterioration in the landscape. Without needed care, the lawn inside the iconic Oval at the end of the Palm Drive axis—once tended by grazing sheep in the early years of the university—became riddled with bare spots and covered in weeds, making its appearance hardly worthy of its high-profile location. With funds diverted elsewhere, restoring the lawn fell to an anonymous $150,000 gift from a former Stanford employee and his wife, who walked past the area every morning and understood it deserved better.[60] Their gift provided new sod, irrigation, and a raised annual flower bed that remains a frequently photographed element on the campus.

At the same time, conversations and controversies over how to improve the conditions of the Inner Quad were continuing. Plans to resurface the crumbling pavement and uneven gravel with a fresh coat of asphalt were met with a counterproposal by two urban design students who advocated a finer treatment using bricks, tile, or pavers.[61] They also recommended adding two additional circles planted with turf to provide "comfortable and relaxing places to sit."[62] In the students' minds, contemporary function appeared to have taken precedence over preservation. This suspicion is supported in an editorial in the *Stanford Daily* that questioned whether Olmsted "would have us bind ourselves to his original concepts more than 70 years after they were conceived" and argued that satisfying students' wishes was more important than a "dogmatic and inconsistent loyalty to 'historical accuracy.'"[63]

As with the lawn in the Oval, anonymous donors spurred the Inner Quad's eventual restoration, this time with a $1 million gift. At the donors' request, the university hired the San Francisco landscape architecture firm Anthony Guzzardo and Associates. Potential solutions included restoring the space to its original design or adapting it to meet the changing needs of the dynamic campus. The initial plan supported a change approach by adding red brick pavers, expanding the diameters of the circular planters from fifty-six feet to nearly ninety feet, bordering them with three-foot-wide seat walls, and filling them with turf and new plantings.[64] A member of the planning staff reported that although the plan was not historically accurate, it was "true at least in concept to the original Olmsted idea."[65]

Interested constituencies within university communities can be complex and challenging to fully understand. In addition to faculty, administrators, and staff with varying degrees of applicable expertise, transient study bodies often provide a constantly changing set of interests, issues, and pet projects current in their time. What to do with the Inner Quad generated heated debate during campus meetings and in a series of newspaper editorials, letters, and articles. There were differing opinions about how closely to emulate the space's original design. In March 1983, on the same day that the university welcomed Queen Elizabeth II to campus, the *Stanford Daily* published a letter from Philip Williams, the director of planning, about the controversy. In it, he confirms a compromise approach, writing, "We begin with strong respect for history and the original plan concept of Frederick Law Olmsted, openness to enhancement, and a directive to study a wide range of design and materials options."[66] In other words, the design would be modified as a respectful reflection of history that met contemporary needs. Ultimately, after decades of discussions, the project was completed. The quadrangle was resurfaced with reddish brown concrete unit pavers, and the circular planting beds were retained at approximately the same size, with low concrete seat walls. Existing plants were pruned, shaped, and supplemented with new plant material and irrigation. Although still not what Olmsted had envisioned, the space benefited from the open debate and dialogue within the university community.

Focus on the Core

During the relative quiet of the 1980s, planners balanced the intimate scale of the Inner Quad restoration with a macro effort focused on correcting decades' worth of suburban sprawl. Consolidating academic programs geographically emerged as the preferred way to unite the disjointed campus while fostering interdisciplinary collaboration and densifying the campus as was originally intended. In 1985 the planning office led the master planning for an ambitious $250 million initiative referred to as Near West Campus Redevelopment. Stanford's eighth president, Donald Kennedy, called the project "the greatest single privately supported building program for science and engineering . . . that has been undertaken by any university in the world."[67] William Hewlett pledged $50 million to the initiative as part of an aggressive, campuswide billion-dollar fundraising effort, the Centennial Campaign, launched to commemorate the university's upcoming anniversary. The plan proposed tearing down existing buildings to the west of the Main Quad and creating a forty-one-acre collection of carefully and deliberately arranged new buildings. It framed open spaces per the original master plan yet failed to restore the axes envisioned by Olmsted.

Momentum for the project was interrupted on October 17, 1989, when the 7.1-magnitude Loma Prieta earthquake struck the Bay Area, killing sixty-seven people and inflicting over $5 billion in damage. As was the case in 1906, the quake's impact on Stanford was also significant. The direct cost to repair damage to more than two hundred buildings climbed to $160 million and subsequently spurred around $100 million more to be spent strengthening seismically unsound buildings across campus.[68] Comprehensive planning for recovery took precedence over erecting new buildings, and although several structures already under construction were completed, the Near West plan stalled.

Figuring out how to proceed within these evolving conditions fell under the purview of David Neuman, who had been hired a few months earlier as the university's first resident university architect.[69] After decades of placing planners in charge of the physical campus, the university's trustees now wanted a more design-focused architect to ensure the aesthetic merit of new buildings. Neuman, who had previously worked at the University of California, Irvine, was known for both his planning and his architectural expertise. Like Sanders before him, he recognized and promoted the value of the campus landscape, once commenting, "If you took away the landscape and you took away the order of [the campus] and just kept the same buildings, it wouldn't be Stanford."[70] Over the next fifteen years, he would be the leading force behind the university's renewed commitment to its historical plan.

Plan for the Second Century

The earthquake altered university priorities and triggered the need for a new master plan. The original plan by Olmsted and Leland Stanford garnered renewed attention as a result of the university's centennial and prompted the board of trustees to ask Neuman to study whether the plan could be resurrected in a meaningful form. Upon considering this question, he realized, as he later recounted, "it was a damn good plan and it was strong enough that we could recover it if we wanted to."[71] It was not about just Olmsted or Stanford, however. Regardless of its cultural significance or whose name was attached to it, the plan's fundamental premise that expanding within a logical and ordered framework could reinforce the clarity and quality of the campus environment worked just as well in the twenty-first century as it did in 1891.

Out of this basic understanding came the Plan for the Second Century, which, unlike the complex intricacies found in Church's master plan, distilled the campus core into a simplified diagram. Harking back to Olmsted, the plan conveys horizontal and vertical axes defining developable quads encircled by the now familiar loop road ringing the campus. With its simple geometries and evident intentions, the plan is a perfect example of the landscape architect William Johnson's tenet, "If your president can draw it on a napkin, you can achieve it."[72] The board of trustees was convinced and voted to approve the plan in March 1991. History, it seems, was in style again.

An early project that emerged from the plan was the reconstruction of the monumental and iconic tree-lined Palm Drive, which had evolved as a functional piece of infrastructure but never achieved its intended quality. As it transitioned from a route for horse-drawn buggies to one carrying thousands of cars per day, it received only minor repairs, primarily the filling and repairing of endless cracks and potholes that drew the ire of the university community. In 1994 the university invested more than $3.5 million to improve the drainage system, evaluate the 152 historic Canary Island palm trees that flanked the drive, and add sidewalks, bike lanes, new paving, native plantings, and the granite curbs specified in the 1890s.

During this same time, a renewed interest in building a new precinct for sciences and engineering led to the design of the rebranded William R. Hewlett and David Packard Science and Engineering Quad, named in honor of the two men, who gifted Stanford nearly $80 million. The architectural firm of Pei Cobb Freed & Partners joined the Olin Partnership landscape architecture firm in a collaboration with Neuman and his staff to produce a quad-based plan similar to the Near West plan, though significantly reduced in size. The design replaced five demolished buildings with four postmodernist versions arranged around an open space framed by a contemporary arcade made of steel, stone, and fabric. The complex mimics the scale and organization created by Olmsted while continuing the habit of introducing architecture specific to its period. The new open space was designed as a response to the adjacent Inner Quad, with a palm-lined corridor connecting the two related landscapes. Several spaces designed by Church were demolished to build the complex, which seemed to prove that Church's and Olmsted's visions could not comfortably coexist.

During the planning for the new quad, Neuman hired the landscape architect Cathy Blake to join his staff. She came to Stanford from private practice and had previously worked on the campus as a consultant. Like Neuman, she brought design expertise to an office that had, for most of its history, been populated with planners. Her role involved directing the work of consulting landscape architects on individual projects, as well as managing the character of the campus from a more comprehensive perspective.

A Sustainable Campus

When the Science and Engineering Quad opened in 1999, the Stanford community could see firsthand the Second Century Plan's strategy for rebuilding Olmsted's framework and begin to understand just how far the campus had strayed from its original intent. This was supported by the first phase of what would be a decades-long project to transform one of the primary east-west axes from an open roadway for cars and other vehicles into a controlled, bicycle- and pedestrian-oriented transitway. In addition to reinforcing one of Olmsted's long vistas as a meaningful open space, the incremental project provided a safer and more attractive route for pedestrians, shuttle buses, and bicycles, which increasingly traversed the campus grounds. Bikes had been fixtures since at least the 1940s and their numbers grew exponentially thanks to Palo Alto's near-perfect climate, the flat terrain, and a lack of available or affordable vehicular parking spaces in the center of campus. Between 1995 and 1999, an aggressive program facilitated the installation of three thousand bike racks—more than a mile's worth—across campus to meet the growing demand.[73]

In addition to the popularity of nonpolluting transit, the university's environmental efforts benefited from a sustainable landscape ethos that dated back to Olmsted, who argued strongly for native plants and against the use of lawn. This contradicted the proclivities of Leland Stanford. According to Olmsted, Stanford was "bent on giving his University New England scenery, New England trees and turf, to be obtained only by lavish use of water."[74] The landscape architect believed the rough and unmanicured aesthetic of the arboretum and surrounding foothills was just as important to the character of the campus as the more formal landscapes associated with the gridded core. Although there were surely exceptions, the campus evolved accordingly, embodying a general dichotomy between opposing landscape types: organic and geometric, rough and polished, or, as Mumford had said, urban and rural.

Even with this long-standing preference for drought-tolerant plantings, numerous ornamental landscapes, some composed with large lawns, were installed on campus for decades. In 2001 the university developed a Water Conservation, Reuse and Recycling Master Plan to document its ability to meet municipal water mandates and "feasibly mitigate the impact from increases in water consumption associated with Stanford's land use and development activity."[75] In the 1880s, Leland built a 115-million-gallon reservoir, Lake Lagunita, as a source of water for his orchards and horse farm. The lake now serves as the main source of irrigation for the campus. The conservation plan proposes the use of water-efficient plant material in new landscapes, which the university had been doing for decades by that point. It also recommends retrofitting certain existing landscapes, particularly decorative, nonprogrammable lawns, using the same criteria. Turf was not completely discouraged, however, as Blake encourages the planting of lawns that provide functional and flexible space used by students and others in the university community for recreation and events. Most of the campus, however, remains of the rural type, designed as casual masses of native and drought-tolerant shrubs and trees. These coarse and rustic spaces weave through the campus, tying together disparate spaces and establishing the aesthetic juxtaposition that is fundamental to the campus's character.

Contemporary Issues

Between 2000 and 2020, Stanford invested around $7 billion in its campus.[76] Neuman led the office until 2003, when he left Stanford for a similar position at the University of Virginia. On his departure, the university's board of trustees formally commended him for, among other things, "creating a place of beauty" and "returning to the planning principles of Frederick Olmstead's physical plan."[77] After an eight-month search, the university hired his replacement, the architect David Lenox, who, on embarking on the job, said he welcomed the opportunity to reinforce Stanford's "rich tradition of integrating the exterior spaces and the architecture of the buildings."[78]

Lenox, Blake, and others on the staff remain closely involved in the planning and design of buildings and landscapes, sometimes spending at least a year studying specific projects before engaging consultants.[79] They adhere to established planning principles while allowing consulting designers to interpret precedent and define parameters with a degree of aesthetic freedom. This freedom within guidelines is particularly evident in the numerous contemporary urban landscapes added to the campus since the early 2000s, each a space unique unto itself with respect to detailing yet adapted in scale and character to the greater campus. Blake facilitated this work with a white paper she produced as the guiding philosophy for the design and stewardship of the landscape. The document, which builds on campus history, establishes a baseline for expectations and bridges the discordant processes of design and maintenance. It also encourages creativity while emphasizing the need to respond to the "understated casual elegance that is emblematic" of Stanford's campus and consistent with the spirit of Olmsted's plan.[80]

Among the most ambitious projects undertaken since the adoption of the Second Century Plan was an expansion and partial redo of the Science and Engineering Quad, the first phase of which was completed in 2015. Two buildings and much of the existing landscape were demolished to accomplish the new plan. In their place came four interconnected buildings totaling 600,000 square feet with a more than eight-acre open space designed by the San Francisco office of Hargreaves Associates (now Hargreaves Jones) landscape architects. As with the previous scheme, the open space conveys a quadrangle with a mix of lawn and a paved plaza to maximize programmed and casual use. It also achieves the previously unrealized goal of extending one of the primary east-west axes critical to the original plan. In another overt nod to Olmsted, a series of circular beds alternating between manicured lawn and drought-tolerant herbaceous plants is scattered across a large plaza as an abstraction of the Inner Quad's circular planters.

Fig. 30. The Science and Engineering Quad (toward the center of the photograph, with white paving) reestablished Olmsted's original plan by removing buildings to open axes and creating open spaces framed by buildings. (Stanford University Libraries, Department of Special Collections and University Archives)

Also in 2015, the university demolished the much-maligned Meyer Library, which did not meet current seismic safety standards. Rather than clog the Olmsted axis with another building, the administration converted the space into Meyer Green, a two-and-a-half-acre, bowl-shaped landscape with a circular paved area (another nod to the Inner Quad), sloped lawns, and trees well adapted to the arid climate. In addition to creating a beautiful and functional new open space for the campus, the move opened Escondido Mall as a visual and practical axial corridor as had been intended in the original plan. Blake and the team of in-house landscape architects and planners that she leads oversaw the design of the new space, as they did with the Science and Engineering Quad and other projects on campus.

In addition to the high-profile work done in conjunction with consultants, staff members lead the design of extensive site improvements across the campus. Many of these projects are funded through the Stanford Infrastructure Program, which is essentially a tax (4.6 percent as of 2020) charged to capital projects to pay for otherwise unfunded enhancements to the public realm. Such projects have included bicycle and pedestrian paths, exterior lighting, public art, transportation infrastructure, and other landscape elements that tie together previously disparate spaces and create a more seamless landscape experience. In its 2019–2020 Capital Plan, the university anticipated $52.9 million to be spent from the program over the following three years on projects, including road infrastructure and "annual SIP program elements."[81]

That the university funds such a program to that magnitude makes it unique among higher education institutions, many of which struggle to allocate adequate resources for the landscape. While this situation is partially attributable to Stanford's relative wealth, it also reflects a commitment to quality dating back to the grand ambition of its original patrons and persevering through generations of leaders and in-house planners and designers. The campus may have evolved, often intentionally, as a collection of individual buildings and landscapes more than as a cohesive whole, but its recent success in reconstructing the historical planning patterns has corrected most major missteps. While the monumental clarity of Olmsted and Stanford's plan deserves a lot of the credit, it would be a relatively meaningless document today without the willingness of university leaders to interrupt the status quo and look to the past to shape a campus well-suited to respond to needs of contemporary society.

Tuskegee University, Tuskegee, Alabama

Tuskegee University was founded in 1881 as the Normal School for Colored Teachers. It was incorporated by Alabama's General Assembly as Tuskegee Normal and Industrial Institute in 1892, renamed Tuskegee Institute in 1937, and, in 1985, acquired its current status as a university. Like Stanford University's, its early history is inextricably tied to a larger-than-life historical figure. Rather than a wealthy benefactor, however, Tuskegee's origin story is linked to its first principal and president, Booker T. Washington, who was born into slavery in Virginia and became a famously celebrated educator, reformer, and one of the most influential thinkers in nineteenth- and twentieth-century America. His commitment to improving the lives of oppressed Americans by increasing access to higher education was one of his most consequential accomplishments. Over time, the school clung to Washington's legacy as it rose from a poor, struggling, ambitious institution to become one of the leading universities in the country. Its history parallels the paths of generations of African Americans striving for equal footing in a country that today still struggles to reconcile with its racial past.

The town of Tuskegee is rural, small, and shrinking. Its reported population in 2023 was around 8,300 people, down nearly 5,000 from its peak in 1980. More than 93 percent of its residents are African American. It sits in Macon County, which was established in 1832 on land taken from the Creek Nation following passage of the Indian Removal Act of 1830.[82] Its name is a derivative of Taskigi, the name of a leader of the Creek. Although it is officially within the Coastal Plain region, the land exhibits the gentle rolling hills and broad ridges typical of the adjacent Piedmont. It is also part of the geographic region known historically as the Black Belt, after the band of dark and fertile soil that runs for around three hundred miles through Alabama and Mississippi and became home to numerous antebellum cotton plantations. After the Civil War, the term took on social and political connotations, referencing those southern counties in which Blacks, predominantly freed enslaved people, outnumbered whites, typically by large percentages.

The university expanded from an initial one hundred acres at its founding to a sprawling five thousand acres made up primarily of forestland. Its campus design heritage includes footnote-level input from the noted landscape architects Warren Manning and John Nolen; it is more significantly marked by a long line of in-house architects and landscape architects who shaped the campus and its landscape for generations. As at Stanford, there is a rich tradition of planning at Tuskegee, although the school has not always had the financial resources to achieve its ambitious goals.

The context of Tuskegee's history traces back to 1862, when President Abraham Lincoln signed the Morrill Act, which granted federal land to individual states for the establishment of public land-grant universities. This legislation led directly to the construction of numerous colleges and universities that provided practical education to help a predominantly rural population adapt to and excel in a rapidly industrializing world. Three years later, the end of the Civil War brought freedom to around four million mostly illiterate enslaved people. At the time, the halls of most educational institutions—land-grant and otherwise—were primarily filled with white men, although there were a few colleges specifically for women. Some private colleges for Blacks were opened with support from the Freedmen's Bureau and white missionaries and philanthropists, but educational opportunities for them were limited.

In former Confederate states such as Alabama, Blacks were not permitted to attend public colleges and universities. To remedy this discrepancy, Congress passed a second Morrill Act in 1890 mandating that states either integrate their existing land-grant institutions or create public ones for African Americans. If they failed to comply, they would no longer receive funds from the federal government. Specifically, the legislation required southern and border states to designate or establish a college to train Black students in agriculture, mechanical arts, and architecture. Out of this came nineteen historically Black colleges and universities (HBCUs) that became known as the Negro Land-Grant Universities and, later, the 1890 Land-Grant Universities. Among these was Tuskegee. Through Reconstruction and the subsequent Jim Crow period, these schools successfully educated and elevated opportunities for many thousands of Black and minority students.

Establishing the Legacy

Washington came to Tuskegee in 1881 as the institute's first principal. This was two years after an uneducated, formerly enslaved man named Lewis Adams rallied support within Tuskegee's community to establish a school for Black students. This effort, which was supported by a white former slave owner named George Campbell, led to House Bill 165, which, once passed by the Alabama senate two years later, allocated $2,000 to fund teacher salaries and free tuition for Black students at what became the Normal School for Colored Teachers. In contrast to this arrangement, federal land-grant funds created the white-only Auburn Normal and Agricultural Institute (Auburn University today) twenty miles away with a large enough endowment to generate more than $20,000 in annual income.[83] At Tuskegee, the $2,000 supported the operations of the new school yet provided no funding to construct a campus on which to teach.

Washington attended Virginia's Hampton Normal and Agricultural Institute, entering as a sixteen-year-old freshman in 1872. He worked as a janitor to help pay expenses before graduating with honors three years later. In 1879, he returned to

Hampton to teach a program for American Indians. On accepting the job as principal of Tuskegee, he taught his first class, with fewer than forty students. Because there was no campus at the time, he held classes in a run-down local chapel. Washington's commitment to the school was evident from the beginning. He sought and received a personal loan from a Hampton colleague totaling $200, which enabled the school to make the down payment on the $500 needed to purchase an abandoned, hundred-acre former cotton plantation known as "the Old Burnt Place."[84] As reported by Washington in June 1881: "The land has several old buildings on it which we could make answer for school purposes for the present, besides there is a young orchard on the place which is worth at least $75 itself."[85] Much of the land had been left infertile by decades of cotton farming and was in the early stages of a transition back to its native forest ecosystem. Washington believed that practical labor would provide Tuskegee's students with hands-on training and industrial education, and instructed them to clear the brush in the emerging woodlands to prepare the land for the cultivation of fruits and vegetables. In addition to such agricultural chores, students managed the construction and, along with Black masons and carpenters, provided the labor for the erection of campus buildings.

Unlike at Stanford, where the Olmsted plan retains its status as an exalted and relevant artifact, there is no famous drawing cementing Tuskegee's original design intent. If there was an early master plan—and research suggests there was—it appears to no longer exist. This lack of physical record does not make Tuskegee's legacy less significant, however. Washington was a force, and his potent combination of ambition, passion, and vision fueled the successful establishment of the institute and became instilled in its culture moving forward. The campus quickly embodied Washington's goal of teaching practical trades to students. In addition to fabricating thousands of bricks every year, students made and then sold furniture, clothing, and other agricultural products. This work was educational and instilled pride in the students and helped to create a campus that would attract new students. As observed by the historian Kenrick Ian Grandison, a lack of funds would be expected for an African American institution so soon after emancipation, and it surely "motivated a self-help approach to the school's goals of black education, campus development, and community building."[86]

Although there is not a formal master plan to refer to, there is a rendered bird's-eye view from 1899 that depicts the campus as likely envisioned by Washington. It shows existing and proposed buildings towering over the vast property. The landscape is shown as being relatively flat, with formal walkways and ornate plantings creating an imagery more consistent with a traditional campus aesthetic than with the reality of what could be found at Tuskegee. Like many master plans, it reflects lofty aspirations that would never be fully realized. Without the means to make monumental moves, buildings and landscapes emerged from an ongoing series of incremental decisions, which was typical for American colleges and universities at the time. From the beginning, the landscape was simple, consistently largely of grass and trees, with shrubs and flowers planted in high-profile spots and around the major buildings. Concrete paths and asphalt roads moved people through the campus on foot, in carriages, and, later, in cars and buses.

One of Washington's most significant moves was to organize new buildings by programmatic use and string them in clusters along the prominent ridgeline known as Tuskegee Ridge and the three smaller ridges that emerge from it perpendicularly, like fingers. Campus Avenue, the early designed landscape's most dominant feature, sat atop Tuskegee Ridge as the symbolic corridor along which students moved and the institute held marches and other ceremonial events. Buildings were sited with one façade facing the street and the other overlooking three shallow valleys, the largest of which earned the name "Big Valley." With this pattern in place, the campus grew quickly and gained national distinction in concert with Washington's spreading reputation. To support his view that the physical campus should be a source of pride for the students, he promoted cleanliness "in every corner from morning until night, from the beginning of one season to the end of that season."[87] As he explained in an 1881 letter, "We want to have the grounds beautiful; we want to have the yards beautiful.... We want to have everything beautiful that the students touch here, for in beauty there is always great inspiration."[88] Historical photographs offer proof of this conviction, with markers shown in lawn areas reading, "Don't cross here," in hopes of keeping the turf pristine.

In 1892, Alabama's legislature granted Tuskegee the authority to govern itself outside the control of the state, thus making it a private institution. The campus expanded with the purchase of a thousand-acre plantation that only decades earlier had been operated with the labor of enslaved people. Included on the property was an ornate, Italianate antebellum mansion named Grey Columns that would eventually become the residence for Tuskegee's president. The evolving landscape continued to lack formal order and looked different from other emerging American campuses, most of which adopted the traditional imagery of rectilinear manicured lawns, yards, and quadrangles. Even though Tuskegee followed the proven pattern of intentionally siting campus buildings to shape open spaces, the overall informality and direct response to the rolling topography conveyed the appearance of a pastoral landscape more than that of a traditional campus. Perhaps appropriately, some land in the undeveloped valleys was used to feed and shelter livestock. According to Grandison, "The rambling layout gives the campus a special ambiance that is rooted not in the conventional notion of the college 'campus as a work of art,' but, rather, in the physical geography and cultural circumstances of place."[89]

In addition to Washington, another influential figure on the campus from this era was the soon-to-be world-renowned scientist and educator George Washington Carver, whose inventions and discoveries related to peanuts and sweet potatoes earned him recognition as "one of the foremost agricultural chemists in the country" and helped boost the economy and social progress of the South.[90] Before Washington hired him to establish Tuskegee's agricultural program and to run the school's farm in 1896, he had been the first African American student, graduate, and faculty member at Iowa State University. In addition to teaching and conducting research at Tuskegee, Carver also managed the horticulture department, essentially being responsible for managing and enhancing the landscape, planting trees, and laying out walkways on the grounds. He created areas of manicured turf, which he considered important because, as he once commented, "The people of our race have never had a carpet of grass to walk on."[91] This intrinsic relationship between and among the school's pedagogical mission, cultural history, and physical grounds would be continued for decades.

By the onset of the twentieth century, and with significant monetary assistance from northern philanthropists, industrialists, and businessmen, Tuskegee's campus had grown to more than 2,400 acres and housed forty-six buildings. A 1900 account describes the condition of the landscape as such: "All of the larger buildings have ample grounds around them, shaded by trees and brightened by numerous flower beds. A northern eye is caught at once by the profusion of southern vegetation, the cacti, yuccas and palmettos. The entire care of the grounds, flowers, shrubs and trees devolves upon the department of floriculture and horticulture."[92] This suggests attention paid to the landscape, as well as the role of students as both laborers and stewards.

Much of the campus's early expansion was laid out by the École des Beaux-Arts–trained African American architect, Robert Taylor, who was the first Black graduate of the Massachusetts Institute of Technology's School of Architecture. When he started at Tuskegee in 1892, he was the only academically trained Black architect in the country and, like Carver, was expected by Washington to carry academic and practical responsibilities. In addition to teaching architectural and mechanical drawing, Taylor either designed or directed the design of many early campus buildings. They included the iconic university Chapel and The Oaks, a grand house that became the home of Washington and his third wife, Margaret Murry Washington. Taylor also led much of the planning work and supervised the campus infrastructure for forty years.

In addition to Taylor, whom Washington trusted as an architect and a planner, outside professionals were, on occasion, also sought for their advice and expertise. During an 1894 trip to Massachusetts, for instance, Washington visited Frederick Law Olmsted at Fairsted, Olmsted's home and office in Brookline, to seek assistance from the man he described as "the landscape gardener who laid out the World's Fair Grounds and the Central Park grounds."[93] As recounted by Washington, Olmsted, who was working on Biltmore at the time, said he "would gladly come to Tuskegee to help us lay out our grounds providing we pay his traveling expenses from Asheville to Tuskegee and return."[94] Likely because of his failing health, however, Olmsted never made it to Alabama for the consultation.

Other outside consultants were engaged and appear to have been far less impactful than Taylor and Carver. Manning, for instance, was engaged by Washington in the early 1900s to produce a campus plan. His involvement at Tuskegee—thought to have started in 1901—culminated in a campus plan that recommended various planning strategies, notably plans to develop around a heavily forested gulley running through the campus. The plan, which was little more than a sketch, failed to distinguish new and proposed features or include topographic lines, scale, labels, or other practical details. [95] Even so, the student newspaper endorsed it, writing, "The essential unity of the buildings will be better preserved in the future than heretofore and the grounds will be laid off and kept hereafter with regard to their permanence and beauty."[96] Despite that enthusiasm, the plan remained essentially unrealized.

The Influence of David Williston

In 1902, Washington recruited David Williston to Tuskegee in the dual role of professor of horticulture and landscape gardener. Williston graduated from Cornell University's College of Agriculture and is acknowledged as the first professionally trained African American landscape architect in the country. He taught grading, planting design, landscape maintenance, and horticulture while overseeing the campus landscape. Early in his tenure, Williston collaborated with Taylor on a master plan intended to add a more traditional order to the informally arranged campus. He also designed a series of drives and walkways across the campus and added numerous ornamental plantings, frequently conferring with Carver on plant selections. At The Oaks, which had a more detailed and ornate landscape than the rest of the campus, Williston designed the walks, plantings, and structures around the house for a fee of $5 per day.[97] Flowers and vegetables filled the residential garden tended by Washington, who equated working in the dirt with his hands to touching nature, noting in his autobiography, "I pity the man or woman who has never learned to enjoy nature and to get strength and inspiration out of it."[98]

Williston shared Olmsted's preference for romantic and informal campus landscapes that work with the natural topography.[99] This was easy at Tuskegee, where the rolling hills and native woodlands influenced the campus's development and helped establish its character. He is credited with leading campus planning efforts after 1906 and achieved a level of influence that grew in 1910, when he transitioned into the role of Tuskegee's superintendent of buildings and grounds, a position he retained until 1929. In this new role, he collaborated briefly with Nolen, whom Washington had hired to create a study plan for the campus. However, like Manning's plan before it, this plan was never realized.

Tuskegee administrators enjoyed a mutually beneficial relationship with Williston, who would have found it difficult to gain employment in one of the white landscape architecture firms in the Jim Crow South. As with Taylor and Carver, the school saved money by having him fulfill roles as both teacher and practitioner. His planting designs, which favored native plants yet also used more ornamental species for visual interest, quickly permeated the campus grounds. It did not take long for his positive impact on the campus to be recognized. In his 1905 annual report for the school, Washington commented that "greater progress has been made during the past year in beautifying and putting our grounds immediately around the buildings in shape than ever before; and it is earnestly hoped that our finances will permit us to continue to improve the landscape features of the grounds during the coming year."[100] Funds remained low, however, which surely prevented Williston from accomplishing as much as he would have liked. Most of the landscape installation and maintenance was done by students, who had to find time between classes and other activities to work. In addition to employing student labor, Williston saved money by transplanting trees from local woodlands rather than buying them from nurseries. When a greenhouse was constructed in 1904, Williston began growing ornamental plants from seeds and cuttings.

Using Williston and Taylor's master plan as a guide, the school began siting new buildings to form rectilinear open spaces that introduced a version of collegiate quadrangles to the primarily informal landscape. A uniform planting of American elm trees was introduced along Campus Avenue, establishing a patterned layer of street trees that would be used in other areas as well. These structural moves added a significant cultural and aesthetic layer to the evolving campus. Among Williston's most significant designs was the landscape for what became known as the Historic Quadrangle, one of the most iconic and formal areas on the campus. His planting plan for the space is typical of his designs, with groupings of evergreen shrubs intermixed with native overstory trees. He was careful to preserve and frame the long views across lawns and valleys that are integral to the campus's character. A gazebo and band stage he designed for the space survived for decades before being reconstructed and then partially removed.

Following the death of Washington in 1915, Dr. Robert Russa Morton became Tuskegee's second president. By this time the campus consisted of around eighty-five buildings, with some of the early ones having been demolished and replaced. Growth continued westward from the historic core during Morton's two-decade tenure, with numerous new buildings designed by Taylor and other architects and an expanding landscape led by Williston. Under Morton's leadership, Tuskegee

Fig. 31. As part of a campus master plan drawn up by Robert Taylor and David Williston, traditional and rectilinear quadrangles were built on the campus at Tuskegee, which had previously been laid out informally and organically. Known today as the Historic Quadrangle, the landscape was planted with loosely arranged trees and shrubs and punctuated by the Adler Band Stand (left), designed by Williston. (Library of Congress)

transformed from a secondary trade-focused school into a nationally respected college offering bachelor's degrees in various disciplines. In 1930, he expressed his pleasure with the campus, noting its transformation from "red sand hills to a garden of beauty."[101] There are other accounts of positive reports about Tuskegee's appearance. The historian Ellen Weiss recounts a 1930s report by a Bahamian student named Cleveland Eneas in her biography of Taylor: Eneas "found himself in 'acres of sheer beauty.' The rolling lawns, flowering shrubbery, paved roads, sidewalks, and 'correctly spaced and majestic buildings' made a 'magnificent spectacle.'"[102]

In 1930, Williston moved to Washington, DC, where he opened the first African American–owned landscape architecture firm in the nation. He continued his campus-focused work by contributing plans and designs to dozens of HBCUs, including Fisk University, Alcorn State University, Lane College, and Philander

Smith College. At his alma mater, Howard University, he redesigned the main quadrangle, the symbolic heart of the campus known as "The Yard," with new walkways and trees. He also continued to guide the planning of Tuskegee's campus as a consultant for nearly two decades. In 1942 the local newspaper, the *Tuskegee News,* reported that Williston (still referred to as a "landscape gardener") had returned to campus "after several months away" and stated that his "many friends here are always glad to see him around."[103] The report credited Williston and his genius for the beauty of the campus more than anyone else, and explained that his services were in demand at various colleges and universities, as well as on "many government projects."[104]

Following Morton as president of Tuskegee was Frederick D. Patterson, who took over in 1935, when the country was mired in the Great Depression. Economic troubles continued into the World War II era, when many college-age males in the United States were sent abroad to fight. In a 1943 op-ed for the *Pittsburgh Courier,* Patterson addressed what he saw as a grave financial situation facing "Negro colleges . . . which heretofore have relied largely on gifts from substantial members of the white race for their support and maintenance."[105] This, he knew, was not a sustainable financial strategy. He proposed that a group of ten or so colleges join fundraising efforts to create a collective account accessible by each school. All it would take to establish this new fund was for each donor solicited by each institution to contribute as little as $1. "At least during these critical times, a unified financial campaign for several Negro colleges seems to be an idea worth toying with," he wrote.[106] A year later, he founded the United Negro College Fund, which was the first fundraising venture of its kind.

Like many other American colleges and universities, Tuskegee's campus growth accelerated following passage of the 1944 G.I. Bill. To help support this expansion, the institute engaged Williston as a consultant in 1948 to create a new master plan, which appears to have been his final project there. As was traditional practice, he organized a spate of new buildings in clusters based on programmatic function. His plan also proposed extensive changes to the landscape, including the addition of a lake and a large amphitheater that would take advantage of the ridge and valley topography, tree-lined roads and walkways, formal entrances to the campus, and landscaped courtyards framed by the new buildings. While some of the organizational elements are apparent today, many of the individual features and buildings were not built as intended.

The Influence of Edward Pryce

In 1949, another important landscape steward, Edward Pryce, began work at Tuskegee Institute. Pryce was a Tuskegee graduate, having transferred there from the University of California, Los Angeles, where he had grown tired of being one of only a handful of Black students out of thousands.[107] At Tuskegee he studied under Carver, who taught until his death in 1943. He also met Williston, who instilled in him a passion for the profession of landscape architecture, which, he came to realize, combined engineering, horticulture, and art, all subjects at which he excelled.[108] After receiving his bachelor's degree in agricultural education in 1937, Pryce moved back to California, where he found a job as a landscape gardener at the Huntington Botanical Gardens in San Marino. He worked under the tutelage of the head gardener and developed an interest in effective landscape management. [109] He then switched his career focus and received a bachelor's degree in landscape architecture from Ohio State University and then a master's degree from the University of California, Berkeley. After earning his license, he faced similar employment obstacles as Williston had owing to his race. As he stated in 1993: "I came out of Ohio State with Highest Honors and I was a Merit Scholar from ASLA and I wanted to practice landscape architecture. I had the qualifications but I couldn't get a job. I would interview with people and they would say 'I'd hire you, but as soon as I do we'll start losing business because you are black.' That is simply the way it was. So I came to Tuskegee to teach."[110] He remained there as a teacher, planner, and landscape architect until 1990.

Like Williston before him, Pryce was a trailblazer for the profession. He was the first African American landscape architect licensed in Alabama and for many years the only one practicing in the state.[111] He was appointed superintendent of buildings and grounds at Tuskegee and oversaw the campus as a planner. In 1953 the school's president announced a $1 million expansion plan that was, he said, "a reflection of the growing need for Negro professionals and technically trained persons."[112] Over the next decade, Tuskegee constructed dormitories, laboratories, and classroom buildings, for which Pryce frequently served as the landscape designer. He also advised the famed modernist architect Paul Rudolph on a quirky, period-based master plan in 1958. Although little of the plan was followed, the seriousness of the effort shows how changing design styles can influence a campus's evolution. Rudolph's office did locate and design a new chapel to replace the one designed by Taylor, which had burned down a year earlier. Joining the team on the project was the firm led by John A. Welch and Louis Fry, both Tuskegee graduates. Displaying his talent as a professional artist, Pryce created a mural for an interior wall that, he said, expressed the powerful influence of Africa on Christianity.[113]

Pryce understood that properly preserving the historic campus was beyond the financial capability of the institution and, in 1972, prepared a report recommending that the original campus as laid out by Washington, Taylor, and Williston be designated a National Historic Site.[114] It was already a National Historic Landmark, having become the first educational campus recognized as such in 1966 thanks to Pryce's leadership in the nomination process. This designation, however, brought with it no financial resources. Making the campus a National Historic Site meant the federal government would take partial ownership and assume some responsibility for maintenance, restoration, and preservation. The proposal made its way through the institute and the NPS and was greeted with enthusiasm by Senator Jim Allen of Alabama, who cosponsored the bill making Tuskegee a National Historic Site. Speaking on the Senate floor in 1973, Allen said, "One place that above all others demonstrates black achievement . . . is Tuskegee Institute."[115]

In 1974, around fifty acres of Tuskegee's historical campus core were designated a National Historic Site. The Department of the Interior celebrated the agreement, with the assistant secretary writing, "Tuskegee Institute is an ideal site for interpreting to the American public the history of Negro education and of the economic and social advancement of blacks in the post–Civil War period."[116] As part of the deal with the university, the federal government took ownership of the three main historical sites and structures—The Oaks, Grey Columns, and the George Washington Carver Museum, which had been authorized by Tuskegee's Board of Trustees in 1938 and for which Williston had produced the landscape plan. In addition to the three primary structures, a memorandum of agreement between the institute and the NPS established a comanagement system for the historical campus, for which it mandated a process of preservation. The NPS provided preservation expertise and funding as appropriated by Congress and the institute committed to preserving "in perpetuity the historical character" of the property.[117] While this was before the NPS took an interest in cultural landscapes and focused primarily on buildings, the agreement did preclude changing the appearance or character of the grounds.

Tuskegee's partnership with the NPS was finalized at a 1977 dedication ceremony, during which Luther Foster, Tuskegee's fourth president, handed deeds to the property over to the NPS. The new arrangement boosted the school financially even as many of its HBCU peers struggled. Collectively, the institutions suffered from a persistent second-class image. They also faced increased competition from public universities in the South, which were actively recruiting Black students to, in part, help meet a 1972 federal order to increase minority enrollment.[118] Between 1976 and 1982, African American enrollment in higher education increased 6 percent, while enrollment in HBCUs dropped by 5 percent.[119] During the decade between 1975 and 1985, however, Tuskegee's enrollment jumped from 2,800 to 3,600 and its endowment increased by 50 percent, to $25 million, the second largest of any HBCU at the time.[120] How much of this success reflects its status as a National Historic Site is unknown; it is likely that the rich historical legacy played some role.

Fig. 32. Except for the four I-shaped buildings in the lower left, this aerial photograph shows the general limits of the property ceded to the National Park Service as a historic district. The informal arrangement of buildings along the ridgeline is evident. (Tuskegee University Archives)

As recognized by the NPS in its 1978 *General Management Plan* for Tuskegee: "Historic photographs clearly show that the core campus has changed little in the 70-odd years of growth, and that the spaciousness and openness of the area enhance the beauty and scale of the historic scene."[121] The new partnership essentially transformed this well-preserved campus conceived as a place specifically for learning into one with an interpretive historical park embedded in its core. The arrangement quickly benefited the buildings as a process of extensive restorations ensued. The landscape, however, continued to suffer from a lack of investment. Even with its verdant valleys and abundance of mown lawns, the campus's pedestrian realm was frequently interrupted by cars, roads, and parking spaces. The NPS correctly projected the campus, now formally identified as a cultural attraction, would lure an additional

50,000 visitors to the campus per year.[122] In addition to people arriving in personal cars, there was already an onslaught of school groups in the spring months, most of whom arrived in buses.[123] As stated in the plan, "Traffic congestion on campus today is considerable," yet it still recommends that visitors first drive to Grey Columns and then "either drive their own vehicles or walk to The Oaks and the nearby Carver Museum," thus further clogging roads in the heart of the historic campus.[124] Service areas and surface parking lots invaded the campus core, severing the once significant relationship between buildings and open space and signifying, as reported by the Southeast Regional Office of the NPS, "an erosion of historic character."[125] A primary objective listed in the plan is familiar to most historic landscapes: "[to] obtain the necessary staffing, facilities, equipment, and funds so that the historical resources and facilities . . . can be adequately managed and protected."[126]

In 1987, Pryce joined the Nashville-based architecture firm Ehrenkrantz Group in developing a new campus master plan. Pryce's influence is easily seen in the depth of analysis and insight into the history, character, function, and potential of the landscape. Harking back to Washington's line of thinking, the plan describes how "the aesthetic elements of campus site and landscape planning are positive forces" and observes that "the educational value of educating the student to appreciate nature, cleanliness, and order is a prime objective."[127]

The plan's description of the landscape provides a snapshot of its evolving condition and points to a lack of maintenance as an obvious problem. Regarding the original historic landscape, it reads, "Components such as trees and shrubs . . . have not been properly maintained. Trees have not been properly trimmed or pruned and frequently gaps have been allowed to occur in hedge rows and other low planting groupings."[128] In describing the formal planting beds that had been added on campus, it remarks that they are high-maintenance features and should be limited to the "most important locations and should be attempted only when a good cyclical maintenance program is in place."[129] Such concern for maintenance is certainly not limited to Tuskegee. Many universities, well funded or not, failed to adequately support their campus landscapes during this period. Not every problem can be solved by increasing maintenance dollars, however. The plan defines the vehicular circulation system as a primary challenge, noting that because of "the heavy reliance by students and staff on vehicular transportation, the resulting road system and chaotic series of small parking lots has resulted in the campus having a chopped up rather than cohesive appearance."[130]

Recognition and Resources for HBCUs

As Tuskegee dealt with the positive and negative aspects of its relationship with the NPS, increased recognition of the unique challenges faced by HBCUs triggered help for the broader collection of such institutions. In 1980, then president Jimmy Carter signed Executive Order 12232 establishing the White House Initiative on Historically Black Colleges and Universities as a means to "overcome the effects of discriminatory treatment and to expand the capacity of Black colleges and universities to provide quality education."[131] The legislation, which applies to colleges or universities established before 1964 with a principal mission to educate Black Americans, enables institutions to participate in federally sponsored programs from which they were previously and unfairly barred. The next year, President Ronald Reagan boosted the order with a version (Executive Order 12320) that, among things, initiated the first Annual Federal Plan for Assistance to Historically Black Colleges. The initiative was rebranded the White House Initiative on Advancing Educational Equity, Excellence, and Economic Opportunity through Historically Black Colleges and Universities and has been supported by each subsequent president.

Out of these significant orders emerged a 1988 historic preservation initiative led by the secretary of the interior to identify, restore, and preserve structures on HBCU campuses. Selected projects received federal grants covering half the cost of repairs, restorations, and renovations. While this money was spent almost exclusively on brick-and-mortar projects for buildings rather than on landscapes, the process raised awareness about the need for broader campus improvements. The program eventually granted over $60 million to qualifying institutions.

In 1998, Tuskegee reacquired Grey Columns from the NPS to house the university president. The same year, the university responded to a historic preservation survey issued by the U.S. General Accounting Office to inventory restoration needs and costs on HBCU campuses. According to the report, "All 103 historically black colleges and universities responded to our survey. Respondents identified 712 historic properties that were owned by the schools. Most of these properties were at a small number of these schools and were mostly buildings rather than structures, sites, or objects." [132] A collective need for $755 million was identified, with Tuskegee reporting a need of over $25 million to restore twenty-eight structures. The National Trust for Historic Preservation boosted this recognition by including southern HBCUs on its 1998 list of the country's "most endangered historic places." The trust emerged as a leader in raising awareness of the struggles facing the institutions and eventually partnered in a grant program established to "empower HBCUs with the resources to protect, preserve and leverage their historic campuses, buildings, and landscapes."[133]

It was not only the HBCU institutions that experienced decline during the second half of the twentieth century. Many colleges and universities focused on expansion left older buildings and landscapes without necessary resources for repairs and improvements. The common practice of deferring maintenance spending to pay for new facilities created a persistent and challenging problem to solve. Some schools did begin addressing their most historically significant buildings as early as the 1960s, though few paid similar attention to their equally valuable landscapes.[134] As campuses met the evolving needs of modern education and research with new construction, acres of associated designed landscapes stretched the work of maintenance crews that were likely already inadequately funded. Heading toward the end of the twentieth century, even celebrated campus spaces were displaying the effects of aging tree canopies, denuded lawns beaten down by overuse, and undermaintained landscapes in general need of restoration.[135]

In response to the visible deterioration of historic campus facilities, the Getty Foundation introduced the Campus Heritage Grant program in 2000 with the intent to assist colleges and universities in managing and preserving the integrity of significant historical buildings and landscapes.[136] It was the most comprehensive funding program directed specifically toward the built environment of academic campuses, ultimately donating more than $13.5 million in support. The director of the Getty Foundation cited as motivation the fact that "American colleges and universities are frequently unique repositories of some of the country's finest historic architecture and designed landscapes."[137] Rather than target specific repairs and restorations, however, the Getty grantmakers took a more holistic approach, encouraging the careful documentation of existing conditions and the production of preservation plans as parts of longer-term stewardship strategies.

In 2006 the Getty Center awarded Tuskegee a $115,00 Campus Heritage Grant specifically to produce a comprehensive conservation plan for the campus. While the completed document pays expected attention to the historic buildings, it also emphasizes the significance of the cultural landscape and acknowledges that spatial patterns and open space are important contributors to the campus's historical character.[138] The plan breaks the campus into landscape preservation zones, including the Big Valley, the main campus drives, and one of the quadrangles designed by Taylor and Williston. Along with a series of implementable strategies and guidelines, the plan recommends reestablishing the practice started by Taylor and Williston and continued by Pryce whereby the university would employ an in-house architect and landscape architect to "maintain the cultural heritage and sense of continuity as new development comes online."[139]

Around the same time as the Getty grant award, a new ceremonial element was added to the campus landscape with the Legacy Walkway. This attractive brick path weaves through the Historic Quad and extends into the broader historic district.

Fig. 33. Running down the middle of this contemporary view of the Historic Quadrangle is the Legacy Walkway, one of the most notable changes in the core of the campus. To the left is what remains of the Adler Band Stand. (Photo by author)

People may purchase and engrave bricks to recognize and honor alumni, friends of the university, and organizations that have contributed to Tuskegee. The walkway enhances the campus's pedestrian realm while adding another layer to the university's rich cultural landscape.

The success of such initiatives seems to have established a broader ethos of historic preservation that once again merges practice and pedagogy. In 2018, Kwesi Daniels, head of the architecture department at Tuskegee, began developing a historic preservation program in response to the increased interest in the campus's history. A year later, six of his students participated in a national program named Touching History Preservation in Practice, which was developed by a partnership consisting of the NPS, the National Trust for Historic Preservation, and the Advisory Council on Historic Preservation. Its purpose is to connect young people to preservation and conservation fields by, as its name suggests, having them tackle projects with their own hands as a means of inspiring interest and achieving results, much as Washington originally intended with his hands-on learning philosophy. The NPS's deputy director for operations said about the synergies involved in the project, "Investing in our future leaders is one of the most important things the National Park Service can do to further its mission."[140]

Contemporary Issues

By 2015, when Tuskegee celebrated the centennial of the death of Booker T. Washington, three thousand students were enrolled. The institution's $120 million endowment was one of the largest among the HBCUs. Even as it looked back in remembrance of its founding president, the university looked to the future by embarking on yet another campus master plan to guide its development for the next ten years. The plan presents a comprehensive road map for the ongoing development, enhancement, and preservation of buildings and grounds. It recommends making the campus more pedestrian-friendly by reorganizing academic functions within the 240-acre core and moving vehicular circulation to the perimeter. The plan also prioritizes creating a significant open space and revitalizing existing ones "central to the legacy and traditions of the campus."[141] It endorses environmental sustainability by using some of the campus's undeveloped acreage for generating wind and solar power and establishing an urban farming program to produce food for campus dining facilities. Such moves enhance Tuskegee's twenty-first-century pedagogical mission while reinforcing the innovative spirit established by Washington in 1890.

In 2020, a year in which American colleges and universities struggled to accommodate remote learning as the fallout from the coronavirus pandemic led to layoffs, budget cuts, and reduced enrollment at many institutions, Tuskegee's fortunes changed dramatically. As part of a nearly $6 billion charitable program targeting social inequity, public health, and climate change, the philanthropist MacKenzie Scott gave Tuskegee $20 million, the largest donation in the school's history. In acknowledgment, Tuskegee University's president Lily D. McNair said, "This gift comes at an opportune time for us. We are about to launch a five-year strategic planning effort and, with this investment from Ms. Scott, we will be able to unfold the next phase of our mission in ways we simply could not have considered before."[142] One year later, Tuskegee was one of eight HBCUs to share in a $650,000 grant sponsored by the National Trust's African American Cultural Heritage Action Fund to assist with targeted restorations or the production of cultural heritage stewardship plans. The university used its portion to develop a stewardship plan for two of its historical buildings.

Tuskegee's success as an educational institution was never dependent on its partnership with the NPS. As Pryce predicted, the condition of the historic campus core improved once the two entities sealed their relationship. The NPS introduced an ethic of preservation and provided peerless expertise when the university faced rapidly aging infrastructure and the deterioration of cultural and historical resources. The partnership also raised the profile of Tuskegee, made Americans more aware of the contributions of Booker T. Washington and George Washington Carver, and boosted the image of the HBCU system. The NPS fostered a sense of pride in the university by documenting and interpreting its rich history. As one student who worked on the preservation project in 2019 said, "I've developed a great appreciation for [Tuskegee's] legacy. Being here at this moment, I feel like I'm following in the footsteps of the students who built the school. I'm touching something that students built, and I am here making it better."[143] This student later applied her interest in historic preservation as a cultural resource intern at the NPS, thus fulfilling the agency's mission to inspire a new generation.

Walking Tuskegee's grounds today makes apparent the ongoing challenge associated with managing a campus landscape with limited funding. Whereas Stanford's relative wealth and comprehensive investment in its campus are obvious throughout its buildings and landscapes, Tuskegee's aspirations for its physical environment appear largely unfulfilled. An example of this is the pedestrian realm, which has been prioritized at Stanford by implementing the long-planned perimeter ring road and moving vehicles into numerous above- and belowground parking garages. At Tuskegee, however, surface parking lots and on-street parking dominate the landscape, often at the expense of people on foot or bike. There is still a lot to appreciate about the campus. Meticulously mown turf, cleanly pruned trees and shrubs, and

Fig. 34. The Big Valley is one of the most prominent and defining features in Tuskegee's landscape and makes evident the campus's natural beauty and significant topographical challenges. (Photo by author)

litter-free grounds comply with Washington's mandate for a well-tended campus. The nearly ten-acre Big Valley, with its long vistas, rounded slopes, mature trees, and open lawn, is one of the most impressive collegiate landscapes in the country and the meaningful heart of the Tuskegee campus.

Despite the challenges they face, both Tuskegee and Stanford excel at respecting the ideals and vision of their respective founders. Leland and Jane Stanford's hopes of creating a university to educate Californians and building a physical campus to serve as a meaningful memorial to their beloved son have been fulfilled, with little qualification. Investing billions to rebuild Olmsted's master plan required consolidated decision-making at the leadership level. At the same time, implementing the work relied on decades of effective collaboration in planning and design between staff, faculty, administrators, and consultants. Finding money is one challenge; spending it well is another.

Even without the same deep pockets, Tuskegee has achieved Booker T. Washington's goal of improving the lives of generations of Black Americans by increasing access to quality higher education. His aspirations for the campus landscape were more modest. He wanted to ensure that it would be clean and beautiful and serve as an inspiration to the students. In this regard, the university has continually succeeded. While the campus may not display the most beautiful or innovative built architecture or landscape architecture around, the well-maintained grounds embody the legacies of Washington, Carver, Williston, and Pryce while providing Tuskegee's students with an abundance of memories and something to be proud of.

4

COMMUNITIES

Make big plans, aim high in hope and work, remembering that a noble, logical diagram once recorded will never die, but long after we are gone will be a living thing, asserting itself with ever-growing insistency.

—Daniel Burnham, architect and planner, 1909

In a discussion of designed landscapes, *communities* can mean different things to different people and cultures while encompassing a wide variety of scales and programs. Fundamentally, the term can refer to places and spaces that serve the cultural and practical needs of a particular community. These can be specific spaces, such as memorials and sites of remembrance, or public squares and civic plazas in which groups of people joined by a common nationality, ethnicity, religion, geography, or other cultural identifier gather in shared protest, celebration, or mourning. The focus of this chapter, however, is on the communities themselves—designed towns, neighborhoods, or suburbs in which people live as a collective unit under an organized governance system. Such communities are contained places with defined boundaries and, typically, some level of contract or agreement establishing consistent standards for behavior and aesthetics.

Recognizing a basic history of community design in America helps to understand and explain contemporary issues. People were creating residential communities within the geographic boundary of the present-day United States centuries before the arrival of European settlers. Notably, Cahokia, located near St. Louis, Missouri, was the largest pre-Columbian community north of Mexico during the Mississippian Period, between 800 and 1400 CE. It was a planned community, deliberately and precisely laid out with the use of astronomical alignments. Its estimated population at its peak, around 1100, was between 15,000 and 20,000, or larger than London's population at the time.

Europeans brought to the New World their own versions of urban planning. In the sixteenth century, Spain's Philip II created detailed ordinances to dictate the site selection and spatial arrangement of settlements. The Laws of the Indies, as they are called, describe appropriate settlement sites as those not at high elevations or in lowlands, "which tend to be unhealthy," near water, and "where it would be possible to demolish neighbouring towns and properties in order to take advantage of the materials which are essential for building."[1] Per the ordinances, each new town was required to have a main plaza, either at its center or at the main port if on the coast, that was "square or rectangular . . . not less than two hundred feet wide and three hundred feet long," depending on the population of the town.[2] From the plaza's four corners would emerge the grid of streets that created the pattern found in the plans of early Spanish settlements such as St. Augustine, Florida (1565), Santa Fe, New Mexico (1607), and San Antonio, Texas (1718). In addition to the physical layout of the town, the dictates established a town's governance and social and religious structure, assigned property, and set rules for relating to native communities.

For English colonists settling along the East Coast, land ownership was the lure for moving to the New World, and the formal gridiron provided the most efficient structure for dividing and assigning lots. This is evident in the seventeenth-century plan for Philadelphia drawn up by William Penn and Thomas Holme, who instituted a rigorous pattern of regular streets along which buildings were spaced in a uniform manner. As described by Penn, "Let every house be placed, if the person pleases, in the middle of its plat, as to the breadth way of it, that so there may be ground on each side for gardens or orchards, or fields, that it may be a green country town, which will never be burnt, and always be wholesome."[3]

Other planned cities quickly emerged as the New World's population grew and migrated. James Oglethorpe, who founded and planned Savannah, Georgia, in the 1730s, stands out as a worthy successor to Penn as a forerunner of the modern urban planner. Like Penn, Oglethorpe offered three plots of land to each family settling in Savannah—fifty acres, consisting of a city lot, a five-acre garden parcel, and a forty-five-acre farm. Smaller landholdings created a more egalitarian society made up of small farms rather than Penn's community, which was defined by class distinctions.[4] Oglethorpe inserted twenty-four open, landscaped squares into the center of square wards, earning him the distinction of providing more open space than, according to the historian Mark Reinberger, "probably any town founder in history."[5]

Towns and villages soon proliferated across the country in some variation on these early gridded communities. Among the unique plans developed was that of Washington, DC, which took shape as the result of a site selection process in 1790. President George Washington selected Pierre Charles L'Enfant to lay out the new federal city. L'Enfant's famous axial plan imported the drama, symmetry, and monumentality of European Baroque planning and applied a heightened sense of design to America's emerging urban form. Westward expansion and the transcontinental railroad led to the creation of hundreds of new towns and cities from coast to coast. The Industrial Revolution and European immigration overcrowded the nation's largest cities, sparking rapid suburbanization. New technologies and innovations in materials and construction changed the scale of buildings and the rate at which they could be built. By 1850, more people lived in cities, with a declining 64 percent of America's workforce of nearly eight million people working as farmers.[6]

At the same time that urban planning was defining the look of communities, new methods for managing them as designed landscapes emerged. An early promulgator of designed urban landscapes was the New York City lawyer and real estate speculator Samuel Ruggles, who in 1831 purchased twenty-three parcels of swampy rural marshland in growing Manhattan to create a uniquely independent community. He arranged sixty-six lots surrounding and overlooking a two-acre, rectilinear "private ornamental park," for which he "laid out the grounds with paths and adorned them with trees, shrubs, flowers and a central fountain, thus making a charming retreat for the enjoyment of the neighboring property owners and tenants, who have access by means of keys to the gateways."[7] Ruggles correctly believed the central open space, which, along with the surrounding residences, became known as Gramercy Park, would increase the development's desirability. In anticipation of this success, he fenced off the landscape and established through covenants that property owners would be jointly responsible for its care and maintenance. In addition to establishing rules for the park, the originating deed contained covenants prohibiting buildings used for a variety of undesirable things. Prohibited were slaughterhouses, industrial factories, breweries, distilleries, museums, theaters, circuses, or "any other trade or business dangerous or offensive to the neighboring inhabitants."[8] Lots were sold, and, as one sales advertisement from 1833 states, "A covenant against nuisances will be inserted into the bond."[9]

Residents became advocates for the landscape they shared, and largely through their efforts, the park survived attempts by the city in 1890 and 1912 to run transit routes through the center of the park. As the rapidly expanding city grew and encroached on the edges of the park with larger buildings, the relatively quaint neighborhood and its park became more coveted. The *New York Times* referred to Gramercy Park's property owners as "the principal defenders of the park," and claimed that as long as they supported the restrictions, the neighborhood was "likely to remain as it is for all time because of its common interests."[10]

The notion of assigning ownership to common spaces while restricting access to the property by outsiders proved popular. Radburn, a 160-acre, unincorporated neighborhood in the Fair Lawn Borough of New Jersey, around twelve miles west of Midtown Manhattan, presented a suburban planned community designed in the late 1920s as a self-sustaining town accommodating the rapidly growing influence of automobiles. Using the model of the "superblock," groups of houses were laid out facing the street while their backs were oriented toward shared gardens that collectively formed long linear parks. To maintain the plan's integrity and protect the character and quality of the common space, the community established detailed covenants that restricted setbacks, open space, and architecture. These were enforced by the Radburn Association, which was formed with a manager-director, a board of trustees, and elected residents; collectively, these individuals wielded extensive power over the community's appearance. As stated in the 1929 "Declaration of Restrictions": "No building, fence, hedge, wall, sign, billboard, awning, pole, radio antennae or other structure . . . shall be commenced, erected, or maintained upon The Property" until it was "approved in writing by The Radburn Association."[11] Owners were charged a mandated annual fee payable to the Radburn Fund to maintain public spaces. These fees also funded community amenities and services not covered by the borough, such as street lighting, snow removal, and park and open space upkeep.

As at Gramercy, unwanted nuisances were banned. Along with a list resembling the one drafted by Ruggles, no cemetery, asylum, saloon, or reform school would be allowed in Radburn. Only people and uses deemed neighborly could reside there. Although there was no formal restriction of ownership based on race, the community was almost, if not wholly, white. This was true for much of the postwar sprawl that began in the late 1940s and 1950s and exploded in the 1960s. Community interest developments, planned urban developments, and other such communities proliferated across the landscape, along with homeowners associations. While such organized groups eventually earned a reputation for being meddlesome and domineering, they started out with positive goals, at least from the perspective of the physical landscape.

The following case studies represent two types of designed communities, a village and a resort. Each has a respected design heritage based on a particular landscape aesthetic vision that is consistently challenged by, among other things, the potent forces of the real estate market. They remain desirable places to live, enjoying an ongoing popularity that proves success has a downside. As housing prices have soared to match demand, access to the communities has become limited to an increasingly elite demographic. The ensuing influx of wealthier buyers typically leads to the construction of large houses or mansions that disrupt the intended character of the landscape. In both communities, engaged residents play a prominent role in preserving a guiding vision and retaining an accepted character.

These examples are by no means intended to be representative of the many types of communities found across America. Riverside represents the homogenous, nearly exclusively white communities that emerged as rejections of the increasingly diverse urbanization found in northern and southern cities in the late nineteenth and early twentieth centuries. Sea Pines, which was developed a half century later, exhibits a similar lack of diversity among property owners. Each grew in a turbulent century during which federally mandated housing practices, including those that led to redlining, promoted the needs of whites over people of color, enabling and encouraging the creation of segregated communities.[12] This aspect of American history is fundamental to understanding community development and governance and is being increasingly explored by historians and journalists. Because this book takes only a brief look at a few representative landscape types, the discussion in this chapter focuses on a narrow theme, the impact of market forces and climate change on designed communities.

In both designed villages and resorts, the two types of landscape considered in this chapter, landscape stewardship falls, at least in part, to residents and homeowners, who may or may not care about their community's design heritage. Therefore the legacy of Olmsted and Vaux at Riverside or of Charles Fraser at Sea Pines matters not only because of the perceived importance of the men who established the defining character of each place but because the patterns they established translate into higher property values and a desirable quality of life. The design history acts as a tool for advancing property owners' interests and investments at least as much as it functions as a guide for preservation activities.

Village of Riverside, Illinois

Of the early planned residential communities in the United States, probably none has a more revered design legacy than Riverside, Illinois. Carrying the names of Olmsted and Vaux as its original designers and featuring numerous parks, tree-lined drives, and large lawns around houses designed by notable architects, the 1,600-acre community has a character and a heritage that set it apart from most other communities. Driving through it today is like experiencing an idealized version of classic American suburbia. Even as its character has evolved somewhat over time, its original vision remains primarily intact. This continuity can be attributed to a highly engaged residential population as well as involvement from organized groups such as the Frederick Law Olmsted Society of Riverside, which has advised on aesthetic issues and challenges since its founding in 1968.

Toward the end of the last ice age, the area in which modern-day Riverside is located was covered by a giant body of water, Lake Chicago, of glacial origin. As the glacier retreated, less than 15,000 years ago, newly formed streams and rivers prompted the massive lake to drain southward into the Gulf of Mexico, leaving behind today's Great Lakes. Prominent among the outlets was the Des Plaines River, which helped establish the current form of Lake Michigan. That three-hundred-mile lake was mapped by seventeenth-century settlers and given a variant of the Ojibwa word *mishigami,* meaning large lake. Members of the Potawatomi tribe were the area's primary Indigenous residents, with the Ottawa and Chippewa (Ojibwe) also residing in the region. White settlers began claiming the land in and around Riverside in the 1830s. In 1833 a group of tribal chiefs signed a treaty ceding five million acres in Illinois and Michigan to the U.S. government.

By the mid-1800s, family farms occupied most of the land. A group of Chicago-area business professionals formed the Riverside Improvement Company and began buying up tracts of farmland and forests in Cook County on both sides of the winding Des Plaines River around ten miles from downtown Chicago. They sought to create a model suburban village that would serve as a commuter community and an escape from an increasingly urbanizing Chicago, which was plagued by unchecked development and unsanitary living conditions.

Establishing the Legacy

The Riverside Improvement Company hired Olmsted and Vaux to create a community that would combine, as Olmsted described it, "the conveniences peculiar to the finest modern town, with the domestic advantages of the most charming country, in a degree never before realized."[13] The designers' plan conveys a system of organic, cell-shaped neighborhood pods filled with broad residential lots, all contained within a system of meandering pleasure drives that forwent the clarity of the ubiquitous urban grid while conforming to the topography and mimicking the sinuous nature of the Des Plaines River. The intersections of drives form dozens of awkward wedges of landscape that became neighborhood parks, interspersed with swaths of open spaces, remnant forestland, and expanses of lawns. In total, around seven hundred acres were set aside for the public realm. Olmsted and Vaux were not involved in the completion of their design for Riverside, however. After the Great

Chicago Fire of 1871, which destroyed most of the city, the development company went bankrupt. Shortly thereafter, the architect William LeBaron Jenney oversaw the completion of the original plan and designed many of the community's first homes.

Although definitive numbers are difficult to obtain, around two thousand people lived in Riverside by the end of the nineteenth century.[14] The *Chicago Chronicle* recognized the village as one of Chicago's "most ultraexclusive" suburbs and a "place of beautiful homes, of wide-spreading, well-kept lawns and grand old trees."[15] The landscape was a source of pride for residents, who collectively spent a reported $10,000 per year, half on the shared parkland and half on their private yards.[16] By maintaining manicured, grass-covered yards, homeowners in Riverside endorsed the concept that blurring the lines between personal property and public open space contributed to a sense of shared community.

Without the enforcement of explicit covenants, promoting the parklike landscape aesthetic established by Olmsted and Vaux fell to the voluntary efforts of like-minded people willing to sacrifice personal taste in favor of community cohesion.

Fig. 35. The sweeping drives, extensive tree plantings, and ample open space help establish the overall character of Riverside's landscape per the original plan created by Olmsted and Vaux. (Chicago History Museum; ICHi-097786; Charles R. Childs)

These early residents seem to have been primarily well-to-do, with one newspaper account describing houses built in Riverside as "large spacious country homes" set back from the street and surrounded by lawns and landscaped gardens that "give an appearance of fairyland which people living in Chicago can find only in parks."[17] Even Olmsted knew that suburbanization was not a particularly democratic phenomenon, writing about Riverside that it would be part of a "counter-tide of migration [away from cities], especially affecting the more intelligent and more fortunate classes."[18]

Early Development

Riverside grew slowly in its first decades, owing in part to restrictions on housing prices to maintain a certain standard and ensure "nothing of cheap or flimsy construction might find a place in the village."[19] It continued to attract people away from an increasingly crowded Chicago, which had grown to 2.7 million people by 1920. A building boom the following year added around 250 houses displaying a variety of styles and materials. According to the *Chicago Tribune,* "A few will be brick, others stucco, some hollow tile, and others frame."[20] Such was the intention of Olmsted and Vaux, who seemed less concerned about the buildings than the landscape. As expressed by Olmsted: "We cannot judiciously attempt to control the form of houses which men shall build, we can only, at most, take care that if they build very ugly and inappropriate house, they shall not be allowed to force them disagreeably upon our attention when we desire to pass along the road upon which they stand."[21]

The community, which the *Chicago Tribune* called "the most aristocratic suburb of Chicago," gradually opened to a growing middle class. Single-story bungalows gradually mixed in with larger homes, with prices for all houses lying between $3,000 and $11,000.[22] An advertisement from the time glorifies Riverside as having "no factories, no smoke, no dirt" and as being "the most beautiful of Chicago's many charming residential communities."[23] Without prescriptive guidelines or covenants, the village board wielded building restrictions dictating what could be built in the community. This governance model was solidified in the 1910s. Like many other self-proclaimed suburban "villages," Riverside would be run as a business or institution. The village board functioned as a board of trustees or directors; the president acted as the chief executive officer, and the village manager, who dealt with daily operations, was essentially the general manager. In this scenario, residents formed the stockholder base.

As Riverside grew, its sylvan goals were boosted by an ongoing initiative to create forest preserve districts in the urbanizing region. The movement dates to the early 1900s and was led by the architect Dwight Perkins and supported by the landscape architect Jens Jensen, who produced an ecological analysis of the region, while Perkins took the lead on advocating for a conservation approach. This was the era of the famed naturalist and conservationist John Muir, whose writings on the threatened American wilderness were widely read. Perkins understood that preserving the landscape, whether a native forest or a reclaimed successional woodland, was a municipal urban planning strategy that benefited, did not thwart, regional growth.[24] After a decade of controversies and court battles, civic leaders formally proposed a forest preserve district to Cook County residents on November 3, 1914. Thanks in large part to the research and advocacy of Perkins and Jensen, voters approved the bill, and the district was formally established later that month. The forested land, which grew to 33,000 acres by 1940 and ultimately nearly 70,000 acres, was to be owned by the people in perpetuity.

Although the forest preserve was not included in Riverside proper, it did provide a wooded greenbelt around a portion of the village that further enhanced the idyllic sense of nature and separation espoused by Olmsted and Vaux. While beneficial from certain perspectives, the forested area attracted outsiders to Riverside, which disturbed the isolation that some village residents enjoyed. In 1919, Edith Rockefeller McCormick, the daughter of the oil magnate John D. Rockefeller, brought added attention when she gifted to the Forest Preserve District land along the southwestern edge of Riverside on which to build "one of the most extensive zoological gardens in the world."[25] The prospect of an estimated two and a half million annual visitors driving on the quiet neighborhood streets worried some people in Riverside.[26] While it did signify intensifying encroachment, the presence of the Brookfield Zoo, which opened to the public in 1934, does not seem to have adversely affected the village.

Fighting Encroachment

Riverside continued to build out lots through the 1940s, and by 1950, more than nine thousand people lived there. Despite the booming postwar suburban growth of Riverside, which followed national trends, there is little indication that the character of the landscape in the sleepy community changed much, even as newly designed houses exhibited current architectural styles. The village board continued to review requests for new construction. This included everything from garage remodels, house additions, and new pools to a "one-story brick single-family residence with two car attached garage [for] $19,000" and a restaurant with parking for ten cars in the commercial area, as was the case over two weeks in 1955.[27] While minutiae concerning architecture were evaluated, the open space faced no such formal oversight.

With respect to the landscape, residents appeared eager to maintain the status quo and became more anxious and vocal about preserving the character they valued. In 1957, around one thousand residents signed a petition protesting the sale of adjacent land owned by the railroad company to an industrial firm that wanted to erect a 36,000-square-foot warehouse and distribution center on an abandoned coal yard. The company owned the land but needed approval from the village board to get a building permit for any structure. Locals complained that the facility would bring large trucks onto their slow, curving streets and create bottlenecks and traffic congestion.[28] The protest was successful, and the railroad quickly scrapped its plans.

Controversies continued, and two years later, residents wrote letters to their state senators in hopes of preventing the University of Illinois from building a permanent campus on the adjacent, one-hundred-acre Riverside Golf Course, which several Riverside residents founded in 1893. The university had opened a temporary campus on Chicago's Navy Pier in 1946 to accommodate the influx of veterans taking advantage of the postwar G.I. Bill. University trustees selected the suburban site against the wishes of Riverside's board, which formally opposed the move because, its members stated, "the majority of citizens are against that location."[29] Worries over congested streets, increased taxes, and lower property values prompted residents to send a petition underscored with five thousand signatures opposing the move. A newly formed committee of property owners led the charge.[30] Whether directly influenced by this or persuaded by the knowledge that most students preferred an urban campus in downtown Chicago over a commuter campus in the suburbs, the university scrapped its plans for the golf course site.[31]

An Emerging Ethic for Preservation

Such perceived attacks on Riverside's tranquility aroused the preservation instincts of its residents. In the early 1960s, a desire to save the past was manifested when a beloved house designed by the architect Louis Sullivan, the Babson House, was razed to make way for subdivided lots. This followed by a decade the near destruction of Frank Lloyd Wright's famous, nine-thousand-square-foot Coonley House, which a developer ultimately agreed to split into four separate units to preserve the architectural integrity of the building, which Wright considered to be among his best. Losing the Babson House coincided with an emerging renaissance for the legacy of Olmsted. In 1964 the American Society of Landscape Architects and Harvard University's Graduate School of Design joined forces for a traveling exhibition celebrating the one hundred years since the establishment of the profession of landscape architecture.[32] Since Riverside was featured in the exhibition and most of the existing village was designed, in part, by Olmsted, the village manager lobbied to bring the display to town.[33] His effort was successful, and the exhibit was displayed in Riverside's Village Hall in the summer of 1966. In addition to broadening awareness of Olmsted, it established an earned sense of gravitas for residents, particularly those concerned with keeping things the way they had always been. Sadly, Vaux, whom landscape architects never adopted as one of their own, remained largely forgotten in the process.

The same year that Olmsted's works were on display in Riverside, the U.S. Congress passed the National Historic Preservation Act with the intent to preserve significant and irreplaceable historic properties "in the face of ever-increasing extensions of urban centers, highways, and residential, commercial, and industrial development." There emerged a tide of support in favor of places like Riverside. To capitalize on the momentum, a group of village residents founded the Frederick Law Olmsted Society in 1968 to educate citizens about Riverside's history and advocate for preservation. In an early initiative, the society's first president, Robert W. Heidrich, acted as both plaintiff and attorney in a lawsuit against the village board. He charged that the playground equipment and athletic fields installed in the

Longcommon, one of Riverside's premier landscapes, destroyed the concept of open space critical to the community's character.[34]

In 1970 the Olmsted Society took the lead in nominating the 1,500-acre Riverside Landscape Architectural District to the National Register of Historic Places. The nomination pointed out that Riverside offered a dramatic contrast to the "typically unplanned and unsightly urban sprawl" that surrounded it and that, aside from minor changes, the century-old village had "preserved its original landscaped plan almost intact."[35] The nomination was successful, and Riverside earned National Landmark designation on September 4 that same year. The status instilled pride in the community's design legacy, yet there remained no enforceable means to preserve it. Fortunately, the original patterns of the streetscapes and open spaces were unmistakable, and people readily adapted their tastes to meet those of Olmsted, Vaux, and the community. As with many other suburban Americans in the twentieth century, they seemed to have little interest in exploring alternatives to clipped green lawns.

Over the next decade, Riverside reached its peak population, with over ten thousand residents. Its publicized landmark status became a signal to curious outsiders that the community was worth exploring. Students and scholars in planning and design disciplines visited to discern what Olmsted and Vaux got so right, while the Chicago School of Architecture Foundation led tours for people wanting to see the work of such celebrated architects as Sullivan and Wright. Preserving history seems to have become integral to residents' way of life, and in 1971 the village board formed the Riverside Historical Commission to complement the role of the Olmsted Society by, among other things, advising on historical matters, researching and documenting the village's history, and running a historical museum. The commission also took to heart its role as an advocate for the community's legacy. In a fiery letter to a local paper, the chair of the commission protested the encroachment of a proposed expressway. He said that in addition to eroding the community's hundred-year-old landscape and property values, it would "be equivalent to destroying a mile square gigantic memorial."[36]

Over time, a dozen boards and commissions formed in Riverside to foster citizen engagement and involvement. Among these bodies was the Forestry Advisory Commission, which was established to advise the Village Forestry Department on landscape-related issues. This commission evolved into the Landscape Advisory

Fig. 36. Many of Riverside's early homes were large, traditional structures set on wide and deep lots with manicured lawns and trees. Over time, a variety of housing types changed the character of the community. (Flickr: David Wilson, @davidwilson1949; Attribution 2.0 Generic)

Commission in the late 1990s after an internal feud that became known as the petunia wars. The conflict began when a well-intentioned board member planted vibrant red, white, and blue petunias in the central business district without regard for a town ordinance dictating that plants in public areas were not to call attention to themselves.[37] Some traditionalists considered the showy flowers an affront to Olmsted's legacy because of his acknowledged preference for naturalistic and understated landscapes. As one member of the Olmsted Society wrote, "We struggled long and hard to get [the preservation] ordinance, and wanted it to conform to Olmstedian principles."[38] Following uproars from those both in favor of and against the flowers, the board rescinded the ordinance, then appointed a seven-member Landscape Advisory Committee to produce a list of acceptable plants with flowers more in keeping with Olmsted's preferred palette.

Natural Forces

Like many other mature designed American landscapes, Riverside's tree-dependent character changed in the twentieth century owing to the devastating impact of Dutch elm disease. As early as the 1940s, Chicago-area residents were urged to "take up the battle" against the elm bark beetle by spraying their trees with dichloro-diphenyltrichloroethane (DDT) twice a year.[39] In 1976, more than one hundred of Riverside's elm trees were cut down just between May and July, with nearly fifty new trees discovered to have the disease each week.[40] In a move that was later regretted as the negative effects of the poison were better understood, the village supplied DDT to residents, who then, through a volunteer program, inoculated trees on private property and in the parkways.[41] By the 1980s, a less objectionable, injectable fungicide had proved effective in slowing the disease, which gradually subsided, leaving few elms behind. The Olmsted Society rose to the challenge by providing information to residents, planting new trees, and purchasing Arbotect, the tree industry's standard fungicide, to fight the disease in the community.

The aesthetic change wrought by the loss of the elms seems to have inspired a renewed focus on the landscape. Olmsted's star had re-risen and, as reported in the *Chicago Tribune* in 1982: "There are thousands of 'Olmstedians' in the world . . . and many reside in Riverside."[42] Calls for preservation were typically louder than any voices advocating changes to the status quo. The village board recognized the need for more direction regarding landscape stewardship and engaged a consultant team to set forth guidelines for rehabilitating and protecting Riverside's open-space system.[43] The plan's writers acknowledged that the landscape "appears and functions in many ways as historically intended" and offered strategies for planting per Olmsted's vision. These strategies included spatially defining outdoor rooms, accentuating views, reinforcing the curving streets, highlighting entry points, and emphasizing the landscape "spine" running through the community.[44] A presentation to the board stimulated extensive discussion among the commissions and concerned villagers, yet they found no consensus on managing the public lands.[45]

The loss of trees was not the only challenge facing Riverside's landscape. In the 1980s and 1990s, Riverside's public works staff worked to remove rapidly spreading invasive and aggressive plants that threatened the aesthetic character and ecological health of the community. Because the task was so daunting, the Village applied for and received grant money from the Illinois Department of Natural Resources (IDNR) in 2000 to train citizen volunteers in natural resource management and weed control to assist in the efforts. A steady group of around a dozen people passed the training and began removing buckthorn, a highly aggressive shrub brought to America from Europe for a hedging material, from the banks of the Des Plaines River. Beginning in 2009, the Olmsted Society organized regular volunteer work-days to help clear the banks of the invasive plants. In 2014 alone, Riverside spent around $10,000 removing buckthorn; five years later, a thriving population of native milkweed, perennials, and sedges was established in its place.[46]

While residents and leaders managed what they could control with citizen-based stewardship, natural forces in the form of flooding increasingly took a toll on Riverside. The entire village is located within the Lower Des Plaines River Watershed and therefore is at the mercy of upstream conditions. As Chicago continued to grow rapidly, so did the broader region. Much of what was once vast forests and wetlands transitioned into farmland before widespread development took hold and turned cultivated area into many thousands of acres of impervious asphalt, concrete, and buildings. Understandably, millions of gallons of stormwater that once soaked harmlessly into the ground before reaching waterways began surging across parking lots and rooftops into drainpipes and then into local streams and rivers, eroding banks and inundating low-lying properties. By the 1980s the urbanization of the region was blamed for nearly $23 million in annual flood damage along the Des Plaines and other waterways that drain metropolitan Chicago.[47]

One particularly severe storm moved through the region in August 1987, dumping nearly ten inches of rain in and around Riverside in a twenty-four-hour period. At one point, several feet of standing water filled roadbeds and poured into basements, businesses, and houses. Power was out for more than a week as more than one hundred Riverside residents were left temporarily homeless. Federal engineers later predicted worsening flooding owing in part to ongoing urbanization. Conditions remained relatively calm, with only minor flood events, until September 2008, when hundreds of residents were affected by three days of rain caused by the remnants of Hurricane Ike, which made landfall near Galveston, Texas, before moving northward. The Des Plaines River swelled to within a half inch of the record set in 1987.

Because the Village of Riverside has no jurisdiction over the river, concerned residents had to look beyond local officials for potential solutions to what they knew would continue to be a problem. In search of answers, individuals contacted county, state, and federal agencies and elected officials.[48] At the same time, around thirty neighbors met, conferred, and eventually formed a group called Riverside Residents for Flood Prevention in a proactive move to help tackle the ongoing problem. The group met multiple times with officials from the Army Corps of Engineers and the IDNR in 2009 with the goal of collecting pertinent information regarding causes of the flooding, tracking the progress of the various studies already underway, and informing their community about the issues and potential solutions.[49]

In June 2010, severe storms in the Chicago area prompted President Barack Obama to declare Cook County a disaster area. Riverside and surrounding communities were hit hard, with serious wind damage, widespread power outages, and flooding inflicting havoc. A deluge of rain—two inches in one hour and nearly seven inches overnight—overwhelmed sewer systems, flooded streets, and, in some cases, left up to five feet of water in basements. By this time, the connection between intense storms, heavy rains, and climate change was difficult for many people to deny. That same year, following the completion of engineering studies of the river and the watershed, the Army Corps of Engineers and the IDNR proceeded with federal- and state-funded work the village president called the "single biggest Riverside project in our lifetimes" to demolish three dams on the Des Plaines River and regrade and comprehensively restore Swan Pond Park, a small, picturesque low-lying landscape adjacent to a prominent bend in the Des Plaines.[50]

Not long after these 2010 storms, an emerald ash borer—the destructive, ash-tree-eating beetle that made its way to the United States from Asia—was found inside a trap attached to a tree in Riverside. Arborists and residents who had been working for decades to stem damage from the Dutch elm disease now faced a new impending threat. This was not a surprise, however, as the borer had been discovered four years earlier in the state. After the first beetle was found in Chicago in 2008, the president of the Olmsted Society acknowledged this and other perils in the community newsletter: "With the loss of many of our beautiful trees due to storm damage, the looming advance of the Emerald Ash Borer, and the state of our Village Economy, we are all concerned with the future of our piece of Eden."[51] In 2014 the village board approved spending $200,000 from its capital projects fund to remove nearly all the

Fig. 37. The curved roads running through the community created triangular landscaped islands that continue to contribute to the landscape's open and lawn-centric character. (Flickr: David Wilson, @davidwilson1949; Attribution 2.0 Generic)

remaining 566 ash trees located on public land within the community, which was about half of the 1,100 ash trees reported in 2006. An aggressive tree planting program added back hundreds of trees representing more than a dozen species in preparation for a future made more unknown by climate change.

Even as nature threatened the integrity of Riverside's design heritage, it does not seem as if the character of the landscape has ever been seriously threatened. By the 2000s, the appearance of the village was likely not too different from what Olmsted and Vaux had envisioned: a well-maintained parklike community defined by spreading lawns and spectacular trees. Residents, who are permitted to do essentially whatever they want on their property, seem willing to play by the rules set for the broader community. The public landscapes, which account for around 40 percent of Riverside's acreage, are generally well-kept. These parkways, parks, and over sixty discrete landscaped spaces—islands, triangles, and commons—remain somewhat protected by the requirement of a permit from the Public Works Department and review by the Landscape Advisory Commission to make changes.

In 2015 the Landscape Advisory Commission fulfilled its mandated charge and produced a Master Landscape Plan for the public portions of the landscape. The document establishes design guidelines and planting schemes for all parks, parkways, and other public planting areas "in accordance with the landscape planting theories, concepts and principles of Frederick Law Olmsted."[52] Because there are no historical planting plans providing direction, oversight involves interpreting an intended character rather than adhering to specific plant species or planting layouts. Lacking a precise historicist mandate grants the flexibility to adapt new designs to address contemporary issues such as changing recreational and public safety needs, potential pathogens or damaging infestations, evolving ecological best practices, and emerging threats from climate change.

Also included in the master plan is a specific plan for the Arboretum, a status established for the entire village in 2015 in recognition of "its unique tree plantings and for its stewardship of the landscape."[53] By reaching such accreditation, Riverside's landscape enjoyed renewed gravitas within the community as a significant cultural and natural resource. Care of the Arboretum as of all the public landscapes falls under the purview of the village forester. The Landscape Advisory Commission offers recommendations and support as needed in conjunction with its broader oversight. It also periodically reviews the implementation of the master plan because, even with Riverside's lofty design pedigree, "Landmark status does not . . . supplant the day-to-day work that is involved in land stewardship."[54]

Contemporary Issues

In 2018, when nearly nine thousand people lived in Riverside, the average home value was $412,000. That figure had soared to $650,000 by February 2021, when the chaos and uncertainty wrought by the coronavirus pandemic led many urban Chicagoans to flee to the relative calm of suburbia and small-town life.[55] Riverside was one hundred fifty years old by this time, and its visual character remained remarkably intact.

Even with improvements along the Des Plaines River, however, flooding remained a serious issue as storm events in Illinois and across the globe became significantly wetter and more frequent. Spring storms brought a new round of floods and damage to suburban Chicago in 2013, prompting a local resident to comment about the

Fig. 38. A significant and growing challenge for Riverside is continued flooding in the community. Images like this of drowned streets and yards will likely become more prevalent in the face of climate change. (Allen Goodcase)

situation: "Every year it's getting worse and worse, and it's coming more often and coming higher every time."[56] Updated proposals for the Army Corps of Engineers emerged with the goal of spending $400 million along the river for floodwalls, levees, landscape restorations, a new reservoir, and floodproofing initiatives for homes and businesses in thirteen communities, including Riverside.[57]

Plans presented in 2018 outlined the construction of a large floodwall separating the Des Plaines River from an adjacent residential neighborhood. Two years later, following a series of negotiations and revisions, Riverside's village board voted unanimously to approve the plan. While the functional aspect of the wall garnered relatively easy consensus, what it might look like stoked controversy. Looking to Olmsted precedent, options for stone walls similar to ones found in Prospect Park and Boston's Franklin Park were considered. Despite the focus on historical examples, the Washington, DC–based nonprofit, National Association for Olmsted Parks (now the Olmsted Network), penned a letter to the village board in 2021 expressing "serious concerns about plans by the Army Corps of Engineers to proceed with design and construction of a massive floodwall," calling it "contrary to the plans of Frederick Law Olmsted and Calvert Vaux."[58] This opinion, while fundamentally accurate, points to the challenges of balancing the desire to respect established design heritage with the need to respond to contemporary realities. As of 2022, final plans for the floodwall had not been agreed on.

Recognizing that its future was directly tied to the changing climate led Riverside to participate in the Cross-Community Climate Collaboration (C4), in which around a dozen local suburban Chicago communities joined together in a pledge to reduce greenhouse gas emissions while promoting equity and sustainability. This effort started in 2022 and, coincidentally, came shortly after a "supercell" storm blew through the area and wreaked havoc with eighty-mile-an-hour winds that damaged homes and trees and spurred Riverside's president to declare a local state of emergency. This was in the summer of 2022, when areas of Missouri, Kentucky, and Illinois experienced thousand-year rain events in one week, highlighting the pending threats from climate change.[59]

Along with initiatives addressing resiliency, connectivity, natural resource management, infrastructure, and healthy environments, C4 promotes tackling equity issues such as affordable housing, inclusion, diversity, and food insecurity, none of which has risen to a mandate in the long history of Riverside, thanks in part to its stubbornly static demography. Unlike Chicago, which was predominantly white before rapidly diversifying in the twentieth century, the suburbs have changed relatively little. By 2020, Chicago was thoroughly integrated, with whites, Blacks, and Hispanics each making up around 30 percent of the core city's nearly three million people. In Riverside, on the other hand, over 90 percent of residents self-identified as white, a figure likely common in many predominantly white and affluent communities.[60] While the village's trees and lawns remain beautifully tended, the community's cultural homogeneity keeps it tethered to an archaic past that many Americans are eager to move beyond.

In most considerations, Riverside has been remarkably resilient when it comes to maintaining the integrity of its Olmsted and Vaux design. As individual houses come and go and trees mature and die, the landscape's overall structure survives. Whether such consistency is the result of an active preservation ethic generated by a reverence for its design legacy or of a desire to maintain a set quality of life is an interesting point of debate, yet essentially immaterial. Either way, Riverside's community has worked together for generations to manage both a significant heritage and a desirable aesthetic.

Sea Pines Resort, Hilton Head Island, South Carolina

Along with a proliferation of subdivisions and planned suburbs that carpeted millions of acres of abandoned farmland outside many American cities beginning in the mid-twentieth century, a new type of place emerged, the resort community. Rather than targeting local commuters as in Riverside, these places were typically designed to attract retirees in search of new locations to live, people with means looking to purchase second homes, and, to a lesser extent, short-term vacationers. At the same time, as automobiles became ubiquitous features in the American landscape, families packed into cars, now able to travel further afield on vacations. Desirable destinations with good weather, often along a coast, were increasingly available to the masses. People fled to quaint beach towns, most of which had evolved incrementally and organically, with rows of cottages, small commercial areas, and no comprehensive plan or community design. This changed dramatically in the 1950s and 1960s when the nascent environmental movement had people thinking differently about large-scale development. A new, ecologically sensitive mentality led to the creation of places like Northern California's Sea Ranch, which, beginning in the mid-1960s, became a pilgrimage destination for architects and landscape architects because of its beautiful setting overlooking the Pacific Ocean and its conservation- and nature-focused design. Even before Sea Ranch, however, there was Sea Pines Plantation on the southern end of Hilton Head Island in South Carolina.

Sea Pines emerged in the 1950s as the dream of the visionary developer Charles Fraser, whose innovative planning helped spur Hilton Head's rapid and expansive growth. Since its early years as a modest business venture, the resort has struggled to meet the dual and contradictory goals of conserving and monetizing the island's natural beauty. Through the ups and downs of its evolution, the community's lasting success makes it a reliable case study in how a clearly articulated vision can provide a tool for ongoing stewardship.

Hilton Head lies just off the coast in South Carolina's Lowcountry, approximately forty-five miles north of Savannah and about twice that distance south of Charleston. It is an oblong landform shaped a bit like a foot that encompasses around forty-two square miles. It averages over five miles in width and about twice that in length, making it the largest of the more than one hundred appropriately named Sea Islands, which stretch along three hundred miles of coastline from South Carolina to the northern tip of Florida. The island's landscape is lush, subtropical, and filled with massive live oaks, palmettos, and loblolly and longleaf pines buffering winding creeks, sounds, and twelve miles of wide, white-sand beaches. Indigenous tribes explored the island for several thousands of years before Spanish explorers discovered it in the 1500s. In 1663 the British captain William Hilton sailed past the bluffs at the northern end of the island eventually named in his honor. After exploring for several days, Hilton observed in a letter, "The Lands are laden with large tall Oaks, Walnut and Bayes, except facing the Sea, it is most Pines, tall and good."[61] No nation formally claimed the island for itself until the British did so in the 1720s.

The island's landscape proved adaptable for the successful production of indigo and the luxurious Sea Island cotton, attracting wealthy plantation owners, who employed the forced labor of enslaved people imported from the Caribbean and West Africa. Prior to the Civil War, more than twenty plantations divided the island. Because it was accessible only by boat, Hilton Head remained relatively isolated despite its growing economic prosperity. The mild climate was ideal in colder months but could be brutally hot and humid through the warmer months. Even as the densely vegetated natural habitat bred an abundance of poisonous snakes, alligators, and swarms of mosquitos and other disease-carrying pests, the island's population continued to grow, thanks to the slave-driven economy. These threats and the oppressive heat kept most landowners living on the mainland while enslaved people lived in relative isolation on the island. They formed insular communities, which allowed them to retain and adapt their specific African heritage across generations more readily than those in more populated inland areas in the South. Following the war, formerly enslaved people joined a larger culture of freed people along the southeastern coast, known as the Gullah in North and South Carolina and the Geechee in Florida and Georgia. Gullah families on Hilton Head tended their crops and purchased thousands of acres, primarily on the island's northern end. Communities once united by shared bondage remained close, living in continued isolation as Union forces left the island and large stands of pine trees covered abandoned cotton fields and erased much of the island's antebellum past.

The quiet was slowly erased beginning in 1949 when a group of associates in the lumber business purchased close to 20,000 acres of pine forest on Hilton Head for about $60 an acre. By this time the island had around three hundred residents, mostly Gullahs who made their living harvesting shrimp, produce, livestock, and oysters, which were then sold in Beaufort, Savannah, and other nearby communities.[62] There was no electricity or telephone service, and only ten homes had indoor plumbing.[63] The southern end of the island retained its native wildness, while most residents lived on the northern end. The partnership, which was named the Hilton Head Company and included Charles Fraser's father, Joseph, built three lumber mills on the island. Although cutting and shipping lumber was the company's primary business effort, the eventual development of the island was always part of the plan. At the time, the only way to get goods and cars on and off the island was on a six-car ferry run by a local Gullah man named Charlie Simmons. Even while exhibiting signs of change, Hilton Head remained essentially unknown to the world.

Establishing the Legacy

When Charles Fraser was twenty-one, he spent a summer on Hilton Head working for his father, and was captivated by the island. In 1952, while a law student at Yale, he used advice from professors to conceive a master plan for the island that would capitalize on the landscape's unspoiled beauty.[64] He wanted to do something different from what had been done at such popular East Coast resorts as Daytona, Myrtle, and Atlantic Beaches, where developers flattened dunes and filled wetlands to maximize tourists in hotels and houses were crammed in parallel rows extending from coveted oceanfront lots. Fraser considered such places visual pollution and knew that if Hilton Head adopted that approach, the qualities he appreciated most about the island would be lost forever.[65] He believed it would be better to, as he put it, "leave [the island] alone entirely" than let it be "chopped up in a quick-buck, real estate operation."[66] His plan involved building villas and single-family residences in a manner that celebrated rather than destroyed the native ecology.

Developing the island became viable in 1956 when the construction of a swing bridge provided the first permanent connection to the mainland.[67] The Hilton Head Company partnership formally split up that same year and Joseph Fraser received 3,500 acres of land on the southern end of the island as part of the deal. A year later, he and his two sons, Charles and Joseph Jr., formed the Sea Pines Plantation Company. Through a second business, the Sea Pines Land Company, they purchased an additional 1,200 acres of forested beachfront on the southern tip of the island. This took the total property to 5,200 acres, which was ample space to implement their plan for a high-end leisure community buried in nature.

The men named the Sea Pines Plantation in unabashed reference to the island's cultural history. Although such a move was not rare for the time, it surely prompted different responses from their targeted affluent white consumer and the local Gullah population, which was only a few generations removed from slavery. Charles Fraser quickly became the visionary and public face of the community. His ambition was bold: to have the "best land planning and the best architecture and the best deed covenants of any community in existence."[68] He wanted buildings designed to be "unobtrusive in form and color in order to complement their natural setting" and "the entire community [to] be homogenous in feeling in a park-like setting and free from the discordant architectural shapes and colors, which vie for attention and attempt to create greater visual impact than a neighbor's."[69] In his view, a million-dollar home should stand next to one costing a fraction of that and not "over-power the street, or nature, or each other."[70] He was convinced this concept would work only if the developer—in this case, himself—controlled every aspect of planning and setting design standards.[71]

By all accounts, Fraser cared deeply about preserving the island's natural quality. He was also a businessperson and needed consumers to buy into his vision. The densifying Atlantic coastline was rapidly developing, making unspoiled tracts of land like those on Hilton Head increasingly rare and valuable. "Hindsight will be worthless because there won't be any place to start over," Fraser predicted.[72] He was confident his ideas would work, and he possessed the charisma to sell his vision to whoever would listen. He just needed help turning it into reality.

Fig. 39. Even with consistent development, Sea Pines Resort on the southern tip of Hilton Head Island retains a lush natural landscape appearance and wide sandy beaches. (Alex MacLean/Landslides, courtesy of Sasaki)

Fraser selected the Massachusetts-based landscape architecture firm Sasaki & Associates to produce a master plan, even though it had no previous experience in large-scale resort planning.[73] He flew to Boston and dropped in on the firm's office in nearby Watertown, unannounced and without an appointment.[74] According to Stuart Dawson, a practitioner in the firm in his early twenties at the time, Fraser entered the office carrying rolls of maps and drawings and proclaimed, "My daddy has given me this Yale graduation present and wants me to do something good with six thousand acres on Hilton Head Island. . . . Will you guys be willing to help?"[75] They were, of course, interested, and Dawson soon traveled to Hilton Head, which he later recalled as a "semi-tropical jungle" and "the best example of raw America I had ever seen."[76]

Fraser shared his vision, and the landscape architects helped flesh out a land-use plan for the entire property and a more detailed plan focused on a tract close to the ocean. Although the landscape architecture firm was critical to the process, no one could legitimately say the plan for Sea Pines was Sasaki's to claim. News of Fraser's

Fig. 40. As in Riverside, a system of sweeping sinuous roads casually moves people through Sea Pines Resort and remains integral to the community's character. (Photo by author)

plan for the residential resort spread quickly. In 1957, a newspaper reported that "the long term development of Sea Pines will include five major residential areas, providing home sites ranging from one-quarter acre to five acres, supplemented by a yacht club and Inn, golf course, and other recreational facilities."[77] Toronto's *National Post* reported a year later that "many Canadians have already discovered Hilton Head," and recommended Sea Pines as a place to vacation and invest.[78]

Because the intent was for the community to evolve in response to the needs of property owners, the master plan became a guide rather than a prescriptive or detailed design.[79] It used innovative strategies that distinguished the community from typical coastal developments, including shifting the main beach access road several hundred feet back from the dunes and building T-shaped streets that connected to oceanfront lots. Houses sat behind a wide setback filled with protected dunes and stands of native pine trees that formed a natural buffer and preserved uncompromised views along the beach. Fifty-foot landscape buffers running perpendicular to the coastline between each cluster of homes reinforced the natural landscape while also providing generous public access points to those living on interior lots. These moves ensured that the beaches were democratic places for residents and visitors and not the perceived domain of the elite beachfront property owners. Saving trees was identified as a critical goal, and developers strived to preserve the centuries-old live oak trees found throughout the property.[80] As in Riverside, streets snaked through the landscape, tightly hugging the edges of marshes and lagoons and avoiding conflicts with those large trees deemed worth saving. Many such conservation planning moves soon to become common practice were innovative for the time.

Codifying Stewardship through Covenants

While the master plan laid out neighborhoods, roads, and infrastructure, a set of restrictive covenants provided the most effective means of control. In August 1957, Fraser signed the "Declaration of Rights, Restrictions, Conditions, Etc" as a tool for defining what residents were allowed to do with and on their property. Since much of the design for Sea Pines' acreage would ultimately revert to individual property owners, the covenants provided a way to establish and maintain the desired character of the buildings and the landscape. In theory, this would preserve aesthetic consistency through and beyond eventual expansion and inevitable redevelopment. The covenants restricted property owners in, among other matters, which permanent and temporary structures they could build, how their garbage containers were to be kept, and how tall their plants could grow. The goal was to require owners "to prevent the development of any unclean, unsightly, or unkempt conditions of buildings or grounds on such lot which shall tend to destroy the beauty of the neighborhood as a whole or the specific area."[81] Buying into Sea Pines was deemed a privilege, and creating uniformity throughout the community meant property values were less dependent on the personal tastes of others. The company, which in the early years essentially meant Fraser, reviewed and approved owners' design plans, which could be rejected for any reason, including the ambiguously formulated "purely aesthetic grounds."[82]

Although the early versions of the covenants were not legally binding, Fraser was able, at least in the beginning, to effectively control the look of houses and lots. This was his world, and he was seemingly involved in all decisions made. As was reported in the *New Yorker,* "[Fraser] is not merely mayor and the zoning board, he is the living ark of the deed covenant."[83] In his own words: "There is democracy of communication here but autocracy of decision making."[84]

The Early Evolution of Sea Pines

In 1960 the Sea Pines Company built Hilton Head's first modern hotel, the million-dollar William Hilton Inn, just outside one of its entry gates. By this time fewer than twenty homes, mostly beachfront cottages, had been constructed in the community.[85] In-house architects developed the unique "Sea Pines look," which presented a fusion of vernacular Lowcountry architecture mixed with modernist and Japanese influences. The style attracted the attention of the design world and influenced other developers. A growing assortment of modest houses with deep overhangs, screened porches, and low-slung roofs slowly and quietly penetrated the landscape. Clad in muted earth-tone shingles from local cedar, pine, and cypress trees, they blended into the environment, which was intentionally left unadorned and natural. Fraser discouraged residential lawns and favored homes set deep within lots planted with scrubby native plants in swaths of pine straw. The goal was to create a community nestled in the natural world rather than suburban yards designed according to the aesthetic of individual property owners.

Four hundred lots were sold to buyers from over thirty states and seven countries by 1962, the year Fraser built the island's first golf course in Sea Pines.[86] Two years later, Fraser's on-staff architects and landscape architects revised Sasaki's master plan with more focus on the interior land. This was one of nine plans created for the resort in its first fifteen years. The new plan allocated about half of Sea Pines for residential use, leaving most of the remaining land to be used for beaches, golf courses, tennis facilities, parks, marshland, bike paths, and natural buffer areas. Amenities and open spaces were considered at least as important as the residential lots because they contributed to the quality and desirability of the entire community. The intent was that for every acre sold for residential lots, another acre would be dedicated to common open space. To add appeal to interior lots without easy beach access, homes

Fig. 41. This early 1970s house exemplifies the Sea Pines–style architecture created by Charles Fraser and his staff of architects. Fraser believed the landscape was paramount and favored modest, low buildings that blended in with nature. Such homes are becoming increasingly rare as residents and investors continue to build exponentially larger homes. (Photo by author)

were typically sited with views to either a body of water, a natural woodland, or a golf course.

In 1968, with an influx of money from the Travelers Insurance Company, Sea Pines announced a ten-year, $30 million expansion plan that would include retail centers designed in conjunction with two boat harbors. The first phase was the quaint residential and commercial area called Harbour Town, which was advertised as a Lowcountry version of the "old port towns in Barbados, St. Lucia, and the warm Mediterranean."[87] Fraser once again hired Sasaki's firm (which had undergone a name change to Sasaki, Dawson & DeMay), this time to design the landscape and the signature red-and-white-striped lighthouse, which soon became the iconic image of the resort. In a move that epitomized Fraser's insistence on saving trees, he instructed the landscape architects to redesign the circular yacht basin in such a way that would preserve a centuries-old live oak—a decision that cost the project a reported $250,000. After he died in a boating accident in 2002, Fraser was buried at the base of the tree, which he had dubbed the "Liberty Oak."

Consumers continued to justify Fraser's confidence that people would pay to live embedded in nature, and Sea Pines grew rapidly as a residential community throughout the 1960s and into the 1970s. The first legally enforceable covenants were adopted and recorded in Beaufort County in 1970. The company still absorbed all expenses for shared community spaces, while property owners paid a $20 fee for security.[88] The covenants set building standards and established an architectural board to review designs. They also set aside six hundred acres as a forest preserve to "protect selected natural and scenic resources, [and] to promote the conservation of swamps, wildlife, game and migratory birds."[89]

The community's population grew along with Hilton Head's, which saw its full-time population jump from around 2,500 in 1969 to 6,500 in 1975.[90] Marketing the island was in high gear, with a full-page advertisement for the island appearing in newspapers across the country in 1975 with the tagline, "Whether you make it your holiday or your home, you belong on Hilton Head Island." Prospective residents could choose to live in Sea Pines or in one of the other five gated communities that opened subsequently, all promoting a similar nature-centric style and character. Barely two decades removed from isolation, Hilton Head attracted an emerging tourist and resident population that looked starkly different from the local Gullahs, who, for the first time in more than a century, constituted a minority on the island.[91] In 1958, a nearby newspaper reported, "The first people you see on entering the island probably will be the true natives of Hilton Head—Negroes whose people once worked in the . . . fields."[92] As developments for vacationers, renters, and property owners steadily smothered the island's natural landscape, however, evidence of the Gullahs' culture faded from public view.[93]

Empowering Property Owners

The Sea Pines Company's focus remained on attracting property owners even as the resort's beaches and expanding golf and tennis facilities lured an increasing number of tourists. With new interest came soaring housing prices. Half-acre oceanfront lots that sold for around $5,000 in the early 1960s were selling for up to $200,000 a decade later.[94] Such success made international news, established Fraser as a star in the resort-planning world, and led the company on a path of rapid expansion. Fraser aggressively pursued new resort developments in the Carolinas, Florida, Virginia, and Puerto Rico with enthusiasm, confidence, and borrowed money. This forward momentum was halted by the severe economic recession in the early to mid-1970s when debilitating gas shortages and double-digit inflation plagued the nation. Housing starts across the country dropped by half as the real estate market tanked. When interest rates soared from under 6 percent to 13 percent, the combination of excessive loans and minimal capital held by Fraser's company proved lethal, and its upward trajectory quickly and dramatically reversed.

Another impact of the downturn was that fewer people could afford vacation property. Those who were able to buy in Sea Pines enjoyed the upper hand when negotiating purchase prices and conditions. Such consumers were coveted, and making them happy was key to the resort's future. If they found Fraser's aesthetic restrictions too draconian, they could propose designs that only months earlier would likely have been rejected. Control was loosened, which led to the introduction of previously disallowed façade materials, such as stucco and brick, that directly challenged the carefully crafted aesthetic.[95] Allowing individual design expression at the expense of cohesion set a precedent that would forever alter Sea Pines' character.

The persistently poor economy stoked fears about the solvency of the Sea Pines Company, and some property owners worried about the effect a potential bankruptcy could have on the community.[96] Since most of the land remained undeveloped, the nature-first vision they bought into could be threatened if the company folded or was forced to sell to a new owner or, worse, multiple owners. In a proactive move, a group of residents united to form the Association of Sea Pines Plantation Property Owners (ASPPPO) as a negotiating body acting on behalf of all owners. In 1974, in an agreement that signaled the first meaningful change in the governance and stewardship of Sea Pines, ASPPPO and Fraser drafted and codified a new set of covenants that better protected the interests of individual property owners.

Among the significant changes in the revised covenants was the addition of a property owner seated on Fraser's architectural review board. To slow the pace of development, the number of permitted residential units and annual housing starts was capped. This was likely fine with Fraser, who had already recognized rapid development was a problem, saying in 1971, "In my own personal preference and taste, there has been too much building at one time within Sea Pines Plantation, and we hope that we can bring the pace of construction and the volume of construction to a more orderly level."[97] The new rules also doubled the amount of land legally restricted to development to 2,250 acres, essentially codifying the previously understood though unenforceable edict that for each acre developed, one would be preserved.

The fundamental challenge inherent in Fraser's vision was becoming clear: How can you simultaneously develop and conserve land? Reducing the amount of developable land meant less money generated, and to mitigate this loss, the ASPPPO agreed to increase its stake as a steward of the property by mandating contributions to a community fund that would finance the maintenance of common open space. To further boost the power held by property owners, the new covenants established a process by which formal alterations to the document would, in the future, require passage of a referendum by a daunting 75 percent of eligible voters.

Leadership Turmoil

Money troubles continued for Sea Pines throughout the 1970s and 1980s. By 1975 the company had amassed liabilities exceeding $280 million and was essentially broke. Fraser was forced to relinquish his control to others in the company. Even though he was no longer in charge, he remained active as the company's public face. Seven years later, thirty years after first conceiving Sea Pines, Fraser, worn out from several brushes with bankruptcy, sold the Sea Pines Company to the Heizer Corporation for $10 million, a fraction of its value.[98] He then stayed on as a consultant for the company, reaping a healthy annual salary of $150,000 as he pursued other interests.[99] This arrangement was short-lived, however. Heizer liquidated its assets in 1984 and put Harbour Town, the resort's golf courses, one hundred tennis courts, swimming pools, all unsold lots, and other community amenities and open spaces up for sale. A local developer bought the company in 1985, only to go bankrupt a year later. Sea Pines was then tossed in the middle of a highly public, nearly yearlong legal circus one newspaper called "one of the most complex and contentious bankruptcy cases that South Carolina has ever witnessed."[100]

In 1987, 670 Sea Pines property owners joined the fray and purchased the company for $24.3 million to keep the property intact and prevent it from being sold off in pieces. As one owner said, "Prior to this, we were dependent on the ability and character of ownership, whose interests were not always directed to the proper care and management of our plantation. Now we have control."[101] As part of a mandated reorganization, the courts and creditors split the company into two entities: Sea Pines Associates, made up of a board of twenty-five owners, would manage the business and real estate of the resort, and Sea Pines Community Services Associates (CSA), a nonprofit owned by all the property owners, would "protect, maintain, and enhance the resources . . . for the benefit of the Sea Pines Community."[102] The CSA took title to the roads, leisure pathways, open spaces, lagoons, ditches, and other common spaces, while the Sea Pines Associates owned the remainder of noncommercial assets and all of the unsold property.

Despite all this turmoil and uncertainty, skyrocketing property values evinced Sea Pines' ongoing popularity. By the end of the 1980s, increasingly scarce and desirable oceanfront lots sold for over $1 million. The entire island was thriving, as the value of property across Hilton Head rose to $232 million, up from just $14 million in 1973.[103] Not all people benefited from the prosperity, however. As the primarily white developers and new property owners spent and earned millions in the real estate gold rush, many in the local Gullah population were left behind. The proliferation of hotels and gated communities divvied up the island into numerous parcels under private ownership and led to soaring property taxes that poor islanders often could not pay. Those who could not afford to settle back taxes were sometimes forced to give up their homes, and the growing tragedy remained primarily hidden. As stated about the locals by a community organizer at the time: "People hold the land real dear around here. The only thing that they've known is the fish and the farm, and through the land they have a sense of pride and self-sufficiency. Once you take that away from them, they have nothing."[104]

Changing Character and Evolving Habitats

As time went on, Fraser's passive, eco-sensitive aesthetic transitioned from being an understood imperative into a nostalgic idealism that quickly lost steam against the brute force of the real estate market. Even as property owners remained effective advocates for controlled growth, as a group they seemed less interested in policing architectural restraint. In 1984, a revision to the covenants awarded most seats on the architectural review board to property owners. As they and their peers had paid exponentially more to buy into Sea Pines, it became challenging to refuse design proposals that ran counter to the established character. Houses displaying noncontextual architecture were approved as people capable of affording the ever-escalating prices began tearing down the classic cottages erected between 1960 and 1974 to build exponentially larger homes reflecting various styles. Multimillion-dollar oceanfront homes increasingly conveyed a bloated brand of architectural eclecticism that showed little interest in blending into context.

Throughout the community, large houses engulfed much of the residential landscape, and the intended proportion between built form and open space flipped. People understandably wanted to create suitable living environments, and Fraser's predilection for mulch and scrubby natural plantings was increasingly forfeited in favor of lushly planted, suburban-style yards with mown lawns and intensive horticulture. He believed such landscapes were not appropriate in his development and once remarked, "We'll never have azaleas growing in Sea Pines."[105] He was wrong. As early as the 1960s, masses of azaleas and other flowering shrubs began spreading across the community.[106] The covenants did little to discourage this habit. Although there were always specific criteria for saving trees and special requirements for oceanfront lots and those adjacent to golf courses, they were vague about horticulture and landscape design.[107]

Losing natural areas to development had an impact on more than just the visual appearance of the resort. Providing proximity to wildlife was from the beginning integral to the allure of Sea Pines and something easily accomplished before the bulldozers and crowds showed up. In the 1950s, fox, brown bear, wild turkey, quail, boar, and bobcat roamed the island freely, contributing to the wild quality that Fraser both loved and promoted. For instance, an early marketing piece printed in newspapers worldwide showed a female model holding a golf club as an alligator

Fig. 42. One of the early "nature-blending" beach houses (left) sits next to a much larger contemporary home. As the original homes were razed and replaced, the intended dominance of natural open space over built form was gradually lost. (Photo by author)

approached, with the cutline "Sea Pines Course Has Natural Hazards."[108] Another, now iconic image showing a dapper Fraser wearing a suit and striding across a golf course beside a giant alligator on a leash appeared in a 1962 edition of the *Saturday Evening Post* and was seen by millions.

Alligators were particularly important players in the choreographed mystique. As a 1971 article in the local newspaper reported, "The alligator went on to work as a sort of assistant salesman for Sea Pines Plantation."[109] Fraser knew the reptiles' presence would attract an adventurous type of consumer and, when first constructing the roads and infrastructure, he reportedly instructed crews to throw hunks of raw beef to the most threatening alligators to pacify and prepare them for inevitable encounters with people.[110] By this time the species had been hunted to near extinction. After being saved by the Endangered Species Act in 1967, their numbers grew, and by the 1970s, reports of alligators approaching homes and killing family pets in Sea Pines were prevalent.[111] The first reported attack on a person occurred in 1985 when a twelve-year-old girl was dragged into a lagoon by one and shaken "the way a dog plays with a toy."[112] She survived, and the traumatic story brought to the forefront the need to balance the romantic ideal of living in nature with legitimate safety concerns.

Not surprisingly, the increasing rate of development triggered the systematic degradation of Sea Pines' native ecosystem. There remained a seemingly endless variety of birds even as the wild mammal population was gradually reduced to those adapted to urban and suburban neighborhoods: squirrels, rabbits, raccoons, and the ubiquitous whitetail deer. Interactions with deer proved far more common and even more problematic than those with alligators. Their numbers increased dramatically as natural predators disappeared and human residents provided an abundant food source in their lush residential gardens. By the late 1990s, as many as six hundred deer lived within the gates of Sea Pines, a figure five times the natural capacity of the property based on size and environmental conditions.[113] While some residents appreciated them as valued wildlife, others complained loudly about deer devouring their expensive plantings and sometimes crashing through glass doors into their homes. Collisions between deer and automobiles became more frequent, reaching more than sixty per year in the small community by 1998.[114]

That same year, the CSA put forth a plan to reduce the deer population through a program of managed culling, meaning they would be intentionally slaughtered to reduce their numbers. Some residents endorsed the proposal while others protested and united in opposition to form the Sea Pines Association for the Protection of Wildlife. Citing Fraser's vision for harmony with nature, the group promoted shooting female deer with darts filled with an immunocontraception vaccine, which, they argued, would be more humane than killing them.[115] Internal fighting and subsequent court cases drew national attention. Three years of legal battles followed, and in 2001 the Supreme Court of South Carolina upheld the decision of a lower court that ruled against those opposed to the culling. Over the next fifteen years, sharpshooters killed approximately nine hundred deer in Sea Pines.[116]

As native habitat in Sea Pines dwindled, the protected six-hundred-acre forest preserve became the primary haven for wildlife. The large interior tract, made up predominantly of unbuildable marshy wooded land, appropriately shows up on Colonial-era maps as "Sanctuary." The covenants dedicated a minimum of 65 percent of the preserve's total acreage to a Wildlife Habitat Zone, with the rest divided between an Environmental Protection and Sanitation Zone and an Outdoor Recreation Zone, in which 180 acres were set aside for a potential golf course. Four freshwater lakes were dug out and stocked with fish, while winding wooden boardwalks provided access through brackish swamps and palmetto rocklands.

One of the marsh areas, known as Boggy Gut, was a cultivated rice field in the antebellum period before being used as a site for duck hunting for fifty or so years. In 1983 it was converted into a water storage and treatment mechanism to mitigate the loss of the island's wetlands, which had been drying up as upland species took over. The project restored the marsh as a healthy ecosystem by capturing and storing stormwater while also treating tertiary wastewater.[117] The designed wetland also succeeded in creating a rich and diverse habitat and contributed to the preserve's designation as an Important Bird Area by the National Audubon Society in 1998. Over time, however, the lack of regular management led to a gradual takeover by cattail, sawgrass, willow, and tallow, all of which swallowed up much of the open water and transformed the marsh into a densely forested bog.[118]

Such a lack of regular maintenance proved challenging throughout the preserve. Along with dictating permitted uses, the 1970 covenants encourage management practices that would ensure the long-term health of the landscape and its habitat. These include planting programs, installing and maintaining proper drainage systems, and preventing the spread of wildfires by cutting firebreaks and implementing controlled burns. Such practices were never employed regularly, however, and, as was essentially true for all of Sea Pines, management became merely reactive.[119] Across the community's open spaces, crews successfully preserved existing trees where possible but failed to actively manage them.[120] Over time, the canopy grew together, shading out the ground plane to the point that there was little plant diversity or generation of seedlings.

Changing Ownership, Yet Again

In the early 2000s, after more than a decade of leadership by the property owner–led Sea Pines Associates, Sea Pines Resort once again found itself with financial problems. Although the Associates had accomplished certain things, such as building a new inn at Harbour Town, it never had the resources to do much more than strive to maintain the status quo. By this time, many of the resort's amenities were decades old and deteriorating. Even though the community remained a popular draw for families and golfers, it faced increased competition from similar properties, such as Amelia Island and Ponte Vedra in Florida and Sea Island in Georgia, all of which were investing as much as $100 million in enhancements.[121]

One property owner, a developer named William Goodwin, moved to purchase Sea Pines Associates in 2004 through his company, the Riverstone Group. Recognizing the need for an influx of capital to undertake needed improvements, the Associates' board unanimously approved the $23.4 million offer. Goodwin, whose family had owned property in Sea Pines since the 1970s, alleviated owners' concerns about the potential downside of the deal by saying he had "no hidden agenda" and intended to maintain the established character and keep the resort under single ownership.[122] Once finalized, the new arrangement did not change the basic structure of Sea Pines' management; the CSA still owned and maintained 1,700 acres of community open spaces, infrastructure, and most of the Forest Preserve.[123] As promised, the resort embarked on an ambitious program of improvements and by 2015 had invested $60 million in upgrades. With an expressed goal of achieving coveted five-star resort status, it planned to spend tens of millions more over the next few years.

By this time, Fraser's initial intention of creating a residential community had been supplanted by the reality that Sea Pines was primarily a tourist-focused resort. Only around 30 percent of property owners made Sea Pines their primary residence, meaning most houses and villas were occupied by part-time residents and short-term renters. In the fall of 2016, when planning was underway to spend upward of $100 million to improve Harbour Town, Hurricane Matthew blew through and sparked a period of dramatic change, unlike anything the island had experienced since the early days of Fraser.

Hurricane Matthew and the Aftermath

Because Hilton Head tucks into the arcing coastline between Jacksonville and Charleston, it has historically been spared serious damage from hurricanes, which typically target less protected areas such as the Gulf Coast, South Florida, and North Carolina's Outer Banks. This changed on October 8, 2016, when, after ravaging Haiti and skirting Florida, Hurricane Matthew slammed into the island as a Category 2 storm with nearly ninety-mile-an-hour winds, fourteen inches of rain, and a six-foot storm surge. Although flooding inundated some low areas, the wind inflicted the most damage. Seven hundred thousand cubic feet of sand shifted on the beaches and thirty-three acres of dunes were destroyed. The storm completely flattened six thousand feet of dunes on the south end of Sea Pines. An estimated 120,000 trees fell across the island, leaving behind mounds of debris requiring an entire year and over $40 million to clean up.

Sea Pines was the hardest-hit community on the island.[124] Over 435,000 cubic yards of debris were removed from the resort. Existing maintenance budgets forced the CSA to double property owners' annual fees to fund the cleanup. The raging wind had substantially ripped and thinned the tree canopy throughout the property, exposing to sunlight yards and natural areas that had been submerged in deep shade for decades. This exposure triggered seemingly instant intrusion from invasive and aggressive vines and weeds, particularly in natural areas. At the same time, residential yards were now substantially more open. Although some residents lamented the loss of privacy that came with cleared views into their property, others relished the opportunity to have healthier lawns and gardens filled with sun-loving ornamental plants they could not grow before.[125]

Coincidentally, this damage and subsequent change to Sea Pines' landscape occurred the same year the CSA earned a Green Star Award from the Professional Grounds Management Society. As a testament to how the designed landscape's character had evolved from its scrubby, naturalistic origins, the award recognized "grounds manicured with a high degree of excellence" that displayed a "high quality appearance."[126] In celebration of the award, the CSA's president proclaimed, "Our beautiful grounds have long been a matter of pride to our property owners and those who work here, as well as a source of admiration by people who visit."[127] While this

Fig. 43. In 2016, Hurricane Matthew tore through Hilton Head and left behind extraordinary damage. Felled trees and stripped foliage opened the once dense tree canopy in the Sea Pines Forest Preserve to sunlight, triggering encroachment by aggressive vines and invasive plants. (Photo by author)

seems to acknowledge the departure from the resort's initial vision, it is not necessarily a negative reflection on Fraser's legacy. The more manicured appearance responded to changing expectations for how an exclusive resort should look. It had been a long time since Sea Pines was alone as a leader in a new market; it was now one among a field of competitors vying for limited tourist dollars, and the designed landscape was part of the appeal.

Contemporary Issues

The community was more than sixty years old by the time Hurricane Matthew hit, and trees reflected only one part of an infrastructure system failing because of a lack of adequate or consistent attention. The stormwater system, consisting of nearly two hundred acres of drainage lagoons, canals, ponds, and lakes; thousands of catch basins; and over thirty miles of piping, required extensive repair, restoration, and replacement.[128] Ongoing paving repairs to the sixty-seven miles of roads and seventeen miles of asphalt leisure trails also needed funding. The CSA estimated the work on this alone would cost around $5.5 million, of which it had less than half on hand.

As the resort aggressively spent millions of dollars on improvements, the CSA remained hampered by a stagnant budget. Lax enforcement of the covenants had contributed to aesthetic changes over time, yet the document effectively limited how much property owners had to pay for maintenance. Incremental increases in the daily gate fee boosted coffers somewhat, but since most of the resort's funding for maintenance came from owners' annual assessments, only a referendum with 75 percent support would permanently boost the CSA's income. Owners therefore would have to vote to increase their own taxes. With a plea to these voters—"We are all in this together"—the CSA issued a Critical Needs Referendum in 2019 to acquire needed money for the infrastructure.[129] It failed, with only 72 percent of the vote in favor.[130] With no good options to consider, a second referendum passed a year later with 87 percent of the vote.

Much of what continues to attract buyers to Sea Pines is the quality of life it provides and the seemingly unstoppable return on financial investment. In 2020 and 2021, during the coronavirus pandemic, real estate on Hilton Head boomed as Americans fled high-density urban areas in search of new places in which to live and work remotely. Reports of bidding wars, a lack of housing inventory, and homes and villas selling sight unseen, quickly and above asking price, spread rapidly from realtors and between neighbors.[131] Trends changed as more people sought to buy permanent homes rather than vacation or rental property, as had been the norm.[132] In June 2021 the median sales price for detached homes in Sea Pines was $1.4 million, more than 72 percent higher than the year before.[133] Of the twenty-six homes on the market in the resort, only four were listed at under $1 million.

Sea Pines appears today as a fairly typical high-end coastal resort community with increasingly few reminders of the wildness that once attracted Fraser to the island. Houses and villas designed in a distinctively quirky architectural style continue to proliferate, along with expanses of golf courses, tennis courts, and bike trails in a well-tended landscape filled with palm trees and live oaks dripping with Spanish moss. Such imagery has become cliché for vacation resorts and coastal residential communities, many of which target a limited, high-end audience. While Sea Pines, and places like it, earn criticism for fostering a closed-off realm of privilege, there is a lot to learn from its story as an innovative planned and designed community.

It is difficult to determine whether Sea Pines' planning and design heritage plays a role in its ongoing appeal. The same is true for Riverside. Whether or not the names Fraser, Sasaki, Olmsted, or Vaux matter or are even recognized today is less important than the sustained quality of the places they helped create. Each community faces a constant tension between public land and private space, and collaboration and cooperation between all involved parties help them thrive. How long those in charge use historical precedent to inform decision-making will help determine how long their original visions remain recognizable.

5

PARKS

In another sense Olmsted's parks exist before they are finished, which means in fact they are never finished; they remain carriers of the unexpected and of contradiction on all levels of human activity, be it social, political, or natural.

—Robert Smithson, artist and writer, 1973

The history of American parks has been well documented and requires little detailed discussion here. It is, however, important to understand why the parks developed and how they evolved as critical contributors to cities' cultural and ecological infrastructure. Insofar as *park* is an amorphous term that, just in regard to landscapes, can refer to anything from a vast protected wilderness to a tiny undeveloped parcel in a downtown area, it helps to establish parameters for reference. As used in this chapter, *park* refers to a designed landscape or open space created for public use. Obvious examples are Central Park in New York City, Overton Park in Memphis, and many similar versions created in their wake and at various scales in communities across the country. Not all such places carry the label of park. The National Mall in Washington, DC, for instance, fits the criteria, as do the historic squares set within the grid of

Savannah, Georgia, the High Line in Manhattan, and the series of modernist open spaces designed by Lawrence Halprin in Portland, Oregon.[1] Some, such as Dumbarton Oaks Park, which is discussed in this chapter, were designed as a different type of landscape before being converted into a park. What they are called or what they look like is less important than that they are discrete landscaped open spaces created specifically to meet the needs of the communities in which they are located.

From the mid- to late 1800s, as the Industrial Revolution took hold, forward-thinking leaders imported the model of English pleasure grounds into growing American cities in the form of large landscaped parks. Many of these large parks created idyllic pastoral spaces that introduced nature and a sense of wilderness into increasingly crowded and dirty cities. They were instantly popular and filled with throngs of people seeking to escape the ills of urban life by strolling shaded paths, boating on lakes, riding horses, and recreating and picnicking on generous lawns. Although promoted as democratic spaces intended for the masses, their typical placement at the edge of city centers frequently put them out of reach for poor, immigrant, and working-class residents.

Social reformers concerned about this inequity successfully lobbied for small parks in inner-city neighborhoods and tenement districts beginning in the early 1900s. Transporting the recognized benefits of large pastoral parks into these small neighborhood spaces was seen as a way to help disparate immigrant communities assimilate to life in America through a positive shared experience. Around the same time, progressive educators and psychologists pressured local governments to build playgrounds in which urban children could play under the supervision of adults. These were generally sparse, functional spaces encompassing a few blocks at most in the middle of dense neighborhoods. Play equipment and courts were the focus, with virtually no pretense at providing a naturalistic surrounding.

Pastoral landscapes were not entirely a thing of the past, however. The Olmsted Brothers firm continued to design city parks and park systems that clung to the picturesque idiom. In Chicago, for example, the firm designed fourteen parks at various scales, some filled with children's play equipment, running tracks, wading pools, ballfields, and fieldhouses, all within a designed landscape. In the case of Marquette Park, the firm had a three-hundred-acre site on which to design numerous facilities, including an eighteen-hole golf course. In these cases, programmatic use trumped scenic quality. In the late 1910s, Warren Manning designed Agassiz Park in Calumet, Michigan, with a playground, tennis courts, an amphitheater, a picnic shelter, and baseball, basketball, and football fields, intending to provide the community "the greatest service and greatest pleasure to the greatest number of its citizens."[2] Improving the lives of everyday people increasingly took precedence over the artistic preference of landscape architects and other designers.

By the 1930s, new public parks accommodated the mass of people fleeing cities for the suburbs with utilitarian facility-oriented sites that further prioritized function over aesthetics. Design still mattered, though the bar for quality was significantly lower. Park management that once depended on horticultural and arboricultural expertise increasingly involved menial tasks such as mowing grass and removing trash. Over the next several decades, many once celebrated urban parks suffered from the effects of suburbanization, economic downturns, rising crime rates, and changing municipal priorities. Concerned park lovers rallied support for many declining landscapes and formed grassroots organizations such as friends groups to raise awareness and money in support.

Such advocacy has always played an important role in public landscapes. New Yorkers, for instance, supported Central Park from its inception. Threats continued as conditions fluctuated and led to a much-publicized downward trajectory resulting from severe overuse and underfunded maintenance. An eventual low point in the 1970s generated publicized images of deteriorating buildings, graffiti-strewn walls, and barren lawns that advertised Central Park as a place to avoid. Conditions deteriorated to the point that then parks commissioner Richard Clurman claimed in 1973 that Central Park "is not in need of restoration, it is in need of saving."[3] Two years later, eight hundred parks department employees were laid off as part of drastic budget cuts, challenging park maintenance across New York City's five boroughs. Wealthy, civic-minded Manhattanites who recognized that proper care of Central Park was now beyond the capacity of the beleaguered parks department stepped in and organized two groups, the Central Park Task Force and the Central Park Community Fund.[4] Working separately, yet with a common goal, the groups' effectiveness in planning, management, and fundraising offered new hope for the park. Because of their success, the functions of the two groups were combined into a single entity, the Central Park Conservancy.

During the 1980s, other cities, including Dallas, Houston, Louisville, St. Louis, Portland, and Atlanta, emulated Central Park's famous example and created their versions of private conservancies to support public landscapes.[5] The Trust for Public Land defines conservancies as "private, nonprofit park-benefit organizations that raise money independent of the city and spend it under a plan of action mutually agreed upon with the government."[6] Typically established to both raise money and actively manage landscapes, these organizations form the now common public-private partnerships found in cities across the country to aid and preserve parks.

The following case studies present two distinctive landscapes that capture the history, ambition, and complexities associated with managing public parks: Dumbarton Oaks Park in Washington, DC, and Prospect Park in Brooklyn, New York. Although they differ from one another in size, context, and appearance, they share celebrated design legacies central to the history of American landscape architecture. Their respective evolutions under the oversight and jurisdiction of federal and municipal agencies and the influence of citizen advocates and public-private partnerships illustrate the unique challenges and common solutions associated with balancing public demands and cultural heritage. While their paths to today were extremely different, the ways in which staff and volunteers handle the present and anticipate the future convey consistent and sometimes surprising similarities.

Dumbarton Oaks Park, Washington, DC

The terraced landscape at Dumbarton Oaks, the former estate of Robert and Mildred Bliss in the upscale Georgetown neighborhood of Washington, DC, is among the most cherished landscapes in the United States. The sixteen-acre garden, designed in the 1920s by Beatrix Farrand in collaboration with the Blisses, is celebrated for its horticulture, design features, spatial layout, and expert manipulation of the site's significant topography. It is a popular destination for tourists and garden lovers and is considered by many historians and designers to be Farrand's masterpiece. As visually satisfying as Dumbarton Oaks Garden is on its own, however, much of the landscape's success as a designed place is dependent on the borrowed views to the surrounding adjacent woodlands that form the backdrop that remains a critical aspect of its character. The view to the east is cut short by the massive trees in neighboring Montrose Park, a former estate purchased by the U.S. government in 1911 and converted into a public park. A view northward from the terraces and tiered lawn behind the mansion spans the garden to the forested slopes beyond. The vista fosters the impression that the property extends forever and conveys the intended semblance of Dumbarton Oaks as a grand country estate.

Although this visual connection persists, Dumbarton Oaks' landscape is now bisected by a black chain-link fence strung along the southern slope that betrays the sense of unity. Uphill of the fence is the famous and impeccably manicured garden, while downhill, reaching into the valley below, is Dumbarton Oaks Park, a distinct landscape also designed by Farrand, who conceived it as a naturalistic extension of the Blisses' property. The twenty-seven-acre landscape, which is now owned and managed by the National Park Service, slides along a meandering stream corridor and crawls up the narrow valley in which it is contained. Its neighbors include the U.S. Naval Observatory and the National Cathedral to the north, the massive Rock Creek Park to the east, and the crowded street scene of Wisconsin Avenue along its western edge. Other entities populating its perimeter include the embassies of Italy, Denmark, and New Zealand.

The Blisses maintained their property as a unified whole until 1940, when, concerned by the war raging in Europe that threatened the United States, they donated the estate for a purpose greater than their personal use. They deeded the upper gardens, the house, and their extensive art collection, furnishings, and research library to "the President and Fellows of Harvard College."[7] With equally generous intentions, the couple deeded the valley garden to the NPS with the understanding "upon the express covenants that it shall be maintained permanently as a public park."[8]

By making these choices, the couple, perhaps unintentionally, propelled the gardens and the park into dramatically divergent futures and fortunes. Once they were physically separated, the upper gardens instantly began enjoying the benefits of management by a wealthy cultural institution, while the woodland landscape embarked on decades of struggle under the care of the perpetually underfunded federal agency. In a discussion of landscape stewardship, Dumbarton Oaks Park is the more intriguing of the two, particularly when understood in juxtaposition to the other. Making one part public while the other remained private and wealthy invokes distinctions that extend beyond the obvious funding variance and touch on broader cultural issues such as access and advocacy. The park's evolution parallels changing views and policies within the NPS on how it manages designed landscapes, as well as the role citizen-based conservancies and public-private partnerships play in leading the preservation and maintenance of important historical places.

Establishing the Legacy

The main house and museum at Dumbarton Oaks rest prominently atop a steep knoll on a parcel of land with a vista once deemed "the finest view in Georgetown."[9] The Blisses bought the early nineteenth-century federalist-style mansion in 1920, following two decades of living abroad. According to Robert, a career diplomat, the purchase would help fulfill the couple's "dream during twenty years of professional nomadism of having a country house in the city."[10] The house had been built around 1800 on a tract of land England's Queen Anne gave to a Scottish emigrant named Ninian Beall, an important figure in the early history of the Washington, DC, area. He was a farmer and likely the first European American to settle the area. He eventually turned his fifty-acre plot of land into 25,000 acres of real estate, which included a nearly eight-hundred-acre tract he named the Rock of Dumbarton after a famous outcropping in Scotland.

The Blisses chose the name Dumbarton Oaks in reference to both the history of the site and the beautiful mature trees whose canopies spread across the landscape. While the house was important, effectively creating the character of a country estate was dependent on the landscape. The six-acre lot was hardly a garden when they moved in; instead, they encountered a neglected space filled with weeds and overgrown plants. Over time, the couple purchased more land, ultimately acquiring around fifty-four acres, much of which was former farmland on slopes to the north of the house. The considerable topography made it ideally suited for a picturesque landscape yet not particularly conducive to creating a traditional, formally arranged garden defined by ordered geometries.

From visits to Europe studying gardens of various styles and scales, Mildred developed a discerning taste for landscape and art and was both educated and opinionated about design. It was 1921 when she and her husband engaged Farrand to assist in creating their gardens at Dumbarton Oaks. Farrand was already an accomplished designer by this time and had gained recognition as the only woman among the founders of the American Society of Landscape Architects. The two women became friends and worked together on the estate's garden for more than thirty years. Bliss was an active partner and exerted significant influence over the design and evolution of the landscape. After being hired, Farrand wrote to her new client, "What I shall try to do with the Oaks is simply be your gardening pair of hands, carrying out your ideas."[11] Such significant involvement from a client is not uncommon, even though it is not always documented.

The basic concept behind the garden's design is relatively simple: set formal spaces close to the house and then transition into looser, less structured expressions moving deeper into the landscape. Terraced spaces stepping down the slopes become increasingly rustic in detailing and materiality as they move away from the house. For instance, precisely pointed brick walls near the house transition to natural stone walls with crisp mortar joints in the garden and, finally, rustic rubble walls in the valley. This thematic and programmatic progression lends itself well to the physical nature of the site as strong axial and formal elements atop the ridge loosen into less rigid forms mid-slope before a stylized version of nature takes over on the lower slopes and extends into the valley.

Creating wilderness gardens was a fairly common practice at large country estates at the time. In a letter to Mildred in the summer of 1922, Farrand wrote, "It is also hoped that a part of the grounds could be developed as 'Wilderness,' where hollies, yews, ivies and spring flowering Magnolias and winter flowering shrubs would make an attractive walk to be followed in the winter."[12] That statement makes evident she intended it to be a designed landscape rather than truly natural and, although different in style from the upper gardens, the transition between them was intended to be clear, yet subtle. Four pathways constructed of increasingly rusticated materials provide physical connections, while established promontories and vistas unify the landscape visually. Groves of flowering trees, shrub masses, and fields of bulbs running down the slope and into the valley strengthen the relationship. Unlike with the upper gardens, for which there are numerous detailed planting plans, there is little documentation of the specific planting in this area. In the same letter to Mildred, Farrand says of this particular landscape: "The whole scheme for the north slopes of the property should properly be studied from the ground itself rather than from any plan, as the contours and the expressions of the ground will control the plantations more strongly than any other feature."[13]

Farrand created a series of rooms in the valley as she had in the upper gardens. "The brook certainly could be widened and dammed up at various points and used as a mirror in which to reflect large plantations of azaleas and iris, or overhanging dark masses of hemlock, with water-loving plants growing on the still surface, and walks arranged on the different levels so that the plantations could be seen from above as well as from their own level," she wrote.[14] Without definitive drawings, notes such as these functioned as the primary design guidelines for this portion of the property for decades. Most of the constructed design elements in the valley relate to the stream, a small tributary called "The Branch" that feeds into Rock Creek, the

Fig. 44. This early photograph depicts the stream at the heart of the "wilderness garden" envisioned by Beatrix Farrand. The rustic nature of the landscape was intentionally juxtaposed with the more ordered and ornate gardens set closer to the house. (Library of Congress)

Potomac River, and, ultimately, Chesapeake Bay. A series of eighteen small stone dams form small pools and short waterfalls. Intricate features such as footbridges, walls, benches, an arbor, and a stone grotto populate the corridor and exemplify the intent for this to be an integral part of the broader garden.

In contrast to the ornamental horticulture in the upper gardens, the planting along the stream was envisioned to be "rather small in scale and entirely simple in all its arrangements."[15] It was filled with loose arrangements of flowering trees, shrubs, and bulbs that emphasize a spring bloom in sync with the upper garden. On the north side of the stream, the floodplain transitions into a series of five open meadows separated by fingers of woodlands that descend from the upper slope and introduce light, scale, and a sense of prospect into what is otherwise a forested landscape. The wilderness character conveyed in the wooded slopes moving northward was intended to remain managed rather than left to successional evolution.[16] For a while, the original vision for the stream and its landscape was retained, and it remained a beautiful place.

The Garden Becomes a Park

While it is not clear precisely what the Blisses expected from the NPS regarding the management of the landscape, it can be assumed they believed it would fare better than if it were taken over by the District of Columbia. The neighboring Rock Creek Park likely served as inspiration, with its founding 1890 legislation stating it to be "perpetually dedicated and set apart as a public park or pleasure ground for the benefit and enjoyment of the people of the United States."[17]

As with the donation to Harvard, turning the naturalistic garden into a park was not a decision made on a whim. A mid-slope arch, called the Forsythia Arch for its placement within a generous planting of effusive, yellow-blooming forsythia, was designed and erected as a threshold between the upper and lower gardens just two years before the donation. With an inscription on the arch looking up to the house reading "Dumbarton Oaks," and "Dumbarton Oaks Park" written on the opposing face, it is clear the distinction was intended even then. In a 1938 letter, Mildred wrote to Farrand that the "arch-way as entrance to the park from Dumbarton Oaks and vice-versa should carry some weight, both actual and metaphorical."[18] Perhaps understanding the potential implications of turning a garden into a public park,

Farrand unsuccessfully suggested calling it "Dumbarton Oaks Grove" instead. She likely understood that such nuance in nomenclature was important in light of the dichotomy that can exist in the maintenance of gardens versus parks.

Robert Bliss seemed less attuned to this distinction and envisioned the rustic park somehow becoming an American version of Kew Gardens.[19] Farrand had doubts about that notion and questioned "whether either the acreage or topography would lend itself" to that.[20] She diplomatically suggested to her client that it "might be used as an adjunct to 'Kew' where plants of natural stream-side and hillside types would find themselves at home, but there is not room for any considerable variety of shrub planting or herbaceous 'wild' planting."[21] In other words, the park could accommodate plants appropriate to its environmental and aesthetic context, but trying to establish extensive Kew-type collections was not a viable approach. While Farrand appeared to worry about the implications of making it into a park, there seemed to also be optimism that the horticultural integrity of the garden she designed could be preserved and maintained.

Access between the private garden and the now public park was soon closed by Harvard. Tall chain-link fences secured the property from the public, and at some point, the opening within the threshold arch was sealed with brick. All other pedestrian connections between the two landscapes were severed. The NPS seemed happy to get the landscape and promoted it as "an unusual combination of native wildflowers and cultivated plants grown in a setting of miniature waterfalls, ponds, and shaded walks . . . [that] is unique among municipal park areas."[22]

Stewardship after Farrand and the Blisses

Any enthusiasm the NPS felt about the acquisition was tempered by the realization that funding was going to be a major issue. During the first year of ownership, the acting secretary of the interior wrote that "no appropriations for the maintenance or protection [of Dumbarton Oaks Park] have been made," acknowledging the difficulty in providing even limited maintenance service.[23] The budget of the National Capital Parks, the administrative unit for the District of Columbia, had less than $4,000 dedicated to annual maintenance in 1943, which was not much, even for the time. An official wrote to Farrand "it will be exceedingly difficult for National

Capital Parks to adequately maintain the area due to the overall reduced total maintenance budget."[24]

The Blisses and Farrand appear to have cared about the park in its early years. Recognizing that its use as a public landscape distinguished it from its former purpose as a residential garden, they recommended tweaking the design to accommodate more users by widening the path next to the stream and adding a new one across the top of the meadows. They argued for the formation of an advisory committee that would establish maintenance standards and ensure the landscape's future integrity.[25] Although such a group did meet in 1941 and 1942, there is little evidence that it accomplished much beyond overseeing the improvements recommended for the paths and some replacement plantings. Farrand provided a narrative detailing the desired character for the park, but even with this basic aesthetic guidance, the continued lack of funding made the property difficult to effectively manage in relation to its design intent.

During this time, the upper gardens did not suffer the same misfortune. When the director of Harvard's Dumbarton Oaks asked Farrand to document her advice on upkeep and design needs for the formal landscape, she produced in 1944 the detailed *Plant Book for Dumbarton Oaks,* which was far more comprehensive than the narrative she had created a year earlier for the park. The volume became an immediate resource, offering extensive maintenance recommendations and planting suggestions ensuring her design vision would span transitions in ownership and staff.[26] That same year she wrote an article for *Landscape Architecture Magazine* in which she discussed in detail her design for Dumbarton Oaks. While eloquent in her description of the upper gardens, she made no mention of the landscape she had once envisioned as wilderness garden. The closest she came to it was a subtle reference to the "background of trees at its north end."[27] Although she remained actively engaged in the upper gardens for several more years, her involvement in the park seems to have essentially ended by this point.

Photographs and letters show that even with a lack of funding, the landscape was relatively well maintained through the 1960s.[28] For years, the NPS and private groups led tours of the landscape during the spring, when the intent of Farrand's planting scheme was most apparent. In 1962, at a ceremony in the park marking

Fig. 45. A National Park Service park ranger leads visitors on a tour of Dumbarton Oaks Park in the early 1960s. After the lower garden became Dumbarton Oaks Park, the landscape remained relatively intact and well-tended for several years. (National Park Service)

the centennial of Henry David Thoreau's death, the poet Robert Frost remarked to fellow poet Louis Untermeyer, "This is a place Thoreau would have loved."[29] Several years later, the noted historian and preservationist Walter Muir Hill wrote in a book about the Dumbarton Oaks estate that the park was "a complete escapist paradise, particularly when the daffodils carpet the little valley and the dogwood is in bloom."[30]

The Park Reverts to Nature

Even as Dumbarton Oaks Park's quality as a designed landscape began to erode, it retained its value to the public. People in the Georgetown community continued to use it for hiking, exercising, commuting, walking their dogs, and as a practical connector between points in the neighborhood. During the 1960s, when Washington was the geographic and symbolic heart of national demonstrations opposing the Vietnam War and promoting social change, public landscapes such as Rock Creek Park and Dumbarton Oaks Park became resting spots for counterculture participants and protestors. Local college students and others in the neighborhood would scale the perimeter walls and fences to swim in the pool after-hours in the summer.[31] Barbed wire was subsequently added atop the fence in several locations as a deterrent, but, although this slowed the activity, the trespassing continued. "I don't care if you put a guy with a machine gun down there," recalled a former superintendent of the gardens, "you're not going to stop them and you just get kind of tired after a while messing with them."[32] People could also still easily climb through the gate within the Forsythia Arch, prompting Dumbarton Oaks staff to more permanently seal the opening with rubble stone laid in mortar in the late 1960s or early 1970s.[33]

While such modifications were necessary from a security and liability perspective, the visual effect of these moves reinforced the sense of self-imposed separation between the elite private institution and the public realm. The pruned shrubs, mown lawns, and periodically refreshed plantings in the upper gardens could be easily viewed from the valley, while the undermaintained woodland likely looked from above like an extension of Rock Creek Park—a rustic wild amid the urban landscape. As long as the paths remained open and there was a sense of relative safety, the degradation of its original garden design details bore no immediate consequence for visitors. At this point, restoring Farrand's vision was a want more than a need. Soon, however, the look of the park began to change dramatically.

In 1970 the federal government passed the National Environmental Policy Act. This legislation required federal agencies to assess the environmental effects of any proposed actions, whether for the management of scenic natural parks such as Yosemite or Yellowstone or for a small, designed landscape like Dumbarton Oaks Park. With a mandate to "maintain conditions under which man and nature can exist in productive harmony," the NPS intensified its focus on natural resources, often at the expense of design elements.[34] The resource manager overseeing Dumbarton Oaks Park at the time established that it would be managed as a natural area, to the point that some invasive plantings were classified as native and therefore left unchecked for years.[35] A 1972 reorganization of the regional office of the NPS transferred care of Dumbarton Oaks Park from Rock Creek Park to the George

Fig. 46. Over time, a lack of financial resources for maintenance triggered a period of neglect and deterioration in Dumbarton Oaks Park. During this time, volunteer invasive plants and migrating aggressive plantings from the upper gardens took over the natural landscape, while poorly managed stormwater eroded the streambed and left behind piles of debris, as shown in this photograph. (Library of Congress, Historic American Buildings Survey, Jack E. Boucher)

Washington Parkway unit, severing the existing stewardship process and prompting a period of neglect. While the sixteen-acre upper gardens continued to flourish with the aid of Harvard's deep pockets and a staff of at least a dozen workers, maintenance of the public park below was relegated to a single full-time gardener for all twenty-seven acres.[36]

Over time, many of the delicate plant species integral to Farrand's scheme were smothered beneath far more aggressive plants such as wild grape, porcelain berry, and bamboo. Some of the exotic groundcovers she specified in the upper gardens, such as English ivy and periwinkle, that were in fashion at the time, crept through the fence and down the hill, instigating an unwanted takeover commonly found in urban and suburban landscapes. It took only a few decades for the once well-tended garden to degrade into a gnarled mess of vines, weeds, and invasive plants.

Although agency-wide directives and targeted decisions affected the management of the park to the detriment of its design integrity, there is little to suggest that a lack of interest on the part of individuals directly involved with Dumbarton Oaks Park was the problem. Regardless, Maureen Joseph, a landscape architect for the NPS whose instrumental involvement with the park started in the 1990s, said the NPS at the time did lose sight of the landscape and its heritage: "No care was given to it. No one was assigned to look after it. People still visited, but they just didn't have a comprehension of its significance and that it related to what's on the other side of the fence."[37]

Rediscovering Farrand's Naturalistic Garden and the Birth of a Conservancy

The stewardship of Dumbarton Oaks Park forged a new direction when the NPS modified its strategy regarding preserving designed and cultural landscapes beginning with the 1983 meeting that was discussed in chapter 2. The HABS report, which was completed six years later, contained detailed photographs, written historical narratives and descriptions, plant lists, and a set of extensive drawings that included measured plans of the overall property, along with sections, elevations, and isometrics of individual areas, structures, and details. Even though its focus was on the layout and design of built features and it offered little in the way of horticultural analysis, the HABS survey became the most complete accounting and understanding of the landscape since the time of the Blisses. It emphasized the park's status as a designed landscape in need of care beyond that of a native woodland and asserted its significance as "a rare example of an urban estate that is still associated with a 'wilderness' area."[38] For the first time in decades, management of the park could not be undertaken with ignorance, feigned or legitimate, of the intention and legacy of Farrand's design.

Even with the optimism spurred by this renewed interest, maintenance remained hampered by a continued lack of funding. Within the vast scope of land managed by the NPS, Dumbarton Oaks Park existed in relative obscurity, a niche property likely buried beneath a slew of familiar landmarks and popular tourist attractions on a list of priorities. Although the HABS report was a necessary first step, it functioned primarily as a statement of significance and documentation of current conditions, without direct proposals for practical management strategies moving forward. Determining what to do regarding the landscape involved a separate process requiring additional money that would take years of lobbying to secure.

In the meantime, a sixteen-week Historic Landscape Preservation Program Studio from nearby George Washington University added further documentation with a preservation needs assessment. The report, which was completed at the end of the summer of 1993, tracked the progress of invasive plant material and the loss of horticultural diversity. It acknowledged that even with the deterioration of design elements, "Dumbarton Oaks Park retains a high degree of integrity evidenced in the remaining Farrand-designed built features."[39] Articulating a theme that would become repeated in subsequent documents, the authors summarized: "The numerous remaining features are repairable and with proper design guidance as well as a thoughtfully planned long term repair and maintenance program should once again provide the natural beauty and rustic charm that Beatrix Farrand intended."[40]

Not long after this, NPS staff joined the Olmsted Center for Landscape Preservation in a garden archaeology project to dig beneath layers of silt and masses of invasive plants to assess the status of design features. The results of this were documented in the 1997 Landscape Preservation Maintenance Plan, which reiterated that the landscape retained a high degree of integrity despite unchecked damage since the days of the Blisses.[41] The plan detailed emergency fixes and maintenance strategies and emphasized stabilizing the stream, its banks, the associated dams, spillways, and walls. Halting damage on the site while more permanent strategies could be developed became the goal. For the next two summers, students from the national job-training group the Student Conservation Association filled sandbags with silt taken from the stream and piled them on the banks to deter further erosion and to reroute the water to its intended channel.[42]

While these early documents contributed to knowledge about the landscape, neither investigated the design's complexity or its significance in the broader field of landscape architecture. To this end, the production of a more in-depth CLR began in 1996 as a cooperative effort between the NPS's regional Cultural Landscape Program and Rock Creek Park branch, local citizen advocacy groups, and the Trustees of Harvard University and Dumbarton Oaks Garden. There remained no formal agreement connecting Dumbarton Oaks Garden and Park, yet each benefited from the success of the other. Staff representing both landscapes collaborated on the CLR to reveal the original vision of Farrand and the Blisses.[43]

Whereas the Olmsted Center's plan identified immediate repairs, the CLR put forth a comprehensive strategy for the long-term preservation of the landscape resources.[44] It reasserted the park's high degree of integrity, stating: "In fact, the park can be judged to possess greater authenticity as a Farrand landscape than the gardens of Dumbarton Oaks Gardens, most of which have been significantly altered since Farrand's time."[45] Ironically, since there were no mandated controls for preservation, the active management of the upper gardens proved a greater threat to the specificity of Farrand's design than the neglect experienced in the original wilderness garden. As the landscape sat essentially untouched for decades, its details and features buried beneath silt and debris, the upper gardens continued to evolve in response to changing habits, usage, and design tastes. Once Farrand stopped working there, her protégé, the architect and landscape architect Ruth Havey, continued to collaborate with Mildred Bliss into the 1960s, notably on the transformation of the tennis court into the ornate Pebble Garden found today.

For decades, even though her name was no longer on the deed, Bliss retained her role as garden patron. Over time, her taste, along with that of Havey and subsequent consultants and garden staff, translated into work that contrasted with Farrand's patterns and details.[46] Such nuanced change is not uncommon in the subjective practice of landscape stewardship, where someone at every stage of evolution determines what is sacred and what is expendable. Recognizing that meaningful history continued beyond Farrand seems to have enhanced the significance of the park's design features, as the NPS decided restoration was the preferred treatment approach for Dumbarton Oaks Park. The goal became to restore the designed landscape to the condition it enjoyed when made public in 1940 as closely as possible while making necessary alterations to account for changing programs and usage.[47]

Citizen Interaction

Before the HABS report in the late 1980s, it is unlikely that many people in the community were aware of the park's design heritage. In 1992 a group of local citizens, spurred by the NPS's renewed interest, founded Friends of Montrose and Dumbarton Oaks Parks as a nonprofit organization to assist the NPS in its stewardship of the two landscapes, which had been listed together on the National Register of Historic Places in 1967. Both landscapes suffered from similar maintenance issues, and the group sought to enhance their beauty, improve their historical, educational, and recreational qualities, and ensure they were safe and of benefit to the public.[48]

This occurred during the lull before the CLR process when the HABS program had completed its documentation and moved on to other landscapes. Federal money for the Dumbarton Oaks Park flowed no faster. The ensuing lack of meaningful action must have frustrated some people in the neighborhood because they, in coordination with the Friends group, attempted restoration efforts on their own without authorization from NPS staff. In 1995, volunteers used HABS sketches as the basis for an unsanctioned attempt to remove built-up silt in one of Farrand's delicate pools and restore it, along with one of the stone dams.[49] Unaware of the sensitivity required for such a task, the crew used a backhoe, thus burying the path and adjacent plantings beneath the discarded silt and inflicting further damage on the stream banks in the process.[50] Later in the decade, other efforts undertaken with NPS oversight had a positive impact on the stream's health and structure. The Friends group proved most productive in fundraising and engaging volunteers. In the early 2000s, members recruited and organized assistance from the Youth Conservation Corps, AmeriCorps, the Georgetown Garden Club, and the NPS Historic Training Center on even more stabilization and related improvement projects.[51] Although these were neither historically accurate restorations nor permanent solutions to complicated hydrological problems, they successfully stemmed further damage.

Following the example of the Central Park Conservancy and other organizations like it, a more organized iteration of the Friends group emerged in 2010 with the founding Dumbarton Oaks Park Conservancy (DOPC) as a nonprofit partner to the NPS with a stated mission to rehabilitate, promote, and maintain the Farrand-designed landscape. At the public event announcing the conservancy, its first president, Rebecca Trafton, told the gathered crowd, "Dumbarton Oaks is one of the greatest gardens in America, and Dumbarton Oaks Park is an inextricable piece of the whole. It is perilously close to losing its design and ecological integrity, but I think we are right at the point we can bring it back to life."[52]

Typical of conservancies during their infancy, the DOPC was set up with a large advisory board and no paid staff. While Georgetown has its share of wealthy residents, the immediate pool of moneyed donors is relatively small, meaning fundraising would take longer and volunteer labor would be needed to get much accomplished. The DOPC increased engagement with the local community and was able to assemble a consistently growing network of volunteers. By 2020, the end of its first decade, an estimated 75,000 hours of volunteer work had been logged improving the park's landscape.[53]

Restoring Dumbarton Oaks Park

Implementing recommendations laid out in the CLR began in the early 2000s. Along with the DOPC, the NPS chose as an initial project the restoration of the northern end of the Forsythia Dell, one of the landscapes that were bisected when the original garden was divided. As designed, the steps and limestone arch were part of an orchestrated transition within a large and vivid mass of forsythia shrubs that crawled into the valley. Over time, plants on the garden side thrived and successfully displayed a burst of yellow in the early spring while those on NPS property faded owing to neglect. What was conceived as one landscape connecting the manicured and wild garden evolved into two distinct places. Because Forsythia Dell is a popular draw to the upper gardens in the springtime, the NPS recognized this project could engage interest and spur support for future efforts. The steps were rebuilt, and thousands of cuttings from healthy forsythia plants were rooted, propagated, and planted on the slope to recreate the original imagery.

It is important to note that this work was accomplished in coordination with staff from Harvard's Dumbarton Oaks. Joseph, who was actively involved in the NPS work in Dumbarton Oaks Park for fifteen years and led both CLR processes, acknowledges the institution's support in this and other collaborative projects that reconnected the two properties.[54] She referred to Gail Griffin, the director of gardens and grounds at the time, as her "garden twin for the upper gardens," signifying a positive and productive working relationship.[55] Such cooperation was needed since the Forsythia Dell project involved restoring the stone archway, which engages both landscapes. The original metal gate was taken from storage and rehung, meaning that, minus the distraction imposed by the chain-link fence, Farrand's intention for

Fig. 47. The stone archway designed by Farrand as a symbol of the transition between the formal upper gardens and the wilderness garden below became part of the physical barrier separating the two landscapes. (Photo by author)

a consistent landscape was renewed, at least visually, and in this one location. Other significant projects emerged from the CLR, which continued to serve as a guide to restoration work.[56] Trails and paths were reestablished and stabilized, and a program to eradicate invasive plants continued.

Once the DOPC took over as the private partner to the NPS, work began on the restoration of the two-acre landscape named Beech Grove, which sits inside the primary public entrance and runs along the eastern edge of Dumbarton Oaks. The entry path extends through a stand of American beech trees along a slope that, like other parts of the park, was overrun with aggressive Norway maples, English ivy, and other invasive plants. It was denuded and continually washed out by the force of unchecked stormwater runoff. Repairing the eroded landscape became known as the "Signature Project."

To mitigate the damage, engineers and workers constructed a stormwater conveyance system to improve drainage and regenerate groundwater supplies, while teams of volunteers stabilized the steep slopes with matting and biodegradable tubes to slow and spread the flow of water more evenly across the slope. They gradually cleared masses of invasive plants in coordination with a broader, in-progress management program for Rock Creek Park. This took over five years, by which time around six hundred fifty-gallon bags filled primarily with twisted ivy vines had been hauled off the site.[57] Native grasses, groundcovers, shrubs, and trees were then planted to stabilize the slopes and establish a contemporary version of Farrand's horticultural aesthetic for this part of the garden.

Another significant project undertaken involved the restoration of the open meadows that once contributed to the character of the Blisses' wild garden and had deteriorated significantly over the years as invasive plants and natural succession took hold. The NPS and DOPC provided the leadership and organization for the project, while volunteer supporters, neighbors, and students helped eradicate the invasive vines, weeds, and grasses that had taken hold. Following two years of organized events scheduled primarily on weekends, the first phase of the project to establish a healthy meadow ecosystem that aligned visually with Farrand's initial design vision was complete. This, along with the Beech Grove project and a series of smaller initiatives, successfully improved the look and health of the landscape while also building meaningful momentum for the DOPC. All of this laid the groundwork for the daunting challenge of recalibrating the insufficient stormwater management system and repairing the damage it was continuing to inflict on the park's landscape.

The Challenges of Stormwater

Aside from the sloping woodland that envelops much of Dumbarton Oaks Park, the landscape's most significant feature is the stream that runs its length along the bottomland. The moving water adds a dynamic character and a focal point to the naturalistic environment while also enhancing the sense of wildness and mystery by masking much of the surrounding urban noise. Farrand designed the carefully rusticated stone dams and retaining walls to reinforce the rural character desired by the Blisses, thus contributing to the property's design legacy. Although successful both aesthetically and experientially, the redesigned stream had inherent problems from a performance perspective. In 1942, Farrand complained in a letter to Dumbarton Oaks' first director, John Thacher, about the "flooding of the stream and drowning of the pools" and insisted that workers not "remove the plants as well as the mud" when cleaning them out, as that would impinge on the integrity of her design.[58] Routine NPS maintenance was already challenged by a tight budget, and additional periodic work such as repairing storm damage was likely not given the attention it needed. Even so, the fundamental issue was not inadequate maintenance; it was the system's design. While the waterway supported a desired aesthetic, it was not appropriately sized to handle substantial rain events.

A year after complaining to Thacher, Farrand wrote to Harry Thompson, a landscape architect and the chief of the NPS's Planning Division: "Let us hope that the old proverb of much water having run under the bridge will not be true of the Dumbarton Oaks stream, as we certainly had enough floods and washouts last summer and autumn to satisfy the most dry-minded individual."[59] Some of the intricate details were consistently at risk owing to changing conditions. Laurel Pool, for instance, was a small body of water designed to mimic the shape a leaf with outlines that could not easily withstand heavy rains or shifting alignments. Farrand recommended reconstructing the edges "as they were" when the pool migrated over time, as if it would be that easy. "While, of course, the stream is in no way a really natural brook," she wrote, "it should have a certain eighteenth-century quality of the naturalistic, which can be preserved by intelligent management and without much cost of plant material."[60] Her thinking, though logical, proved to be naïve.

Threats to the stream's design integrity amplified as intense runoff poured onto the site from the rapidly developing neighborhoods growing up around Dumbarton Oaks. Land once densely forested before being razed for farmland gradually transitioned into seas of rooftops, roads, and parking lots—hard surfaces that directed accelerating flows of water toward the delicate low-lying stream. Over time, as flooding rains moved more forcefully through the park, washouts and erosion contributed to the degradation of many of the features and details Farrand had designed. With inadequate funding for maintenance, the NPS could do little beyond observe the ongoing degradation.

Hope for the stream emerged along with the renewed focus on the park in the 1990s. In the Olmsted Center's *Landscape Preservation Maintenance Plan, Dumbarton Oaks Park,* which came out in 1997, investigators conducted a systematic analysis of each dam, pool, and waterfall, all of which had experienced damage. Some were less structurally sound than others, and several had silted over or sat buried beneath swaths of bamboo that had migrated down the garden slopes. One description captures typical conditions: "The dam's integrity is unknown due to heavy siltation which completely covers the upper portion of the dam. The stream may have relocated to the left of the channel. Siltation below the dam is minimal. Both downstream banks show extreme signs of erosion."[61] Although it was evident the historic features required significant restoration, it appeared none were lost forever.

Acting on the plan's recommendations, the NPS employed student volunteers to stabilize the banks in hopes of averting further damage. In 1999, a civil engineering firm conducted a hydrologic study of the stream. This was updated a dozen years later with a series of photographs documenting the continued deterioration caused by rushing stormwater that flooded the undersized stream and further eroded its banks, adjacent paths, and surrounding slopes. The plan prescribed removing accumulated stream sediment, eliminating invasive plants, stabilizing surrounding slopes, and separating pedestrian paths from the flow of water.[62] Increasing the stream's holding capacity was identified as being critical. In addition to localized measures, the plan acknowledged that many problems stemmed from adjacent areas and advised cooperation with neighbors to improve stormwater retention and treatment.[63]

Determining an appropriate path forward required negotiating a familiar balance between respecting historical design integrity and adapting to evolving environmental and programmatic dynamics. Over time, several restoration projects rebuilt dams, walls, and stream features in an appropriately rustic style. While these efforts helped locally, increasing capacity to solve greater hydrologic problems would likely predicate a significant alteration of Farrand's design. Slowing the amount of water entering the stream was deemed preferable from a cultural landscape perspective and

Fig. 48. After a series of improvement projects based on the National Park Service's cultural landscape report, Dumbarton Oaks Park's stream corridor, including a series of rustic dams, was restored with a certain degree of historical accuracy. (Photo by author)

would require creating upstream bioswales, rain gardens, or a bioretention pond. It was clear to both the NPS and the DOPC that trying to accomplish this within the park's limited boundaries would be a waste of time unless the flow and force of water could be stemmed before the water entered the site.[64]

Accomplishing any version of the stormwater plan became dependent on garnering support from outside constituencies and acquiring a sizable source of funding. Such help came in 2016, when the DOPC received a $320,000 grant from the National Park Centennial Challenge, one of the initiatives established by the NPS to celebrate its centennial anniversary. The challenge provided the opportunity for conservancies, friends groups, and other partner organizations to match federal dollars for projects that supported the conservation, preservation, and public enjoyment of national parks. The DOPC raised matching funds for a stormwater project and hired a consultant to design a remediation system in the park. A three-and-a-half-acre parcel of adjacent land owned by the NPS sitting just upstream from Dumbarton Oaks was identified as the site for a stormwater feature that would slow, detain, filter, and cleanse water before the water entered the stream with a combination of pools, wetlands, and landforms. Designing any new features to blend with Farrand's historic design was identified as a priority, as was using the new landscape as a tool to support the NPS's focus on environmental education.[65]

Around the same time, in the summer of 2017, while the DOPC and NPS partnership worked on these plans, the upper gardens embarked on a project to address their own above- and belowground water systems. Activities included replacing crumbling, nearly century-old lead and terra cotta pipes and installing an efficient, state-of-the-art irrigation system that responded to the institution's mission for environmental sustainability. Dumbarton Oaks Gardens also added surface features such as berms, swales, and rustic, stone-clad gutters to slow and more effectively direct stormwater flowing into the valley while aesthetically fitting within Farrand's design vocabulary. Although these fixes were undertaken with an awareness of their downhill implications, the process was entirely separate from what was happening in the park. The two, once unified landscapes functioned as friendly neighbors, looking out for one another yet focused on their own interests.

Contemporary Issues

In 2017, following a series of successful projects, the DOPC and the NPS embarked on a master planning process to chart a "prioritized plan of action with measurable goals for the short and longer term time horizons."[66] By this time, twelve of the park's twenty-seven acres had been successfully rehabilitated, extensive stands of bamboo and other invasive plants had been cleared, and more than five hundred new trees had been planted.[67] Although not the pristine version of nature initially designed by Farrand, the informal beauty she envisioned was reemerging. Early skepticism about the potential for the public-private partnership had given way to pride over what the two groups and scores of volunteers had accomplished.[68] The joint planning process further solidified the relationship between the NPS and DOPC and quickly established a tool for building on successful restoration efforts while also establishing strategies for ongoing maintenance and management.

Among the remaining challenges was to reconcile the park's design legacy as a private garden with its current use as a space open to the public (as was mandated by the Blisses). Elevating Farrand's vision and detailing as the highest priority contradicted how many people wanted to experience the landscape. She envisioned it as a place for an aging couple and their friends to stroll and enjoy a semblance of nature, yet it had evolved into a popular spot used by crowds of dog walkers, runners, and cyclists. Generations of people enjoyed the park while unwittingly degrading its delicate path system and fragile stream details. Discussions during the master planning process weighed options for mitigating this damage, including adding perimeter fencing, closing the landscape at certain hours, and restricting activities. Suggestions such as diverting runners and bicycles around its perimeter and prohibiting dog walkers seemed to signal a preference to reclassify the landscape as a garden or, as Farrand had suggested, a grove, rather than a public park.

Such conflict is not unique. Many designed parks struggle to protect their landscape and features while also providing access to the public for which they were created. Central Park, for instance, periodically closes its wildly popular lawns to give them a chance to recover from the abuse inflicted by heavy use. While the system attracts complaints, it is an understood aspect of that park's management policy. Establishing and implementing similar rules of engagement to protect Dumbarton Oaks Park was considered critical to lasting success. In addition to solving this issue, the master planning initiative addressed the need for various management plans and mechanisms, including interpretive signage, communications, accessibility compliance, stormwater management, and routine maintenance. As much progress had been made, there was still much work to do.

After its first decade, the DOPC had matured from a grassroots volunteer organization into one formalized with a full-time staff and an increasingly sophisticated fundraising operation. As beneficial as its partnership with the NPS has been, however, achieving the ambitious goal of reestablishing the Blisses' and Farrand's design vision for creating a unified garden requires cooperation with the upper gardens staff to blur the lines between the two landscapes. With two powerful, bureaucracy-laden property owners focused on their unique issues, fostering collaboration and compromise can be a challenge. Even with their differences, however, progress is being made. As the DOPC and NPS cleaned up the valley side of the fence, staff in the upper gardens embarked on a scheme to create a native landscape edge on Harvard's side to bring the woodland character into the formal garden and help diffuse the long, prominent boundary line.

Whether understood and appreciated as a garden or a park, it is evident in hindsight that Dumbarton Oaks Park would be worse off had the NPS not intervened. Many of Farrand's original design details would have certainly been lost without the agency's developing expertise in documenting and treating designed landscapes. Conversely, the NPS benefited greatly from having Dumbarton Oaks Park on which to learn and test methods and techniques. Its ultimate success, however, would certainly have been tempered without the advocacy and volunteer efforts of neighbors, citizens, and professionals. As it stands today, the landscape may never achieve the visual quality initially envisioned by Farrand or the status of a park for the people, as intended by the Blisses. Even so, the collaboration between the NPS and DOPC is helping to create a landscape that successfully functions somewhere in between.

Prospect Park, Brooklyn, New York

Few landscapes capture the aspirations and fundamental purpose of the American urban park better than Brooklyn's Prospect Park. Rightfully considered by many the crown jewel of Olmsted and Vaux's revered portfolio—their masterpiece, their magnum opus—the landscape embodies the principle that people, wherever they live, need and deserve access to nature in whatever form possible. Unlike its more famous neighbor, Central Park, which suffers somewhat from the constrictions of its rigidly narrow and rectangular site, Prospect Park has a more generous footprint on which to erect the designers' pastoral vision. The park is divided into three aligned landscape types, described by Olmsted as such: "first, a region of open meadow, with large trees singly and in groups; second, a hilly district with groves and shrubbery; and third, a lake district, containing a fine sheet of water, with picturesque shores and islands."[69]

Many parks have such features in the form of open lawns for play and picnics, dense woodlands filled with nature and mystery, and bodies of water upon which the eye can rest and across which boats and, in cold climates, skaters can move. As many versions as there are, however, no other park has a space quite like Prospect Park's

Fig. 49. Prospect Park's Long Meadow is one of Frederick Law Olmsted and Calvert Vaux's most famous landscapes and epitomizes the designers' goal of creating scenic and pastoral urban parks. (Courtesy of the U.S. Department of the Interior, National Park Service, Frederick Law Olmsted National Historic Site)

Long Meadow, which is among the most beautifully conceived, constructed, and preserved designed landscapes in this country. The sublime, mile-long, ninety-acre space exquisitely arcs and rolls along nearly the entire length of the park. Olmsted seemed to know how special it was, writing, "The word park as a common noun, as a descriptive word, should indicate such graceful topography, such open pastoral, inviting hospitable scenery as I have indicated."[70]

Brooklynites made Prospect Park an instant success, pouring in even before the end of construction. Rather than using it passively, in appreciation of the carefully staged scenery, as Olmsted and Vaux had intended, they frequently wandered off-path and engaged in active sports. It quickly became apparent that while the design successfully created a beautiful nineteenth-century landscape, it needed to morph with the changing needs of the rapidly densifying city to retain relevance. Much to the distaste of Olmsted, buildings, structures, and monuments popped up in the park, along with playgrounds, plazas, and playfields that challenged his and Vaux's vision. Even the sacred Long Meadow was lopped off at one end to create play space that still exists today. Such alterations are to be expected and, in hindsight, were generally not as bad as Olmsted perceived. Time and again, the park's vast scale and articulated vision have made it remarkably adept at absorbing change, from both design and cultural perspectives.

Like many urban parks, Prospect Park's condition has paralleled that of its municipal context, rising and falling with the relative health of New York City's economy, and often left struggling on the losing end of government prioritizations and cyclical battles for public funding. In tandem with the more publicized process in Central Park, periods of deterioration triggered citizen-based advocacy that prompted action by those with power and money. Preserving an Olmsted and Vaux treasure may have been the rallying cry, but figuring out how to navigate a complex bureaucracy and establish a functioning management system was always the more significant challenge. While its story has frequently focused on its original designers, the park's enduring quality is a testament to the work of many less-heralded people who followed in their path as stewards.

Establishing the Legacy

A cursory look at New York City's chronological or cultural history might suggest that Brooklyn's desire for a public park was merely an attempt to emulate the success of its Manhattan neighbor, Central Park, which, once built, became an instant source of pride for local New Yorkers. While that example surely spurred more widespread interest in Brooklynites, a call for open space had been around for decades before Olmsted and Vaux's Greensward Plan. As early as 1820, when the town's population was a mere eight thousand people, there was an attempt to transform a popular tree-lined stretch on a prominent bluff in Brooklyn Heights into a formalized promenade. The landscape offered unparalleled panoramic views across to the emerging New York skyline, and neighbors in the wealthy suburban enclave frequently strolled along what was essentially an informal linear park.

People seem to have recognized that open space was an amenity and a means of building civic pride even before the urban park movement took hold, and arguments for and against formalizing the promenade as public space went on for decades. Across the river, residents already enjoyed their promenade on the southern end of Manhattan, while Brooklynites had no suitable equivalent to rival that landscape. While lamenting the lack of quality architecture and open spaces in Brooklyn, a local newspaper proclaimed in 1930: "We have not a single public square, and the only walk that our townsmen can enjoy, is the crumbling margin of the heights, left open by the liberality of some. . . . Even this promenade is totally neglected by the hand of improvement—no path is laid out, no tree planted, and no terrace regulated, either for beauty or safety."[71]

In the 1830s the governor of New York appointed a commission to study the potential for a system of parks in Brooklyn, and in 1839, a plan depicting sites for eleven small parks emerged. A few of these were approved, though none sizable enough to meet the growing needs of city residents. The first wave of European immigrants drove the population of Brooklyn, more than doubling it between 1840 and 1845 to around 80,000. One park that attracted attention was Fort Greene Park, intended for one of the hills on which the Battle of Brooklyn was fought during the Revolutionary War. Walt Whitman, while editor of the *Brooklyn Daily Eagle,* used the paper to advocate for public parks. An 1846 editorial depicts the negative aspects of urban life as drivers for parks: "The air is dead, and dry—in the summer, hot and offensive."[72] It goes on to describe the suffocating aspect of the city before conveying the fictional discovery, some thirty years in the future, of "a dash of nature's greenness—a patch of sward, a tree, a grassy knoll!"[73] Whether because of the paper's influence or not, Fort Greene Park was successfully built in 1850 and proved so popular that it spurred public demand for an even larger park.

In 1859, a year after construction began on Central Park, a fifteen-member commission was formed to "authorize the selection and location of certain grounds for public parks, and also for a parade ground for the city of Brooklyn."[74] Commissioners put forth several proposed sites, the most significant of which was a 320-acre tract referred to as Mount Prospect Park in reference to its location around Prospect Hill, the second-highest spot in Brooklyn, which had functioned as a lookout for the troops during the Battle of Brooklyn. The New York state legislature confirmed this recommendation and, in 1860, authorized the purchase of the land for a park. Another commission—the Board of Commissioners of Prospect Park—was appointed that same year "for the laying out, adornment and management" of a great park in Brooklyn like Central Park in Manhattan.[75] James Stranahan, a U.S. representative, abolitionist, and millionaire businessman, was appointed the board's president. Because of his advocacy and leadership, credit for Prospect Park belongs at least as much to him as to Olmsted and Vaux. Known by history as "the first citizen of Brooklyn" and "the father of Prospect Park," Stranahan was a significant civic leader and a proud Brooklynite who worried that his city "was in danger of sinking into the character of a second-rate suburb" relative to its wealthier and more populous neighbor across the river.[76] Even though the city was still predominantly undeveloped at the time, it was clear to him that a large park would help establish it as a world-class city moving forward.

The proposition of a park was not without public opposition. There was a sense within the community that it would be an unnecessary luxury as there was already plenty of open space in and around the city. Amid ongoing controversy, the commissioners moved forward with plans for Prospect Park in 1860. Thanks to the continued influx of immigrants, Brooklyn was the third-largest city in the United States by this time, with more than half of its over 200,000 residents being foreign-born. For many citizens, the potential benefits of parks extended beyond aesthetic beauty, improved health, recreation, and retreat from urban life; they were also civilizing spaces in which a diverse populace could come together as a unified community with shared experiences.

The commission first engaged the civil engineer Egbert Viele to draft a master plan for the park—a familiar task for him since he earlier had created a plan for Central Park that was subsequently disregarded when the Greensward Plan was adopted. His plan reflects the tastes of the time, a scenic version of nature, with water features, gardens, and wooded areas made accessible by a series of paths and carriageways. The estimated cost for converting the property into the park was within the $300,000 authorized by the legislature. Viele's proposal received support, but the process stalled in 1861 at the outbreak of the Civil War. Once the war ended four years later, the plan was shelved for undisclosed reasons. Whether underwhelmed by Viele's design or put off by his reported bristling personality, the commission chose to go in a different direction. In 1865, the commission selected the plan proposed by Calvert Vaux.

Vaux's plan similarly depicts a scenic landscape taking advantage of existing topography and intact woodlands. His most substantial change was to slide the park's boundary several blocks to the west to avoid being bisected by the busy Flatbush Avenue, as Viele had shown. The shift wisely eliminated vehicular cut-throughs while expanding the park by 228 acres. During this time, Olmsted resigned from Central Park and moved with his family to California. Vaux doubted his ability to complete the project on his own and pleaded with Olmsted to return to New York to work on Prospect Park. In 1865, the two men rekindled their partnership just eight miles from the landscape that had made them famous.

The men presented their final plan in January of 1866, and by May they had been hired as the official landscape architects to design the park. Construction soon began yet was complicated by the decision to open a portion to the public in 1867. Around 100,000 people visited the unfinished park each month during the following summer. Stranahan was an, if not the, integral player in the process from the beginning. He was an engaged collaborator, offering input on all aspects of the planning, design, and construction.[77] When he died in 1898, the *Brooklyn Daily Eagle* wrote in an account of Stranahan's life, "It is all very well for him to say that they had the best landscape architects in the world [working on Prospect Park], but they could not have done anything without him."[78] Even as Olmsted and Vaux moved on to other projects, Stranahan remained intimately involved in the park's evolution.

Prospect Park Opens

Prospect Park was primarily complete by 1873, at which point Olmsted and Vaux once again severed their formal association. An economic depression in Europe and North America marked by history as the Panic of 1873 caused the building of several structures designed by Vaux to be deferred or eliminated, and funding for maintenance was severely constrained. The project's total cost was nearly $10 million—comically outside the range of early estimates. Vocal complaints about accelerating costs were left in the past as the public visited the park in droves and immediately proclaimed its unmitigated success. Around 6.7 million people visited in 1873 alone. The local paper re-created a typical late summer day: "The Park was thronged as usual on Saturday afternoon. A great multitude clustered around the Music Stand, and others found shady comfort under wide-spreading trees or trod the sinuous avenue or were propelled over the glistening waters of the Lake. The croquet parties were numerous, and their sports were witnessed by many."[79] Brooklyn's population was rapidly approaching a half million; and "the crowds came, some in their carriages, others on foot, and the majority by the horse cars from the various sections of the city."[80] The *New York Times* consistently published notices of events such as miniature yacht races on the Lake, musical concerts, and various organized sporting events, including cricket, baseball, tennis, and archery. Tens of thousands of people reportedly showed up for military parades and, when conditions allowed, ice skating on the Lake. Prospect Park was no longer merely a landscape designed by the celebrated Olmsted and Vaux; it was now a public space, one integral to city life and owned, loved, and trodden on by the people of Brooklyn.

Because the park was so popular with the masses, it did not take long for maintenance to become a critical issue. As noted in a commission report from 1880: "It must be obvious that with the increased use of the Park, and consequently, the more rapid wearing out of its numerous improvements, and the requirement at some points of enlarged conveniences to meet a public demand, there is great need that the appropriation for maintenance should be increased."[81] The road, in particular, presented a daunting and consistent challenge. Such was not a surprise to Olmsted and Vaux, who prioritized making the landscape accessible and pleasurable to large crowds as the city's "driving room, riding room, walking room, sitting room, skating, sailing and playing room," even as the practical impact of those people on the landscape could threaten its "rural, natural, tranquilizing and poetic character."[82]

Stranahan, who never accepted compensation for his position, was, along with all the other commissioners, removed by the mayor in 1882, thus ending a period of relative stability. John Culyer, an engineer who had worked with Olmsted and Vaux on the construction of Central Park before joining them on Prospect Park, became a significant figure in managing the landscape. He was promoted to chief engineer in 1871 and, on the departure of the two designers, became the park's superintendent three years later. As of 1885, when the City of Brooklyn was bogged down by a dysfunctional political system, $120,000 was allocated to tend the borough's six hundred acres of urban parks. At the same time, the City of New York across the river provided over four times that amount for maintenance of less than nine hundred acres of parkland.[83]

As part of an 1886 management restructuring undertaken in part to decentralize power, commissioners eliminated the powerful position of chief engineer and superintendent in 1886. This move not only left Culyer without a job, it severed the last direct connection to Olmsted and Vaux and deprived the park of an effective steward. The new organizational structure, which was deemed "business-like and economical," divided leadership among several superintendents in unique specialties.[84] The *Brooklyn Daily Eagle* recounted Culyer's legacy after his death in 1924: "He loved every tree and shrub and blade of grass in Prospect Park, and took as much interest in seeing its beauty grow and develop as his plans were carried out as an artist or sculptor takes in the work of his brush or chisel."[85]

With the departures of Culyer and Stranahan, Prospect Park was without two of its strongest advocates and effective stewards; and the landscape suffered. In 1888, the *New York Times* reported that Brooklyn's mayor, Alfred Chapin, "made a visit to Brooklyn's former pride, Prospect Park, yesterday and discovered that it had fallen into a shameful state of neglect since the Park Commission . . . had discharged Chief Engineer John Y. Culyer."[86] On finding several features, primarily rustic bridges, in disrepair to the point of being unsafe, the mayor complained about the ineffectiveness of the commission in letting the park become run down. He soon proposed spending $100,000 on improvements, signaling a temporary period of revival for the landscape.

When the Rustic Meets the Beaux-Arts

The next few years brought extensive changes. The extent to which Chapin's visit directly influenced this period is difficult to discern, although several weeks after his visit, New York's governor did sign a park bill for $500,000 and Brooklyn's park commissioners requested an additional $540,000, primarily for improvements in Prospect Park.[87] The economy had rebounded from the 1873 collapse, and people and municipalities were spending money more freely. While urban living conditions improved for many, the chasm between the wealthy elite and the expanding masses of impoverished residents deepened. The lower class swelled with immigrants who poured into New York and Brooklyn from Ireland and Germany, along with an increasing number of Russian and Eastern European Jews. Domestic migration brought farmers and rural residents into the cities looking for industrial jobs at the same time the Great Migration was directing African Americans from the Jim Crow–plagued South to Northern states. On the Manhattan side of the river, the Lower East Side became satiated with crowded and unsanitary tenements, while the Gilded Age's super-wealthy, the Astors, Vanderbilts, and Rockefellers among them, comfortably lined Fifth Avenue with grandiose mansions designed by in-demand architects such as McKim, Mead & White, Richard Morris Hunt, and Horace Trumbauer.

Stranahan believed Prospect Park would help to foster an upper class in Brooklyn by offering "strong inducements to the affluent to remain in our city, who are now too often induced to change their residences by the seductive influences of [Central Park]."[88] Brooklyn Heights was prosperous, even as the city remained primarily working class, filled with merchants, dock workers, professionals, and farmers. The city's economy boomed thanks to expanding trade, manufacturing, and the processing of agricultural commodities. Most of its subsequent success resulted from the 1883 opening of the Brooklyn Bridge, which connected Brooklyn and Manhattan far more efficiently than the existing ferry system. In 1890, when Brooklyn's population reached 800,000, over 40 million people crossed the bridge—roughly 250,000 per day.[89] As the port cities of New Orleans, Boston, Baltimore, and Philadelphia gained in population and status, conversations about consolidating Brooklyn and New York into a unified metropolis amplified.[90] In 1898, following decades of discussions, Brooklyn gave up being an independent city and became a borough of New York City, more closely tied to Manhattan than ever before.

Coincident with Brooklyn's ascent was the growing popularity of the Beaux-Arts style, which American designers and their upper-class clients imported from Europe. Architects influenced by the style designed buildings for their clients that embodied wealth, using materials and detailing that emphasized neoclassicism, ornateness, and symmetry. The City Beautiful movement, a broader planning and design initiative that expanded the formal approach and attitude of the Beaux-Arts, took hold around the same time. The movement endorsed a brand of urban reform rooted in artifice, in which aesthetic improvements and monumental planning were tools for inspiring residents' pride in their city. Civic buildings, such as public libraries, train stations, and city halls, adopted the style as cities invested in their public realms with fountains, sculptures, monuments, and other landscape improvements such as grand plazas, tree-lined streets, and dazzling displays of ornamental horticultural. Formality in the landscape became the fashion, presenting a stark visual counterpoint to the picturesque style, which had been popular in the United States for decades.

The aesthetic contradiction between the naturalistic and the monumental began appearing in numerous gardens, educational campuses, and parks. This was, for instance, the era during which Leland Stanford and Olmsted argued over which direction the campus at Stanford University should take. The distinction first became evident in Prospect Park in the 1890s, when ornate metal bridges replaced two of the original rustic wooden bridges. The now iconic Terrace Bridge, designed by Vaux, and the Lullwater Bridge, designed by McKim, Mead & White, introduced a new design vocabulary into what had been an intentionally consistent landscape experience. These structures were followed by other grand architectural gestures, notably classical Greek– and Roman–inspired entrances into the park, also designed by McKim, Mead & White. Marked by granite columns, turrets, statues, pedestals, colonnades, and other formal elements, these features presented a dramatic departure from the simple allée of trees originally designed by Olmsted and Vaux. These new architectural elements were often viewed alongside large, formally laid out, Victorian-style flower beds that had become popular and permeated the landscape. When asked by the *Brooklyn Daily Eagle* what he thought of the new structures, Culyer conveyed his bias for tradition by commenting, "I really haven't thought much about them, but I do not think the designer had carefully studied the harmony of the designs with the surroundings," adding that they were "not at all in keeping with the general natural style of Prospect Park."[91] An editorial comment in the same piece expresses a similar sentiment: "When people go to Prospect Park it is to enjoy themselves and the beauties of the surrounding scenery and not to contemplate incomplete and chilling specimens of classic monumental art."

Even with this criticism and others, these features located at the perimeter read more as extensions of the city fabric than as spaces integral to the park itself. They convey a departure from the initial intent, but their presence hardly threatens the park's design integrity. Potentially more problematic are the insertions of classical buildings into the park's interior. This includes three ornate structures designed by the architectural firm Helmle, Huberty and Hudswell: the Tennis House (1909), the Willink Entrance Comfort Station (1912), and the white terra cotta–clad Boat-

Fig. 50. A Victorian era craze for flowers led to the proliferation of formal annual beds throughout Prospect Park. This was just one of the changing design styles that gradually altered the landscape aesthetic intended by Olmsted and Vaux. (Courtesy of the U.S. Department of the Interior, National Park Service, Frederick Law Olmsted National Historic Site)

house (1905), which reflects a dramatic detour from the modest wooden structure it replaced. Another curiosity is the Peristyle, a Grecian-style shelter framed by a perimeter of Corinthian columns designed by Stanford White of McKim, Mead & White. White was surely aware of the Olmsted and Vaux legacy, as they were contemporaries and his father was friends with both designers. People recognized White for his flamboyant designs, which relied heavily on decoration and ornamentation—a look distinct from Vaux's quaint, rustic structures.

Whether the intrusion of these buildings into the pastoral landscape is considered positive or negative is open to debate and personal taste. Time seems to have proven that the scale of Prospect Park allows it to absorb numerous insertions from various periods without negatively affecting the quality of the whole. Vaux's mostly wooden structures may have been appealing in their context, but many of them could not effectively survive the lasting effects of time, weather, and public use. The Beaux-Arts structures provided a relative permanence while reflecting changing views on urban design. As a landscape still only a few decades old, Prospect Park was likely seen as a malleable public space open to improvement rather than as something sacred or in need of preservation.

Moving beyond Olmsted

As the addition of buildings altered the park's character, other deviations changed the landscape. On Lookout Hill, to which generations of parkgoers had flocked to take in its expansive vistas, the adjacent plants had become so overgrown by the early 1890s that, as was reported to the commissioners, it was "almost impossible to get a good view of the Park and the surrounding country."[92] Throughout the park, crews monitored and cleared shrubs and trees as necessary. Olmsted and Vaux endorsed this practice. On touring the landscape in 1886, they recommended removing half the current growth. They explained that "the present condition was the result of massing trees and shrubs to produce the immediate effect of first planting, and that it was the design to cut out as nature progressed the growth and not to permit the vegetation to kill itself by overcrowding."[93]

Park commissioners were tasked with managing the landscape and had the authority and financial control to legitimately make changes they considered optimal for the benefit of the landscape and the community. Their power became concentrated, reduced from a healthy panel of eleven men in 1870 to eight in 1880 and three in 1888, until by 1892, a single man held the title and decision-making influence. Frank Squier, the lone commissioner as of 1894, brought Olmsted back to Prospect Park that same year to be the "consulting landscape gardener." Olmsted was not impressed with Squier. Following a tour, he wrote that he found the administrator had little interest in trees or landscape and no "higher notion of his duty than to estimate . . . what would best please 'the people,' and move accordingly."[94]

Olmsted's renewed involvement seems to have accomplished little aside from stirring his frustration. He wrote to Vaux that what he encountered in the park "was melancholy and perplexing; wide and erratic variations from our plan" and evidence that "certain radical changes of design have been already made and that the ground lost by them cannot be recovered."[95] He seemed particularly irritated by a redesign of the Children's Playground and Children's Pool, spaces he had envisioned as informal and unified within a bowl-shaped landscape in the northeastern quadrant of the park. McKim, Mead & White had recently transformed the playground into a formal rose garden with three water-filled granite basins and had hardened the edge of the naturalistic Children's Pool at the base of the site with copings topped with marble balustrades. This space was renamed the Vale of Cashmere, a moniker Olmsted considered absurd. He also complained the space compounded the habit of adding formal features, statues, monuments, and buildings that he believed were changing the character of the park. In a subsequent letter to Squier, he wrote: "If this tendency continues, and more and more architectural features are introduced, the time will come when the beautiful, quiet, rural landscapes of the Park will be, to a very great extent, marred, and the Park made to resemble a confused and fussy-looking ornamental garden, or the best of our rural cemeteries."[96] While Olmsted seems to have viewed this change as an affront to his design, Squier's predecessor hints at a justification for the project, having reported years earlier that the playground was not a successful space but an "arid waste" and a place where "the children never go to play."[97]

When Olmsted retired in 1895, the second generation of his firm—first named Olmsted, Olmsted & Eliot and then F. L. & J. C. Olmsted—retained its advisory role at Prospect Park for several more years. Similar to what happened at Stanford around the same time, there is little to suggest the changing relationship produced many positive results. Even though professionals in the firm were without doubt qualified as park designers and experts in the field of landscape architecture, they were perhaps too invested in a literal reading of the original plan to be capable of dealing effectively with its evolution as a community resource. There were legitimate reasons to be concerned that the intended character of the park would change as it accommodated new programs, facilities, buildings, and objects, yet the extent of damage to the overall park experience is difficult to quantify.

After this period, the park continued a course of overuse and undermaintenance. The 1914 annual report of the department of parks relates a familiar refrain: "With the increased use of the parks by the public, their proper maintenance becomes each

Fig. 51. The Beaux-Arts-inspired influence of the McKim, Mead & White architecture firm permeated Prospect Park in various applications. An example is the Vale, a quiet landscape with a small pool and gardens that was redesigned by the architects with a hardened edge, balustrade, and marble coping. (Courtesy of the U.S. Department of the Interior, National Park Service, Frederick Law Olmsted National Historic Site)

year a question requiring closer study, especially so for the reason that the greater intensity of use happens to be coincident with a sharply declining maintenance budget."[98] Perhaps as a result of the onset of World War I, it notes that the budget for 1915 was $90,000 less than that of the previous year, signaling an unfortunate future for the park.

By 1918, nearly a third of New York City's 5.6 million residents lived in Brooklyn. World War I was ending, the Spanish flu was ravaging the world's population, and a particularly cold winter caused extensive damage in Prospect Park. John Hylan, New York's seventh mayor since the consolidation of the boroughs, entered office that year, emphasizing the need to beautify the public parks. Harking back to Olmsted's reference to urban parks as being the lungs of the city, Hylan described New York's parks as the "breathing places of a great city."[99] He had an aggressive agenda for parks, promoting them as places for active recreation, sports, large celebrations, festivals, and concerts. Although the deployment of men to the war effort triggered a loss of skilled labor, a substantial landscape restoration program in the park pruned or removed hundreds of trees and planted thousands of new trees, shrubs, vines, and groundcovers. Crews repaired "miles upon miles of walks and roads," fixed and painted benches, renovated structures, and strengthened whatever rustic benches remained.[100]

A familiar argument was reborn that same year when a spot overlooking the Long Meadow was proposed as a potential site for a large memorial to World War I soldiers. A former commissioner argued against the plan and lamented the apparent lack of concern by the public. He wrote in a letter to the *Brooklyn Daily Eagle* that, whereas in Manhattan, "people and press instantly resent and defeat every project for putting up buildings in Central Park," in Brooklyn, "the press generally favors such proposals as to Prospect Park, and the people are silent, while the governing body that must provide the funds leans favorably toward every trespass."[101] He mocked the lack of expertise of park administrators as "any insurance man or lawyer or man of other business or profession" could be "suddenly given the management of that great and magnificent pleasure ground."[102]

The park commissioners could not make such decisions on their own. In 1898 the state legislature established the Art Commission of the City of New York to provide aesthetic consistency in a chaotic environment filled with numerous, often contrary, civic and political opinions and points of view. Per the approved charter, mayoral-appointed members of the first Art Commission had to include three laymen, an architect, an artist, and a sculptor. Notably absent from the roster was a landscape architect. Seemingly unmoved by concerns over potential loss of the park's natural character, the commission decided to make the "Greek gothic" memorial even taller and wider, craft it out of "the best granite on the American market for monumental work," and situate it prominently on the edge of the Lake.[103]

Even though this series of new structures and monuments diverged from their pastoral context, the park's vast scale allowed its character to remain largely intact. They could be easily seen, but their presence did little to detract from the landscape's scenic intent. At least as disruptive to the scenery was the decision to harden the naturalistic edge of the Lake with a masonry wall in 1927 to stabilize the banks and to save mature trees along the shore that were "being undermined and ready to fall."[104] Beginning a couple of years later, the landscape suffered the fallout of the Great Depression, which affected New York perhaps more than any other city because of the city's direct relation to the financial and banking industries. By 1930, many of the city's seven million people were unemployed or homeless. The inherent benefits of green spaces became apparent as people escaped to the closest version of nature they could find for solace, respite, and fresh air. Urban parks flooded with crowds. Along with this increased usage, a lack of public money for maintenance propelled Prospect Park and many other urban landscapes into a dispiriting period of decline.

By this time, New York City parks had been at the mercy of various park commissioners acting solely or as a group who had, at least according to the *New York Times*, left parks abused and neglected as the officials "lived well and got rich by diverting maintenance funds."[105] Prospect Park embodied a landscape of abandonment, displaying denuded lawns, dead and dying trees, broken pavement, and other signs of abdication. Olmsted and Vaux's legacy faded, even as people cared deeply for the park and its landscape. A significant new chapter began in 1934, when the incoming mayor, Fiorello LaGuardia, formed a unified City Parks Department that would service all five boroughs under the direction of one man, Robert Moses.

The Reign of Moses

By 1934, nearly one-third of New York City's workforce remained unemployed. Many families were homeless, forced to live in shantytowns, shelters, and hotels. Thousands of children without proper care were placed in institutions and foster care. President Franklin Delano Roosevelt's New Deal programs funneled millions of dollars of federal money into communities across the country. LaGuardia capitalized on these funds for his city, frequently traveling to Washington, DC, to lobby Congress or the president for money. Roosevelt once commented about the mayor, "He comes to Washington and tells me a sad story. The tears run down my cheeks and tears run down his cheeks and the first thing I know, he has wrangled another fifty million dollars."[106] At one point, one in every seven dollars made available by federal New Deal programs, such as the Works Progress Administration (WPA), was spent in New York City. This money primarily supported the dramatic and controversial parks improvement campaign that followed.

Moses, frequently described as New York's "Master Builder" out of either respect or scorn, depending on one's perspective, proved the right person to spend this influx of money. He had attended Yale, Oxford, and Columbia Universities and by many accounts was keenly intelligent. His career, which is exhaustively documented in Robert Caro's epic biography, *The Power Broker: Robert Moses and the Fall of New York,* began in 1924, when he was appointed president of the Long Island State Park Commission. Over time, he developed a clear vision of what he believed enabled modern cities to thrive and, wielding the force of public authority, spent decades transforming New York's parks, beaches, suburbs, freeways, and urban cores. What his administrations accomplished is staggering, if not entirely admirable. According to Caro, Moses "built parks and playgrounds with a lavish hand . . . for the rich and the comfortable. Recreational facilities for the poor he doled out like a miser."[107]

By the time he started as New York's parks commissioner, Moses had built a reputation as someone who could get things done, meaning there was legitimate hope that his new consolidated position would be a good thing for the city's parks. Recognizing the significance of Olmsted and Vaux's legacy, the *New York Times* enthusiastically welcomed him in 1934, declaring, "A great heritage has been entrusted to him for safekeeping."[108] With an unintended irony that would become apparent years later, the paper also observed that "some of Moses' predecessors have manifested an abysmal ignorance of what parks are for," and that it is because he was "capable of seeing the park, playground and parkway problem as a whole that friends of the parks are especially glad to welcome Mr. Moses." The *Atlantic* described his role five years into his tenure as such:

> His responsibility covers more than half a billion dollars' worth of new city parks, playgrounds, swimming pools, and other recreational facilities; transformation of old ones; acquiring and developing scores of large park areas throughout the state; reclamation for public use of miles of waterfront around lakes, along rivers, and at the seaside; and the linking of these vast playgrounds and scenic places by a system of parkways, causeways, and bridges that are bold

> examples of engineering skill and are solving traffic problems in the most congested area in the world.[109]

With such an expansive purview, it seems unfair to assign blame or credit to Moses alone for all of the decisions made under his administration. However, he was such an imposing figure, and was so frequently bestowed with dictatorial labels such as *czar, autocrat,* and *authoritarian,* that emphasizing the centrality of his role is justified.

When it comes to historically significant places such as Central Park and Prospect Park, Moses has been regarded as a sort of anti-Olmsted—ignorant of or even antagonistic to the landscapes' uncluttered and picturesque intent. To him, twentieth-century cities had evolved beyond their need for bucolic retreats into nature. He was a pragmatist who believed that public parks are valuable to the community for their practical use more than their scenic quality and that effective management is a compromise. "There are people who hold that parks should be places of quiet and repose where the weary may go to rest . . . and where nature should be disturbed as little as possible," he once wrote. [110] He believed these views needed to be balanced with those of the "so-called playground enthusiasts who feel that parks will serve the greatest number by being turned into places of recreation."[111] His record as a builder makes evident he rested in the latter camp, but at least he acknowledged that both perspectives existed. Moses considered Olmsted and Vaux to be relics who "planned a place with walks and benches" and could not anticipate the needs of the future city.[112]

Moses was by no means alone in his opinion about parks. Municipalities across the country were filling public parks in cities and suburbs with swimming pools, playgrounds, sports fields, tennis courts, and other in-demand recreational facilities, while many naturalistic parks were left in various states of disrepair and neglect. During his first year in office, Moses recounted the condition of New York's parks as he found them:

> The parks were in a disgraceful condition. Fences, playground equipment, iron benches, steel bridges, piers and other iron work had been allowed through neglect to rust. Walks, paths and roadways were acres of broken pavement. Lawn areas were meadows of woods. Thousands of trees were butchered or

hacked down to provide firewood for political favorites. There was not a single park building in the city that did not need immediate repairs and painting.[113]

He reorganized his department to better tackle the work. He fired unskilled and unqualified staff, including white-collar workers who had lost their regular jobs and, according to Moses, been "placed in charge of construction projects on the theory that men unable to do common labor work could at least stand around and watch laborers dig and rake leaves."[114] Led by a team of architects, landscape architects, and engineers, a workforce of 75,000 men funded by the Civil Works Administration (CWA) completed 1,700 projects across the system within a few months.[115] A gardener who had worked in Prospect Park for eighteen years told the *Brooklyn Daily Eagle* the new workforce was making a difference. He said people did not appreciate the park as they used to and just threw trash on the ground. Now, he said, "We have eight CWA men just in this part of the park picking up papers all day."[116]

What had been a period of neglect in the park evolved into several years of intensive action. Extensive improvements to the landscape included reestablishing lawns and repairing walks, roads, comfort stations, and bridle paths. Hundreds of trees, many infected with Dutch elm disease, were pruned, trimmed, and cut down across the park, while crews planted thousands of new trees to reinforce the canopy so critical to the landscape's character. Countering these enhancements were the negative impacts caused by the accelerating influence of the automobile. Moses was notorious for his love of the car, once comparing a city without traffic to a ghost town. He strove to make Prospect Park more automobile-friendly, widening and straightening internal drives designed for horses and carriages to accommodate cars and other vehicles. To further accommodate vehicular access, he paved over open lawns and replaced them with asphalt parking lots.

Moses conveyed little sentimentality for the park's design heritage. He reportedly ordered boxes of original drawings for Prospect Park to be hauled off and deposited in the local dump, and called for the razing of more rustic buildings, shelters, and bridges designed by Vaux. Among the buildings destroyed was the Dairy, a quaint architectural gem that once processed milk supplied by cows grazing on the Long Meadow and provided to picnickers in the park. Although it needed repair, reports make evident it could have been restored and repurposed as had a similar structure in Central Park. Prospect Park's menagerie, which Moses referred to as a fire trap, was torn down to build a new zoo in 1935. The WPA funded the new facility, which took up eight acres along the eastern edge of the park, making it the era's grandest architectural statement. WPA money also funded a band shell that provided modern facilities for concerts, plays, social dancing, and, according to a press release from the parks department, "popular Barber Shop Quartet contests."[117] Both structures reflected the sturdy, unadorned character of WPA architecture, which contrasted with the frail rusticity of Vaux as much as with the ornate classicism of McKim, Mead & White.

Adjacent to the band shell was one of several new playgrounds added around the perimeter and one of the more than six hundred playgrounds built across New York City by Moses's administration. In contrast to the passive space for children conceived by Olmsted and Vaux as part of Prospect Park's rural recreational grounds, the playground had equipment for active uses such as see-saws, slides, swings, and a volleyball court. While out of character for a scenic park, these playgrounds successfully fulfilled a need in the growing metropolis for structured play environments for children—something yet to be considered in the mid-nineteenth century.

Funding from the WPA lasted only a few years, as the federal government redirected funds to the World War II effort. The parks department then focused on maintenance, although reduced staffing as a result of the war led to Prospect Park's renewed deterioration. Moses became a target for criticism as never before, likely owing, at least in part, to his very public persona, which was caustic, belligerent, and wickedly sarcastic. A member of a well-known Manhattan family wrote to him about the conditions in Central Park, complaining: "Having seen the scrupulously clean parks of other Nations of the world maintained at a fraction of the cost of our famed one I say with authority that Central park, in toto is a disgrace to the City and the Nation."[118] Moses, obviously unmoved by the critique, replied: "If you like foreign parks so much, why don't you patronize them?"[119]

The postwar fortunes of Prospect Park aligned with those of Brooklyn, which suffered from a movement away from local manufacturing. The city also suffered from the painful and symbolic losses of the Navy, which closed the Brooklyn Navy Yard, and the Brooklyn Dodgers, who fled to Los Angeles in 1957. The borough's population peaked at nearly 2.74 million people in 1950 and dropped by almost 100,000 over the next decade. Crime rose, as gang violence ignited by an influx of heroin scared away an increasing number of Brooklynites, many of whom were already fleeing to the suburbs. Violence and death by overdose became more routine in Prospect Park. To enhance perceived safety beginning in the late 1950s, parks department crews opened views within the wooded, nearly 150-acre Ravine by ripping out much of the lower plantings, thus creating drainage problems and essentially destroying the woodland's ecology and aesthetic character. This work coincided with the beginning of a six-year, $9 million enhancement initiative meant to spruce up the buildings and landscape, the most significant investment in the park since its initial construction. While this effort offered some relief from persistent deterioration, it was only a short-term reprieve.

In his last big move in Prospect Park, Moses facilitated the design and construction of the Kate Wollman Rink, an outdoor skating facility that expanded the skating season beyond the short winter period when the ice spanning the surface of the Lake froze to the needed depth of four inches. While the new rink quickly became a popular amenity, its design was particularly insensitive to the park's design heritage. Constructing the facility required partially filling in the Lake and destroying Music Island, the intimate landscape created by Olmsted and Vaux as a setting in which musicians would play music intended to carry across the park. Although it soon became apparent the plan would not work since the sound of music did not reach waiting audiences, the island did enhance the park's scenic character. Rather than save the island by erecting the rink on the adjacent, already paved carriage concourse, Moses elected to preserve the pavement as a parking lot for three hundred cars.

In 1960, after twenty-six years, Moses retired from his position as commissioner of parks. Whether admired or vilified—and he certainly was both—he had become far more famous in relation to New York parks than either Olmsted or Vaux, who had retreated from the public's collective memory by this time. Although Moses was controversial and by no means the type of nurturing steward commonly associated with historic landscapes, he did seek to make Prospect Park a more functional and accessible place for the citizens in the ever-changing community.

Fig. 52. Among the monumental changes instituted by Robert Moses in Prospect Park was the Kate Wollman Rink, an ice-skating facility that proved popular for visitors but destroyed a significant portion of the picturesque Lake. (NYC Parks Photo Archive, Rebekah Burgess)

Rising Advocacy amid a Failing Economy

Newbold Morris, a former president of New York's City Council, replaced Moses and remained commissioner until 1966, the year of Prospect Park's centennial. The park's first one hundred years were marked by a general decline, interrupted by minimal landscape restorations and the construction of architectural structures and features that distanced the landscape from its scenic origins. The *New York Times* marked the occasion by referring to the park's original designers as Calvert Vaux and "Frederich Law Olmstead," misspellings that exemplify how far removed the landscape architect was from the fame he had once enjoyed.[120] Perhaps it should not be surprising, therefore, that despite ample evidence that the landscape continued to suffer, the first vocal public cries for preservation addressed something Olmsted and Vaux had nothing to do with: the Beaux-Arts Boathouse.

When Moses entered office, he facilitated a change in legislation enabling him to demolish park structures without review—something he took advantage of numerous times in various settings. The subsequent administration chose not to wield that same authority when, in 1964, it announced it was investigating whether the deteriorating, seventy-year-old Boathouse was worth renovating or whether it should be razed and replaced with something new. The local architectural community, still stinging from the demolition of the McKim, Mead & White–designed Pennsylvania Station in Manhattan a year earlier and firmly in the camp of the urbanist and activist Jane Jacobs and her crusade against urban renewal, rallied to support the building. In a letter to the *Times,* Frank Helmle, the son of the building's architect, acknowledged that although "many architectural gems must disappear to make way for progress," he hoped that "here in the middle of one of the most beautiful parks in the world . . . our city fathers in their wisdom would preserve this rare example of classical architecture not only for its historical value but also for its artistic significance."[121] Calls for preserving architecture were amplifying even beyond New York City, but, as was being experienced with City Hall Plaza in Boston around the same time, few such emphatic pleas were made on behalf of Prospect Park's landscape architecture.

Preservationists found an apparent ally in John Lindsay, New York City's mayor as of January 1, 1966. Often referred to as "the Republican Kennedy" for his good looks and progressive policies, he called the park "one of the most pastoral and naturalistic, man-made parks in the United States" and lamented the process that had led to its current condition, with "historic structures altered or destroyed, the finest trees cut down, its beautiful gardens turned into swamps, magnificent statuary vandalized, walks and bridges shattered, and the erection of ugly and makeshift structures."[122] He hired Thomas Hoving, a fellow member of New York's upper crust, as Morris's replacement as parks commissioner. Hoving's educational background was in art history, and he developed an interest in the history of parks, particularly the works of Olmsted. He once said he considered Prospect Park "the greatest naturalistic manmade park in the world."[123]

Lofty praise from influential leaders might suggest the emergence of a renaissance for Prospect Park. However, thanks in part to chronically limited budgets and a municipal government preoccupied with numerous social and economic crises, that was not the case. Lindsay and Hoving successfully secured over $600,000 from city coffers, which was then matched by a federal grant specific to Prospect Park improvements, including $225,000 to renovate the Boathouse. Crews repaired and added sidewalks and repaved roads while clearing much of the shrubs and understory in the Ravine as a security measure. Other targeted projects for that funding included improvements to playgrounds, picnic areas, the wall around the Lake, the zoo, and a croquet shelter. While these changes helped, most of the features deemed significant enough to preserve had been added after the time of Olmsted and Vaux and therefore addressed practical contemporary needs rather than the less definable decline of the park's original scenic intent.

Exemplifying the dynamic nature of municipal administrations, Hoving left the parks department in 1967 for a position as director of the Metropolitan Museum of Art. Apart from his role in securing funding for restorations, his most significant legacy regarding Prospect Park is probably the appointment of a curator, which is essentially a professional—in this case unpaid—assigned to oversee the development of the park specifically as a work of art. In announcing this position and one like it in Central Park, Hoving said, "Just as one would not take a wire brush to a great masterpiece, one must have a sensitivity and expertise in the care and restoration" of parks, too.[124] The position offered no real authority but did institutionalize a needed advocacy for the landscape.

Prospect Park's first curator was Clay Lancaster, an architectural historian who made a name for himself advocating for the preservation of historic homes in Brooklyn Heights and had argued passionately for saving the Boathouse in 1964. Lancaster hoped the park's centennial celebration would be a turning point. The landscape had experienced a dramatic decline in attendance, from around 15 million visitors per year in the early nineteenth century to a third of that by 1966, and he hoped the centennial would "reawaken the people to the wonderful facilities the park has to offer."[125] In 1967 he wrote *The Prospect Park Handbook,* at the time the definitive history of the park. Part chronology and part commentary, the book details the history of Stranahan, Olmsted, and Vaux and offers critiques on the influence and architecture of the Beaux-Arts period. Surprisingly, Lancaster went easy on Robert Moses and his impact on the park. This could be because although Moses no longer had a formal role in the park system, he still retained considerable political influence.

Lancaster dedicated the handbook to Robert M. Makla, a fellow park advocate who in 1965 cofounded the nonprofit Greensward Foundation with a $1,000 grant to support naturalistic urban parks, specifically Central Park and Prospect Park. In addition to publishing numerous books and maps to promote the parks, the foundation sponsored tours, offered lectures, and provided on-the-ground volunteer maintenance support. A year later the Friends of Prospect Park emerged with Lancaster as chair. It attracted around seven hundred members in its first eight months.

By 1971, five years into Lindsay's million-dollar restoration process, there were mixed responses to the progress. While some citizens seemed pleased with the improvements, Makla complained the money was not being spent to the best effect for the park. He and others who had experienced the park for many years argued that the landscape was moving away from its "wild and beautiful" past.[126] Makla griped, "There's lots of money going for asphalt walks and widening roads, but the things that make the park a park—the horticulture, landscaping, care of trees and shrubs—have largely been ignored."[127] As stated by another civic-minded resident in similar terms: "The city is making it sort of beautiful in a concretized and modernized way, when before it was wild and beautiful."[128]

Lancaster and Makla were instrumental in fostering a meaningful sense of landscape preservation in the community and were joined in their efforts by the poet Marianne Moore, who was a president of the Greensward Foundation and played a significant role in raising awareness about the park with her 1967 poem about one special tree, the Camperdown Elm. Moore was distressed that the tree, a rare weeping elm that had been in the park since 1872, had been neglected and was at imminent risk. Her poem spurred to action the Friends of Prospect Park, who were moved by the final lines: "Props are needed and tree-food. It is still leafing; still there. / *Mortal* though. We must save it. It is / our crowning curio."

The Darkest Years

The rising interest in preservation coincided with a period of social unrest and fiscal catastrophe in New York City. As hundreds of thousands of people fled the city for the suburbs in the 1970s, urban life produced new challenges and increased dangers. Violent crimes soared across the five boroughs, with instances jumping from 58,802 in 1965 to 155,187 a decade later.[129] The crime rate in parks also increased—from vandalism and litter to theft, rape, drug dealing, and assault. With the city government mired in financial straits, funding for parks was reduced, leading to a further deterioration of the landscape.

In 1973, incoming parks commissioner Richard Clurman caused a stir by stating his department would essentially withdraw its services from parks that were repeatedly vandalized and destroyed because of what he said constituted community abandonment. His department, he said, "has not enough men, money, or equipment to win a war against the very population it is trying to serve."[130] This seemingly nihilistic approach to park management angered many people, who assumed he must be anti-park. He claimed in attempted self-defense that his comments reflected a reality more than an ideology. He proclaimed respect for Olmsted and supported the idea that the parks Olmsted (and Vaux) designed were works of art, but believed they were "more works of the philosopher's art than the composer's, painter's, or poet's."[131] Seeming to echo Moses, he said that parks were meant to change with society and were not "immutable works for contemplation."[132] At the same time, Clurman railed against previous changes to Prospect Park—in particular Moses's ice rink, which had required the destruction of Music Island. Local communities, he believed, needed to act as effective partners in the stewardship of parks, once saying, "If no one cares or will take any responsibility, no amount of repair, maintenance or police attention will produce any lasting results."[133]

Clurman held the office only one year and in this respect is more significant for his ideas than for his accomplishments. Although he drew criticism, his insistence that the health of parks depended on organized community involvement was a precursor to what was to come. Despite, or perhaps because of, the park's decline, New York City's Landmarks Preservation Commission designated Prospect Park a Scenic Landmark in September 1975. This was the same year the Central Park Task Force and the Central Park Community Fund formed in Manhattan and one month before New York City, with $34 million in the bank and a $453 million debt payment coming due, found itself on the brink of bankruptcy.[134]

Tupper Thomas and the Birth of the Prospect Park Alliance

Prospect Park's decline in the 1960s and 1970s was a manifestation of New York City's broader economic and cultural breakdown. Systemic funding shortages, consistently ineffective leadership, and violence (real and perceived) perpetuated the cycle of decay and kept people out of the park. Maintenance focused on repairing buildings, structures, and play equipment, while the slow decline of the landscape remained left largely unchecked. By the end of that period, things had gotten so bad that Makla proposed in an editorial in the *New York Daily News* that the only hope for restoring the landscape to the quality it deserved was to turn it over to the NPS.[135]

Following the brief and relatively ineffectual tenures of four successive parks commissioners, new mayor Edward Irving ("Ed") Koch appointed Gordon Davis to the position in 1978. This coincided with Prospect Park's nadir, during which annual visitation dropped to an all-time low of 1.7 million people. There was, however, a positive change in the works. A study of the park's condition was initiated in 1976 after a bronze statue depicting Columbia, the Greek winged goddess of victory, fell from atop Grand Army Plaza's Stanford White–designed Soldiers' and Sailors' Arch. "Had she not fallen out," said Koch, "we would not have been in a position to know how much needed to be done [in Prospect Park]."[136]

In 1980, the same year the National Register of Historic Places added Prospect Park to its roster, Koch announced a $10 million first-phase plan for the park's rehabilitation and preservation. Advocates for Prospect Park supported Koch's plan for what was envisioned as a decade-long, $60 million program that would include the restoration of several arches, bridges, buildings, fields, and roadways. Prominent landscapes such as the Long Meadow and the severely denuded woodlands of the Ravine were soon researched, documented, and planned as subjects of Historic Landscape Reports, which were intended to be the basis for restorations. "I cannot think of a better way for the Mayor to begin a New Year and a new decade," Koch proclaimed.[137]

To oversee this work, Davis took the significant step of using a community development block grant to hire a park administrator, a position parallel to the one held by Elizabeth Barlow Rogers in Central Park. He selected Tupper Thomas, who lived in the Crown Heights neighborhood of Brooklyn and had no specific park-related experience. She knew little about urban parks and even misspelled Olmsted's name on her application for the position.[138] She had, however, previously worked in housing and urban renewal in New York City and understood many of the issues facing the borough. From the beginning, her role involved changing the park's image, attracting visitors, and overseeing capital improvements. "We want the park used not only for recreation and relaxation but for historical, environmental and cultural activities," she stated several months after starting in the position.[139] She was starting from a deep hole, however. While deferred maintenance continued its assault on the park's structures and landscapes, the parks department's workforce fell from 6,000 employees in 1968 to a record low of 3,800 in 1980. In 1981 the first dedicated horticulture crew in twenty years was formed to tend Prospect Park's woodlands, meadows, and lawns.

Thomas and her effusive personality charmed community members, potential donors, local politicians, and even the press. The *New York Daily News* referred to her as "Joan of Park" in 1982 for her efforts championing the cherished landscape.[140] She initiated new programs and events and oversaw the renovation of several architectural structures and significant landscapes. The number of reported crimes in Prospect Park dropped by more than half during her first year on the job. By 1983, more than 4.4 million people were enjoying the park, the most in decades. The *New York Times* acknowledged the improvements made under her leadership in 1984, claiming, "The grass is so green in the Long Meadow in Prospect Park these days and the elm trees so lush that the greensward looks more like a 19th-century English landscape than a very urban park deep in the heart of Brooklyn."[141] Although improvements to the landscape were readily apparent, what differentiated the park at the time was the extent of research and strategizing devoted to planning its ongoing restoration and stewardship.[142]

By 1984, it was clear to Thomas that substantially more money would be needed to restore the park to the quality it and New Yorkers deserved. Rogers's example at Central Park helped her make a case that private donors were interested in contributing to the improvements of a public park. "Every park in the country has benefitted from that," Thomas once said.[143] In 1987, after several years of planning, the

Prospect Park Alliance was officially incorporated as a nonprofit organization. By this time, thanks largely to the efforts of Thomas, the City of New York had spent $17 million on park restorations. "I spent seven years . . . just trying to get the place fixed up enough so people would want to give money to it," said Thomas.[144] She was named the alliance's first president, which allowed her to raise private money while spending the $45 million the city planned to invest in the park over the next five years. Because she retained her position as park administrator, Thomas simultaneously served as a civil servant in the public sector and as a private employee responsible for park management, fundraising, community outreach, and public relations. Henry Christensen, a trust lawyer who led the citizen- and business-based process to create the alliance, was named the organization's chair.

Even though Central Park and Prospect Park share a prestigious design heritage, the fortunes of the alliance and the Central Park Conservancy have by no means been identical. While conservancy fundraisers can target the many Manhattan millionaires and billionaires piled into the high-rise luxury condos and apartments lining the park's perimeter, Prospect Park's surrounding donor pool is generally less prosperous. With limited funds and a desire to spend money directly on rehabilitation projects, the alliance started with a small staff, many of whom performed multiple tasks to accomplish everything needed.[145] Unlike many other such organizations that codify municipal partnerships with formal memoranda of understanding, Thomas chose to keep the arrangement less formal, signing agreements on practicalities such as concessions and licensing only as needed.

Restoring Brooklyn's Last Forest

Among the most significant projects undertaken by the Prospect Park Alliance was the rehabilitation and reforestation of the park's extensive natural areas. Central to this work was the restoration of the nearly 150-acre wild and natural woodland spine that runs through the hilly middle third of the park. Olmsted and Vaux conceived of the landscape as a lushly planted forest in which to escape from the city. "Although we cannot have wild mountain gorges, for instance, in the park, we may have rugged ravines shaded with trees, and picturesque with shrubs, the forms and arrangement of which remind us of mountain scenery," Olmsted wrote in 1866.[146] With its streams, boulders, and rustic bridges and structures, it must have seemed like a miniature version of the Adirondacks. It was not long before the quality of the landscape suffered, however, as it became beaten down and denuded by the millions of people who poured into the park annually as early as the late 1800s. Even as the number of visitors dropped dramatically in the twentieth century, the landscape continued to deteriorate. In the 1960s, a continuation of Moses's program to heighten security by eliminating shrubs and groundcover succeeded in opening up views while also leaving behind severe erosion and a nearly complete loss of topsoil on some slopes. As native trees died and were not replaced, light penetrating the canopy fostered optimal conditions for non-native and invasive tree species to thrive, creating a familiar and depressing lack of species diversity and essentially no ecological integrity.

Recognizing the severity of the problem, city staff engaged Walmsley & Company, a New York City landscape architecture firm specializing in historic preservation, to assist Thomas and her office in preparing the park's first Historic Landscape Report. In addition to the firm's president, Anthony Walmsley, the report's credited authors include Patricia O'Donnell and Charles Birnbaum, both of whom later emerged as leaders in landscape preservation and for whom this was surely formative work. The *First Historic Landscape Report for the Ravine, Prospect Park* is a thoroughly researched document that resembles the CLRs that subsequently became part of everyday practice for the NPS. The report presents a method for historic park preservation using a richly drawn history and existing conditions analyses as the basis for a prioritized list of restoration projects to be accomplished over time. Fundamental to the report's findings was that the original plan by Olmsted and Vaux retained its relevance as a master plan for the park's ongoing design and management. An extensive phase one project emerged from this initial study. Before work started, however, it became evident to alliance staff that the designed landscape had evolved into a natural landscape and required additional study from a purely ecological perspective.[147]

The fortunes of Prospect Park brightened dramatically once the alliance firmly established itself as the park's primary force of stewardship. Thomas created the Landscape Management Office to handle day-to-day operations and maintenance and, in a move that would prove significant, she hired Ed Toth as its director in 1989. He was a botanist from Ohio who had worked in the park as a service worker since 1985, so he knew the landscape well and likely understood the challenges he faced. Although some improvements occurred before this time, the park remained in desperate need of attention. In an early project, he and others in his office surveyed trees and created a database from which they developed a comprehensive tree planting master plan to guide replanting of specimen trees and those located outside the wooded areas.

Toth pushed for the use of native plants and for maintaining the park's woodlands as natural areas. In support of this aim, and with grant money from the Lila Wallace-Reader's Digest Fund,[148] the alliance formed a Natural Resources Crew dedicated to restoring and maintaining the park's woodlands under Toth's direction. In 1991, Toth and others embarked on a collaboration with the landscape architecture firm Andropogon Associates on the production of *A Landscape Management Plan for the Natural Areas of Prospect Park,* which was paid for by a grant from the National Endowment for the Arts. That document, which was completed in 1994, establishes a detailed management program for the reforestation and management of nearly 250 acres of woods—100 along the perimeter and 150 in the interior. "The collapse of these woodland ecosystems is at hand," the report warns, as it makes clear the goal of establishing an ecologically healthy and sustainable landscape rather than an artificial or design-focused "pristine reserve." [149] One recommendation highlights the use of native plants to foster biodiversity and encourage regeneration—a practice at odds with Olmsted's preference for exotic plants that helped achieve his desired aesthetic. By emphasizing a natural resources approach, the report helped establish a broader paradigm for the care of the urban landscape. In the report's acknowledgments, Toth thanks Thomas for her leadership and for providing "the atmosphere within which innovation is not only possible, but nurtured."[150]

To complement the operational aspect of the Landscape Management Office, Thomas formed the Office of Design and Construction to lead the comprehensive restoration envisioned for the park. She was frustrated by the layers of bureaucracy in the parks department and believed Prospect Park deserved targeted focus rather than being one of many within a huge network of parks.[151] The first hire in the design office was Christian Zimmerman, a recent landscape architecture graduate from the University of Idaho who had decided to move to the East Coast while most of his classmates relocated to San Francisco, Portland, or Seattle. With a personal interest in history, he started at Prospect Park in 1990 after working as a designer for the

New York City Parks Department for two years. He soon found himself leading the design for a complex, twenty-five-year, $20 million woodland restoration project. The first phase of the multiphase project was restoring around twenty acres of woodlands and waterways in the Ravine, which Olmsted described as "an irregular course, with numerous small rapids, shoots and eddies, among rocks and ferns."[152] By this time, the landscape looked little like that.

The project tested Thomas's belief in the efficiency and effectiveness of using a focused in-house team rather than hired consultants. As stated by Mary Fox, the alliance's vice president for capital and planning, "You get a chance to see a problem; you go out and raise your own funds; you design solutions, write up the documents, get bids, oversee construction. Then you live with the results."[153] In other words, the same people who conceive a project also design it, build it, and, more important, maintain it and manage its continual evolution. While Zimmerman and others worked on the design and implementation of the Ravine, the Natural Resources Crew planted nearly 14,000 trees, more than 12,000 shrubs, and over 200,000 herbaceous plants in the woodlands beginning in 1994.[154]

In 1996, during the project, crime in Prospect Park re-entered the headlines as reports of an alarming rape, the fourth reported within the park that year, stoked fears about safety in the woodlands and renewed discussions with police about security. Denis Hamill, a columnist for the *Daily News* who had once referred to Prospect Park as "the Sherwood Forest of my youth, where you can jog, walk, picnic, make out, crack a baseball, toss a Frisbee and go boating,"[155] adopted an ominous tone when describing the landscape into which the victim, a female jogger, was dragged. "The dirt path leading into the woods suddenly resembles a corridor to the gallows, something out of a Nathaniel Hawthorne nightmare," he wrote.[156] Citizens in the community expressed concern over the incident and made evident that no matter how potent design legacy and ecological integrity were as motivators, creating a safe environment mattered more.

Without the original working drawings, which had been destroyed during the Moses era, Zimmerman and his team researched historical drawings and photographs, the writings of the two original designers, and numerous other documents to "try to get inside the heads of Olmsted and Vaux, to figure out their intent."[157] Their plan sought to replicate what Olmsted and Vaux had created as closely as pos-

Fig. 53. An area of the park that suffered the most from lack of maintenance was the heavily wooded and naturalistic Ravine, which exhibited erosion and degradation by the 1970s. (Prospect Park Alliance)

sible while balancing authenticity with safety, current environmental factors, and innovative practice. For instance, the once-picturesque watercourse had long ago ceased to function as intended. For all his talents as a designer, Olmsted, like Farrand at Dumbarton Oaks, proved ill-equipped at engineering hydrology. "He did it beautifully, but didn't know how to do it in a way that would hold up over time," remarked Thomas.[158]

Construction lasted several years, but by the early 2000s, the waterways running through the Ravine had been restored and opened to the public. This early phase of the project cost around $6 million, much of which went to stockpiling and relaying massive boulders and removing 12,000 cubic yards of silt from the main pools. The extensive planting program included nearly 200,000 vines, ferns, grasses, and wildflowers; more than 10,000 shrubs; almost 30,000 aquatic plants; and over 9,000 trees.[159] In the process, a long-hidden waterfall named Ambergrill Falls was redis-

Fig. 54. Extensive restorations of the Ravine beginning in the 1990s reestablished the lushness and mystery originally envisioned by Olmsted and Vaux. (Photo by author)

covered and rebuilt using contemporary techniques to recreate the original vision. Fencing was kept in place for more than ten years to protect the fragile plantings through an establishment period, a practice common in the stewardship of heavily used public landscapes.

The project's next phase tackled the stabilization and reforestation of fifty acres of adjacent woodlands. Slopes were rebuilt with imported topsoil held in place by erosion control fabric and log cribbing similar to what was used in Dumbarton Oaks Park's Beech Grove. The Landscape Management Office, now without Toth, who had left in 1998, expanded these reforestation efforts to one hundred acres of woodlands outside the project boundaries. After these efforts to remove invasive plants, stabilize slopes, and plant hundreds of thousands of native trees, shrubs, and other plants, Brooklyn's last remaining forest was a healthy and biodiverse ecosystem for the first time since early American settlers razed the old-growth forest.

Changing of the Guard

In 2010, after thirty years working on behalf of Prospect Park, Tupper Thomas announced her retirement from the Prospect Park Alliance. Under her leadership the organization's annual budget had climbed from $200,000 to around $9 million and the number of annual visitors to the park had more than quadrupled. Adrian Benepe, New York City's commissioner of parks at the time, referred to a walk through those landscapes once lost and restored on her watch, "from Grand Army Plaza through Long Meadow into the Ravine, past the rivers and waterfalls, to the Nethermead, and to the lake and boathouse," as "her masterpiece."[160] This was significant coming from Benepe, who knew New York City parks well. He had worked in various roles since starting as a sixteen-year-old seasonal employee in the parks department, and by 1980, the year Thomas started at Prospect Park, he was a ranger in Central Park. His statement emphasizes that Thomas, like Olmsted, Vaux, Stranahan, Culyer, and Moses before her, was cemented in the legacy of Prospect Park.

Thomas worked for several months on a final goal to raise the last $20 million of a projected $70 million project called Lakeside. She had been fundraising for six years to build a new structure for year-round skating that would replace the Kate Wollman Rink, which, in addition to being used only for a short period during the year, was outdated and required frequent maintenance. Also included in the project was reconstruction of the Lake and its shoreline per Olmsted and Vaux's plan. Without a pool of wealthy donors, Thomas relied heavily on support from elected officials for the first $50 million, notably an $11.5 million contribution from Brooklyn's borough president.

By the time of her January 2011 retirement date, the Lakeside project was more than $15 million short, a reflection of the still struggling economy following the 2008 downturn. Emily Lloyd, a former commissioner of the New York City Department of Environmental Protection, took over for Thomas in the dual role of Prospect Park administrator and Prospect Park Alliance president. Lloyd acknowledged the fundraising challenges and other issues the park faced. Along with raising money for Lakeside, one of her stated priorities was to control the increased and unauthorized pedestrian traffic within the still-fenced, recently restored woodlands.[161] She seems to have recognized yet been undaunted by the magnitude of Thomas's shadow, saying shortly after she started, "We of course want to maintain what Tupper accomplished, and because I'm a different person we have to find different ways of doing that."[162]

An early challenge for Lloyd was overseeing progress on Lakeside, which was Prospect Park's first major construction project in five decades. Costs had jumped to $74 million by this time, amplifying complaints by some that the money would be better spent on other park initiatives, including maintenance. The existing rink was demolished and the site was converted back into the circular cove, and Music Island in the Lake was rebuilt. At the same time, two other historic spaces—the strolling Esplanade along the shoreline and the Concert Grove gathering spot—were rebuilt. Zimmerman once again led the in-house team on the landscape design, creating sweeping curvilinear paths, lush native plantings, and historically accurate or appropriate materials and details that fit into the park's cultural and aesthetic context.

While the revised landscape conveys much of the authentic historic preservation for the twenty-six-acre site, the new skating rink structure, which the architect claimed was "not so much a building as a place in the park," introduced twenty-first-century architecture into the landscape.[163] Much like the Beaux-Arts architecture from a century earlier that was now firmly ingrained in the fabric of the park, this insertion of a contemporary structure reflected changing times and tastes. Ten million people visited Prospect Park in 2012—around twice as many as in the early 1990s—and the need for updated and expanded facilities could not have come as a surprise. In addition to a state-of-the-art ice-skating rink perched on the edge of the Lake, a second rink was built beneath the protective cover of a giant roof, making it available for roller skating even in the hot summer months, during which the outdoor rink became a water park.

Early concept plans drawn up by in-house architects had proposed a historicist design directly referencing the quaint nearby Vaux-designed Concert Grove Pavilion. The jewel-like quality of the pavilion did not work at the scale needed for the rink, however, and an unabashedly contemporary approach was pursued.[164] The ultimate design presents a massive, low-slung rectangular roof required to meet energy requirements set by Leadership in Energy and Environmental Design (LEED). Although it meets sustainability goals, the roof's strict geometric form awkwardly contrasts with the organic curves of the naturalistic landscape. A design critic commented in the *New Yorker* that the cantilevered roof "stands out like a billboard" even as the entire building is partially contained and shielded within a landscaped berm.[165] While the notable departure of the architectural style drew reactions, one

Fig. 55. The LeFrak Center at Lakeside replaced the Kate Wollman Rink with a historically and contextually sensitive facility that provides all-season recreation and restored the original quality of the Lake's shoreline. The rigid rectilinear geometry of the Center's roof contrasts the sinuous paths that hearken back to Olmsted and Vaux and navigate the site's gentle topography. (Michael Moran)

element of the design more in keeping with the park's original intent, massive cut stone blocks, quietly became part of an emerging contemporary landscape design vocabulary. In early 2021, a new entrance to the park on the eastern edge—the first in more than seventy years—incorporated similarly honed granite blocks along with rustic boulders to reinforce the twenty-first-century design layer.

Return to the Vale

Following the construction of Lakeside, Prospect Park Alliance turned its attention to two adjacent landscapes in the little-visited northeastern corner of the park—the Rose Garden and the Vale of Cashmere. Olmsted and Vaux originally conceived of the space as a single naturalistic area for children. The landscape was later divided and transformed with a formal flair by McKim, Mead & White. The area appears to have been well maintained, for a while at least. In 1945, a letter to the editor of the *Brooklyn Daily Eagle* states about the Vale: "Here you see a pool of water surrounded by huge rhododendrons, flowering dogwood and azaleas in full bloom to delight the eye and make glad the heart."[166]

Conditions deteriorated over the next two decades, and by 1965 the garden was described as "a jungle of wild vegetation, with a lake from which the water has been drained, a decorative fountain stolen and carved granite balustrades removed."[167] Mayor Lindsay decried the condition of the area, calling the Vale a "dismal swamp" and complaining that the basins in the Rose Garden had become receptacles for beer cans.[168] Thanks to his push for improvements to Prospect Park and in anticipation of its upcoming centennial celebration in 1966, a $450,000 enhancement to the area was undertaken. In addition to hardening most of the remaining natural edge of the Children's Pool with concrete block retaining walls and installing brick pavers in the Vale of Cashmere—an anomaly in the park—the project replaced the Rose Garden's granite basins with poured concrete and added fountains. The quality of construction was reportedly poor, and within five years the Pool was dry and the fountains no longer worked. The curator of the park at the time blamed a lack of funding and maintenance staff, commenting in 1974, "If Prospect Park were a private estate with lots of money there probably would be little problem with the fountains."[169]

This was, of course, aligned with the neglect and deterioration experienced across the park as the economy languished, crime increased, and the public stayed away. The Vale became a forgotten landscape, surprisingly close to the popular Long Meadow yet concealed within densifying woodlands and enveloping overgrowth. It became known as a spot for making drug deals and cruising for casual sex. Many people avoided the landscape, perceiving it as seedy and dangerous. In the pre-AIDS era of the 1970s, it became a haven for African American and Latino gay and bisexual men. In the forward to a collection of landscape photographs and intimate portraits of men by the photographer Thomas Roma taken between 2008 and 2011 and titled *In the Vale of Cashmere,* the poet and author G. Winston James writes, "Gathering at the fountains and along the benches in the Rose Garden above the Vale of Cashmere during the day is as much (or more) an act of fellowshipping and strengthening of community ties as it is one of cruising for sex."[170]

Even though the Friends of Prospect Park reportedly spent $40,000 improving the Vale's landscape in the 1980s, a general lack of active management preserved the space as a hidden secret until late October 2012, when Hurricane Sandy violently blew through the region. Visceral and widely publicized images depicted the extensive damage inflicted by Sandy on New Jersey and New York coastal communities, yet inland areas also experienced the impact. In Prospect Park, five hundred trees were toppled, including around fifty on the wooded slopes of the Vale of Cashmere. With a $1.2 million grant from the NPS and the State of New York, the alliance undertook a project to clear storm debris, repair damage, and restore the woodlands in the Vale. During the summers of 2016 and 2017, when once shaded slopes faced the onslaught of invasive volunteer plant species, tribes of goats were imported to devour and remove the threat of emerging weeds. The restoration project rejuvenated the landscape's ecology by planting more than 20,000 native trees, shrubs, and herbaceous plants. While this work successfully addressed the woodlands, the formal portions of the Vale and the Rose Garden remained in need of enhancement.

The restoration of the area, waiting in the wings, requires different approaches. Because the pool at the center of the Vale retained much of the integrity of both the original Olmsted and Vaux design and the alterations made by McKim, Mead & White, a historic preservation perspective was considered appropriate. Zimmerman, as lead designer, evaluates such questions and weighs precedent and current conditions before recommending the extent of restoration versus reconstruction or redesign. As with all projects, the design proposal must follow multiple steps and proceed through a complex bureaucracy. It first needs approval from the Brooklyn branch of the parks department, then the deputy commissioner of capital projects, the commissioner of parks, the borough board, and the Landmarks Preservation Commission. Ultimate approval comes from the Public Design Commission, which has jurisdiction over permanent structures, landscape architecture, and art in city-owned properties. Thanks to an earned reputation for quality based on numerous successful projects, design proposals brought forth by the alliance typically carry a gravitas throughout the process.

Unlike the fountain in the Lower Vale, the landscapes of the Upper Vale and the Rose Garden were considered a blank slate, with no detailed original design apart from a location for play equipment. Because of this, it was determined acceptable

Fig. 56. This contemporary view of the Vale conveys a sense of neglect and wildness while depicting the landscape's temporal layering. Design elements as envisioned by both Olmsted and Vaux and McKim, Mead & White are evident. (Photo by author)

for a new design to meet the contemporary needs of, in particular, the communities of color, young people, low-income families, and recent immigrants who live adjacent to that edge of the park. In 2018 the alliance made public the results of an extensive community engagement process that endorsed preserving the secluded and meditative character of the site while incorporating innovative play and educational opportunities. With input from the community, alliance staff generated rendered concept plans to help raise $40 million for the restoration of the Rose Garden and the Vale. Proposed designs as of 2022 include an architecturally understated restroom building with clean lines that reinforce the park's contemporary layer established at Lakeside, which was renamed LeFrak Center at Lakeside in 2011. Plans for the landscape incorporate a focus on ecology, which Zimmerman considers a major part of Prospect Park's twenty-first-century aesthetic.

Criticism of the plan emerged at a 2023 Landmarks Preservation Commission meeting. Many of the complaints focused on the restroom building, which was accused of looking too heavy-handed and being an inappropriate focal point in what was intended to be an open space. Other comments arose about the layout, with one commissioner expressing concern about the unified space being broken down into distinct zones. He asked, "Help me understand why, from an appropriateness perspective, it's okay to break up the space into what basically will become three spaces: the pollinator garden, the discovery area, and the new lawn. Why is that, from an historical preservation perspective, the right choice to make?"[171] A representative of the Victorian Society New York followed with a similar rebuke, commenting, "Olmsted would never put a building for these utilitarian purposes at the most visible point of the landscape of this type."[172] One hundred fifty years later, Olmsted and Vaux continue to rally support.

A more nuanced perspective was voiced at the meeting by Roma, the photographer who had documented the Vale as a cruising space. More concerned about respecting how the landscape had been used historically than with the perceived sanctity of a particular design legacy, he emphasized the importance of ensuring the "socially significant LGBTQ landscape is preserved and interpreted, rather than erased from history."[173] A preservation advocate agreed with Roma's concern about turning the tranquil part of the park into an activity hub, stating that the people representing the alliance "are tactically using a combination of stranger danger and gay panic to [reimagine] this natural geological hollow, a secluded quiet and poetic corner—unlike anything in New York—into a busy, activated, kid-safe hub to chase away 'undesirables' and 'blight.'"[174] While protecting Olmsted and Vaux's designs is being practiced as a discipline unto itself at this point, addressing cultural history can prove divisive and challenging when plotting the future of a public landscape. This is especially true when the people involved—gay men of color, in this case—fall outside accepted or outdated societal norms.

Contemporary Issues

While people struggle to appropriately address the cultural issues associated with Prospect Park as it evolves, climate change persists as an existential threat. Rising seas, stronger storms, and higher temperatures translate into trouble for such tree-filled urban oases, which become increasingly important resources for providing citizens with access to nature while also making the air cleaner and cities cooler.

Olmsted surely understood the health benefits of urban green space, yet his version of naturalistic woodlands and landscapes was based primarily on visual artistry rather than on ecology. The Prospect Park Alliance changed this mentality beginning with the work of Ed Toth and increasingly pursued a more authentic version of nature rooted in science and research. Pollinator gardens, functioning meadows, native woodlands, and habitat corridors fill the less developed interior of the park while more authentically Olmsted features—the Long Meadow, the Lake, and managed woods—sit on the more public perimeter landscape. Evolving management practices include a program to introduce plant species that increase diversity and support wildlife and eliminating the planting of trees predicted to decline as a result of climate change. At the same time, warming temperatures have meant the planting of new species such as the southern magnolia, which was once limited to warmer regions. An elevated threat of invasive plants and destructive pests with rising rates of carbon dioxide emissions, longer growing seasons, and warmer winters has forced adaptation and new methods of monitoring to proactively protect the park's native ecology.

In the fall of 2021, Prospect Park embraced its role as a participant in the fight against climate change by hosting virtual and in-person events as part of NYC Climate Week, during which more than five hundred related events were held across the city. In addition to lectures and tours, the alliance led hands-on activities geared toward children to help to foster a sense of environmental responsibility among a population that would one day inherit the stewardship of this and other parks. In the meantime, after three decades overseeing the restoration of Prospect Park, Zimmerman was keeping an eye out for a more immediate successor to the role Tupper Thomas had conceived, and he had made his own. He sought a landscape architect interested in historic preservation who could appreciate and identify the layers of history throughout the park.[175] Success in such roles requires more than a specific knowledge base. Historical landscape stewardship is neither a course of study nor a recognized discipline. To be effective, professionals working in positions such as Zimmerman's need an inherent passion and an accrued professional ethic, qualities not easily found. When it happens, the transition to his replacement will continue the cycle of leadership that the alliance, along with most other organizations and institutions, periodically experience. Some professionals in leaderships roles thrive as game-changers, while others succeed at perpetuating the status quo. Both types can be effective, depending on the issues faced during their tenure.

A significant leadership change occurred at Prospect Park in the fall of 2022, when Sue Donoghue, who had replaced Lloyd as president eight years earlier, left the alliance for an elevated position as commissioner of the New York City Department of Parks and Recreation. Morgan Monaco, who has a long history with New York's parks, starting with a high school internship under Adrian Benepe, was subsequently introduced as the organization's first Black leader. Monaco, who inherited a hefty $13 million operating budget and a $25 million endowment, said it was "a dream come true to be able to steward one of the most significant parks in our city."[176]

By this time the Prospect Park Alliance was a seasoned organization with a firm grasp on what it takes to effectively manage the park. Monaco carries on the tradition of fundraising, community engagement, and leadership started by Thomas. Although issues such as climate change will continue to challenge people working for and in the park, the incessant turmoil that defines much of its history seems finally to be a thing of the past. Most important, despite everything that has changed—and continues to change—the landscape is as beautiful and as valuable to the community as Stranahan initially envisioned. Many of the elements Olmsted criticized as being assaults on the landscape are now embedded in the park's fabric and have become critical to its use and enjoyment.

Although Central Park Conservancy receives more press, accolades, and, certainly, financial support, Prospect Park Alliance is perhaps a more relatable example for the many nonprofit groups and conservancies that continue to form in support of designed landscapes. Lacking the abundance of Manhattan billionaires lining Central Park's perimeter, the alliance's success has been slower to achieve and has followed a more realistic grassroots path. Not that things were necessarily easy at Central Park, but the process there was surely expedited in comparison to its Brooklyn neighbor. Despite their impressive progress, both organizations face continued fiscal challenges. In 2023, for instance, New York City's mayor, Eric Adams, mandated a 12 percent cut to the budgets assigned to the care of the city's 1,700 parks as a result of a multibillion-dollar budget deficit attributed to, among other things, the cessation of federal pandemic-era aid. While Prospect Park and, particularly, Central Park are in a better position than most of the other, less-well-funded parks, they will likely suffer somewhat from the cuts. Such is the reality facing all public landscapes at some point during their evolution.

Even with such challenges, the alliance's precedent story surely continues to inspire similar groups, the Dumbarton Oaks Park Conservancy among them. While the two parks share little in the way of scale—Prospect Park is incredibly more complex owing to its size and program—they both struggle with the realities of maintaining naturalistic landscapes while holding on to cherished design legacies in the face of constant change. It may be cliché to say such stewardship work requires passion, but it is true, especially for public parks, which face extra layers of financial and management challenges beyond those faced by private or institutional landscapes.

CONCLUSION

In 1891, Frederick Law Olmsted wrote to the Boston architect Henry Van Brunt about the frustrations that come with working on public parks and the inherent differences between designing buildings and designing landscapes:

> Suppose that you had been commissioned to build a really grand opera house; that after the construction work had been nearly completed and your scheme of decoration fully designed, you should be instructed that the building was to be used on Sundays as a Baptist Tabernacle, and that a suitable place must be made for a huge organ, a pulpit and a dipping pool. Then at intervals afterwards, you should be advised that it must be so re-fitted and furnished that parts of it could be used for a court room, a jail, a concert hall, hotel, skating rink, for surgical cliniques, for a circus, dog show, drill room, ball room, railway station and shot tower? What chance would you see for making a fine affair, in any respect, of your building?
>
> Again, suppose that once in three or four years an ordinary house painter and paper hanger, or even a theatrical scene painter, should be called in to

revamp and improve your decoration of the auditorium? Could you think of such a history without indignation and disgust?

But that, more or less, is what is nearly always going on with public parks. Pardon me if I overwhelm you; it is a matter of chronic anger with me.[1]

It is easy to read Olmsted's statement as the product of a dented ego, but his anger likely extended to sites beyond those places he had designed. He passionately believed that urban landscapes such as Central Park and Prospect Park best served their communities by providing undisturbed stretches of natural scenery, and altering that intent by inserting buildings, monuments, or playing fields undercut the landscapes' cultural and aesthetic value. His letter expectedly exalts the designer who created the "really grand" opera house yet surprisingly suggests that those who dare to implement even minor changes—the "ordinary house painter"—are plebeians surely unfit for the task and whose work should inspire "indignation and disgust."

What Olmsted fails to acknowledge (understand?) in this specific situation is that although the pleasure ground, his version of the opera house, appealed to a mid-nineteenth-century audience, its relevance changed along with the needs of its users. By the 1890s, urban dwellers were demanding more from their public spaces, such as land for active recreation and cultural and ceremonial engagement, even if accommodating these new uses challenged a respected design or designer's vision. His suggestion that public landscapes somehow equated to buildings and thus should be treated with a similar hands-off approach is fundamentally wrong. We know that he considered landscape architecture a fine art on par with other arts, including built architecture, which is valid in concept yet forges a false equivalency between static works of art and dynamic landscapes.

This is not just about Olmsted. Generations of landscape architects and designers have complained that their clients alter or fail to adequately maintain landscapes they designed, which in their view threatens the integrity of what they helped create. In an insightful 2013 essay, the celebrated landscape architect Michael Van Valkenburgh explains: "Neglect or underperformance of landscape management is a massive problem for designed landscapes," while seemingly assigning some blame to owners or "caretakers, who, at least partially, become [a landscape's] new designers" once consultants are out of the picture.[2] He says there is a fundamental and detrimental disconnect between the specific tasks landscape architects are hired for on the front end of a project and the extended cycle of change that kicks in once management of the landscape is placed back in the owner's hands. To remedy this, he proposes that designers remain under contract for a longer time so that they can observe a landscape as it becomes established and tweak it as needed. From his perspective, "Both clients and designers who truly care about the long-term quality of their projects will welcome this change."[3] Van Valkenburgh has good company in this line of thinking, as Dan Kiley once wrote, "As architects of a living medium, landscape designers have a mandate to guide their work through its successive evolutionary stages—or relinquish its integrity."[4] This may be true on a certain level, but as we have seen in the various case studies, there are many alternatives to that sometimes unachievable hands-on approach.

The ongoing role Van Valkenburgh supports is not dissimilar to how Paul Chalfin worked at Vizcaya and Thomas Church worked at Stanford, except both men could almost be classified as in-house designers rather than outside consultants because of how embedded they were in their respective landscapes. An even more successful model is exemplified by David Williston and Edward Pryce at Tuskegee and John Noyes at the Missouri Botanical Garden, all three of whom acted both as client and consultant, designer and steward. No outside consultant could gain the same understanding of the intricacies involved in a landscape as it ages, especially when they just visit from time to time. I came to this understanding while working at Duke. While I value the exceptional work performed by consultants on campus, my in-house position gives me a unique perspective and insights into best ways to address alterations as they arise.

I do agree with Van Valkenburgh's assessment that maintaining some relationship between owner and design consultant is a good thing for all involved. Of course, this, as well as hiring in-house design staff, requires financial resources not always available. Such is especially true for public landscapes, which are often underfunded by public monies and increasingly rely on support from private sources. It is not just about money, however. As seen with Brooklyn's Prospect Park and several other landscapes explored in this book, public spaces face many unusual challenges. Communities inherit a sense of ownership of public open spaces that they would never feel for a civic building, no matter how beautifully designed or preserved. There is an inherent sanctity to the public realm, and it should be regarded as a good thing and not an assault when citizens push for change in parks and other spaces. To reiterate what the chapters in this book have repeatedly asserted through examples, parks, gardens, campuses, neighborhoods, and other landscapes inevitably evolve. If any of these places fail to meet the needs of the people for whom they were created, should not striving to remedy that shortcoming take precedence over the potential loss of the stamp of a particular designer?

Of course, there is a massive gray area in which legitimate claims of design integrity deserve serious evaluation, along with the push for potential deviations and transformations. The voices of the Cultural Landscape Foundation, Olmsted enthusiasts (and the like), and historians arguing loudly on the side of preservation are desperately needed, especially because saving or adapting something established can appear less appealing than providing something shiny and new. Equally critical are the opinions and influence of people advocating adaptations of historical and cultural landscapes. Typically, neither side is 100 percent right, and determining how to effectively manage controversies and divergent opinions falls to professional stewards embedded in a specific landscape and uniquely capable of making decisions that simultaneously respect the past and accommodate the future.

This view underscores the need to evaluate designed landscapes as malleable products of evolution rather than more static works of design. What we find today in New York's Central Park, for example, is a far better twenty-first-century urban park than the pastoral place envisioned by Olmsted and Vaux. The Delacourt Theater, the Conservatory Garden, the Loeb Boathouse, the Heckscher Ballfields, and the Shakespeare Garden are all among the landscapes and structures not original to the Greensward Plan but likely to be enjoyed by today's visitors with little concern for their provenance. Each of these beloved features emerged from trajectory-changing decisions to alter or destroy something historically authentic. Proposals to raze of any one of them to restore an original Olmsted and Vaux feature would surely be met with protests as loud as any of those that have favored preservation. And that is as it should be.

This attitude may seem cavalier or even irresponsible to those people who believe the quality of a historic landscape is directly tied to the enduring clarity of its original design or vision. But it is precisely this temporal and experiential dynamism that makes such places special. The richness of the landscapes and the reason we like them so much is not because of the accuracy of the history they embody but rather because of the aggregate effect of modifying tweaks and corrections implemented over time in response to myriad inflection points and the processes of evolution. Incorporating such tweaks and corrections into a historic landscape is rarely easy and does not happen on its own. The challenge comes in effectively managing these landscapes so that they retain or amplify their relevance rather than become so rooted in the past that they become obsolete.

NOTES

Abbreviations

Dumbarton Oaks OHP	Dumbarton Oaks Research Library, Oral History Project. Dumbarton Oaks, Trustees for Harvard University, Washington, DC
Dumbarton Oaks RBC	Dumbarton Oaks Research Library, Rare Book Collection. Dumbarton Oaks, Trustees for Harvard University, Washington, DC
LoC FLOP	Library of Congress, Frederick Law Olmsted Papers, Washington, DC
SUL	Stanford University Libraries, Department of Special Collections and University Archives, Stanford, CA
VMGA	Vizcaya Museum and Gardens Archives, Miami, FL

Epigraphs

Kevin Lynch, *What Time Is This Place?* (Cambridge, MA: MIT Press, 1972), 237.

Dan Kiley and Jane Amidon, *Dan Kiley: The Complete Works of America's Master Landscape Architect* (Boston: Little, Brown, 1999), 96.

Charles Birnbaum, ed., *Preserving Modern Landscape Architecture II* (Washington, DC: Spacemaker Press, 2004), 24.

Introduction

1. The Philadelphia-based architect, Horace Trumbauer, was hired by the industrialist James Duke to design the university's campus and its buildings. In the 1980s, it became evident that much of the design for Duke University's buildings was done by the architect Julian Abele, who was the first African American to graduate from the University of Pennsylvania's school of architecture.

2. Abbott, "Historic Preservation," 153.

3. Olin quoted in Birnbaum, *Preserving Modern Landscape Architecture II,* 16.

4. City Hall was designed by the architectural firm Kallmann, McKinnell & Knowles.

5. Cobb quoted in Coleman, "Wanted: Ideas for City Hall Plaza." The name of the firm changed to Pei, Cobb, Freed & Partners in 1989.

6. Von Eckardt, "Turned to Stone."

7. T. B. Adams, "Boston Achieves a Triumph in Architecture."

8. R. Campbell, "The Trouble at City Hall Plaza."

9. Menzies, "Three 'Spaces' That Weaken the Cityscape."

10. Kay, "Common, Public Garden Are a Mess."

11. Kay, "Boston Common Sprucing Up at a Snail's Pace—but It Still Shows." In this piece, written nearly a year after her previous column, Kay refers to Falck as chief booster. The Boston Redevelopment Authority had been around since 1957, when it was created to oversee all urban renewal efforts in the city.

12. Coleman, "Wanted: Ideas for City Hall Plaza."

13. Black, "A Floral Facelift for City Hall Plaza."

14. R. Campbell, "Giving Purpose to City Hall Plaza."

15. This statement is supported by numerous articles and opinion pieces in the *Boston Globe* and other local newspapers from the 1980s to the 2000s voicing opposition to selling off the property.

16. "Another Central Park Danger." "Friends of Central Park" as the term is used here refers generally to supporters of the park rather than to an organized group calling itself the Friends of Central Park.

17. Birnbaum, "Preservation Brief 36: Protecting Cultural Landscapes." The brief identifies the primary issues that need to be addressed when considering how a particular cultural landscape should be treated. It is intended to give preservation professionals, cultural resource managers, and historic-property owners a process for preserving historic designed landscapes and vernacular landscapes. This process begins with the planning and research of the landscape, during

which its history, significance, and conditions are documented. Based on an analysis and evaluation, a treatment plan can be developed. Preservation Brief 36 provides a framework and guidance for undertaking projects to ensure a successful balance between historic preservation and change.

18. Thompson, "Land Matters," 17.

19. Cultural Landscape Foundation, "Landslide 2008." "Among the many upgrades that might be envisioned are new construction to better define the plaza, the introduction of fine scale elements to add to its visual texture, new landscaping, and regulatory changes that could encourage commercial interventions such as . . . a hotel alongside the plaza."

20. A. Ryan, "City Seeks Out Ideas to Repurpose Brick-bound Plaza."

21. City of Boston, *Rethink City Hall.*

22. While the master plan does identify an area for "play for all ages," it is clearly not meant to be a playground for children.

23. Jost et al., "Charlotte Shouldn't Demolish Its Waterfall Park Uptown."

24. *Preservation* is defined as the act or process of applying measures necessary to sustain the existing form, integrity, and materials of a historic property. Work, including preliminary measures to protect and stabilize the property, generally focuses on the ongoing maintenance and repair of historical materials and features rather than on extensive replacement and new construction. New additions are not within the scope of this treatment; however, the limited and sensitive upgrading of mechanical, electrical, and plumbing systems and other code-required work to make properties functional is appropriate within a preservation project.

Rehabilitation is defined as the act or process of making possible a compatible use for a property through repair, alterations, and additions while preserving those portions or features that convey its historical or cultural values.

Restoration is defined as the act or process of accurately depicting the form, features, and character of a property as it appeared at a particular period of time by means of the removal of features from other periods in its history and the reconstruction of missing features from the restoration period. The limited and sensitive upgrading of mechanical, electrical, and plumbing systems and other code-required work to make properties functional is appropriate in a restoration project.

Reconstruction is defined as the act or process of depicting, by means of new construction, the form, features, and detailing of a nonsurviving site, landscape, building, structure, or object for the purpose of replicating its appearance at a specific period of time and in its historic location. These definitions appear in the Department of the Interior's "Standards for the Treatment of Historic Properties" (https://www.ecfr.gov/current/title-36/chapter-I/part-68/section-68.2).

1. Gardens

Epigraph: Robert Pogue Harrison, *Gardens: An Essay on the Human Condition* (Chicago: University of Chicago Press, 2008), 7. Harrison is a professor of French and Italian literature at Stanford University.

1. American Botanic Gardens Association, "About Us," accessed December 27, 2020, https://www.publicgardens.org/about-us.

2. Findlen, "Anatomy Theaters, Botanical Gardens, and Natural History Collections," 280.

3. Baker, "Rensselaer Polytechnic Institute and the Beginnings of Science in the United States."

4. Rudolph, "History of the Botanical Teaching Laboratory in the United States," 663.

5. Pinckney, "'Still Mindful of the English Way,'" 155. Creating the famous landscape began in 1741 and took about a decade to complete.

6. Downing, *A Treatise on the Theory and Practice of Landscape Gardening,* 18.

7. Le Nôtre was a seventeenth-century landscape architect and the principal gardener of King Louis XIV of France. He designed the Baroque gardens of Versailles, along with numerous other landscapes.

8. Winterthur was designed by DuPont along with the landscape architect Marian Cruger Coffin. Dumbarton Oaks was designed by Mildred Bliss and the landscape architect Beatrix Farrand.

9. The Trustees of Reservations was founded in 1891 by the landscape architect Charles Eliot as a nonprofit organization to preserve, conserve, and maintain landscapes and places for the use of the public.

10. The Ladies Garden Club of Athens was awarded official status in 1936 by the National Council of State Garden Clubs.

11. The Garden Club of America was initially named the Garden Guild.

12. Scribner, "The Garden Club of America," 269.

13. Seale, *The Garden Club of America,* 17.

14. Weller, "National Arboretum to Serve as Plant and Tree Laboratory."

15. Garden Club of America, "National Affairs & Legislation and Conservation Committees." The quotation is attributed to Mrs. Thomas M. Waller.

16. Cabot quoted in Dickey, *Outstanding American Gardens,* 9.

17. Mia Amato, "Getting Rid of White Fly, Rose Rust Is Murder, She Wrote," *San Francisco Examiner,* April 12, 1992. Facilitating preservation efforts through partnerships and with innovative methods would become common practice for the organization.

18. Adezio, "A Garden Conservancy Project." Adezio, the president of the conservancy at the time, lauded the "on the job training" provided by the project.

19. Adezio quoted in Raver, "Human Nature."

20. Adezio, "The Garden Conservancy Mission, Revisited," 5.

21. Ibid.

22. Garden Conservancy, "Garden Preservation Comes of Age."

23. Ibid.

24. Garden Conservancy, "#OpenDays25."

25. Adezio quoted in Raver, "Garden Notebook," 4.

26. Missouri Botanical Garden, "A Brief History of the Missouri Botanical Garden."

27. Dimmock, "Henry Shaw," 12.

28. Ibid., 12. Although Dimmock does not elaborate on the quotation, he does say this was "according to his [Shaw's] own statement," so it was likely written in one of Shaw's journals from the trip.

29. "Some Early History of the Garden."

30. Rudolph, "One Hundred Years of the Missouri Botanical Garden," 2.

31. "An Act to Enable Henry Shaw to Convey of Devise to Trustees Certain Lands."

32. Flaherty, *Henry Shaw,* 123. According to Faherty, Olmsted signed the garden's guestbook on April 3, 1863.

33. Ibid., 181.

34. The figure of $2.5 million comes from an encomium published in the *St. Louis Dispatch* on August 26, 1889, "Passed Away Quietly." The higher figure is from "Henry Shaw Is No More," *Chicago Tribune,* August 26, 1889.

35. MacAdam, *Tower Grove Park of the City of St. Louis,* 7.

36. Shaw, "Henry Shaw's Will," 35.

37. Ibid.

38. Rudolph, "100 Years of the Missouri Botanical Garden," 17. This was taken from a letter cited by Rudolph from Trelease to his colleague, the Harvard professor William Farlow, written February 16, 1912.

39. Ibid., 7.

40. Ibid., 8.

41. Missouri Botanical Garden, "Twenty-Third Annual Report of the Director," 12. This 1912 reference describes the central section of the garden being "reconstructed in accordance with the plans of the landscape architects for the Board" with a "circular plot subdivided into eight sectors, enclosed by low hedges, separated by radiating walks, and given over to flower beds parted by turf."

42. "Sulfur Dioxide in Air Menace to Human Life and Vegetation Here."

43. The ordinance forced industries and large businesses to burn only clean coal. It also estab-

lished regulations and inspections and set standards for emissions. In just one year, emissions from commercial smokestacks were reduced by two-thirds.

44. "Shaw's Garden Shuts Station in Move for Economy."

45. "Shaw's Garden Needs Additional Support."

46. Ibid.

47. "Rare Trees Lost, Flowers Perish at Shaw's Garden in Sunday Storm."

48. These figures represent the city of St. Louis only, as the broader metro region continued to grow. Today more than 2.2 million people reside in the region, while fewer than 300,000 live within the city limits.

49. Rudolph, "100 Years of the Missouri Botanical Garden," 12.

50. McNulty, *Missouri Botanical Garden,* 58.

51. Rudolph, "100 Years of the Missouri Botanical Garden," 12.

52. Missouri Botanical Garden, "Annual Report 1967," 16.

53. McNulty, *Missouri Botanical Garden,* 59.

54. Gates, "From the Director," 2.

55. Stock, "Saving the World the Ecologist's Way."

56. Ibid.

57. Elsley quoted in Homan, "Go to the Top."

58. Crosby, letter to the editor, *St. Louis Post-Dispatch.* Crosby was the garden's director of research at the time of the letter.

59. The most famous American version of a Japanese garden, the Portland Botanical Garden, had opened to the public a decade earlier, in 1967.

60. Tobar, "Architect Built Bridge between 2 Cultures."

61. UPI, "Japanese Garden Is Pride of St. Louis."

62. M. King, "Garden Says No to Di Suvero."

63. Duffy, "Sculpture in the Parks."

64. Rosenblatt, "The World Is His Garden."

65. Donald, "From the Chairman," 3.

66. Missouri Botanical Garden Board of Trustees, "Executive Summary," 2.

67. "Cultural Institution."

68. Missouri Botanical Garden, "Gateway to the Garden." The quotation is attributed to Peter Wyse Jackson, the garden's president at the time.

69. K. McGuire, "New Garden President Says He'll Stick to What Works."

70. Hollo, "From the Chair."

71. *Missouri Botanical Garden Consolidated Financial Statements and Supplementary Informa-*

tion, Years Ending September 30, 2021 and 2020. According to the report, the garden had net assets totaling $374,818,141.

72. James Deering to Paul Chalfin, April 7, 1914, VMGA. James Deering to Burrall Hoffman, April 7, 1914, VMGA. In his letter to Chalfin, Deering said the name was pretty and easy to say and captured the mystique and glamour he sought for the property by reflecting both the tradition of early Spanish exploration in the region and the natural history of the site.

73. The Deering family had been vacationing in Florida for decades, first in St. Augustine and then in the Miami area. William and his wife, Clara, began wintering in a home they had built near Miami in Coconut Grove following his retirement from Deering Harvester. Like other wealthy families, they appreciated the warmth and beauty of what was still mostly exotic wilderness. In 1908, James's brother, Charles, who remained as chairman of the board of International Harvester until 1910, bought over two hundred acres of land in nearby Buena Vista. He hired the Chicago landscape architect Ossian Cole (O. C.) Simonds to lay out the property. A talented artist himself, he devoted much of his time following retirement to collecting art.

74. When Miami incorporated in 1896, it had five hundred residents, a population made up of large landholders, farmers, sharecroppers, and merchants—an adventurous group whose primarily white families contrasted with the area's diverse cultural history.

75. Deering quoted in Gallagher, "The Garden's Guardian."

76. James Deering to Paul Chalfin, February 27, 1914, VMGA.

77. It was on another trip to Italy that Deering and Chalfin met the young Colombian, who was trained as an architect and had been studying Italian Renaissance gardens. He gave them a tour of private gardens in and around Florence, including several he had designed. Once back in New York, Suarez was engaged to design the formal gardens, an extensive lagoon on the southern side of the property, and the iconic barge situated just offshore from the main façade of the villa that remains one of the most recognizable features of Vizcaya.

78. Rybczynski and Olin, *Vizcaya,* 140.

79. Maher, *The Twilight of Splendor,* 210.

80. "James Deering Dies on Ocean Coming Home."

81. "James Deering Gives Hospital $500,000 Fund."

82. Ibid.

83. "James Deering Dies on Ocean Coming Home."

84. Bell, "Told in Miami Homes." This piece, although surely a gossip-style feature, states that Charles Deering's two daughters had inherited their grandparents' home in Coconut Grove and were taking turns spending winters there. In 1926 specifically, Marion was staying in Coconut Grove and Barbara was at Vizcaya.

85. Chauncey McCormick to Chalfin, March 24, 1953, VMGA.

86. McCormick to Miller, December 4, 1933, VMGA.

87. Miller to McCormick, undated but likely in November 1933, VMGA.

88. R. P. Barfield to Miller, January 4, 1934, VMGA.

89. E. G. Sewell to Miller, January 4, 1934, VMGA.

90. McCormick to Miller, December 4, 1933, VMGA.

91. McCormick to Chalfin, November 1, 1934, VMGA.

92. Ibid.

93. Chalfin to McCormick, February 2, 1935, VMGA.

94. Ibid. Chalfin claims in the letter that "objects whose value in November was nothing" now were worth $13,000.

95. Rybczynski and Olin, *Vizcaya,* 217.

96. Riley, "America's 10," 181. Others on the list include Doris Duke, Mrs. Edsel Ford, Mrs. Merriweather Post, and Mrs. Alfred duPont.

97. Marion McCormick to Burall Hoffman, March 27, 1953, VMGA.

98. McCormick to Chalfin, March 24, 1953, VMGA

99. Grizzle, "How William Cecil Made the Biltmore Estate into Asheville's Biggest Tourist Attraction."

100. Ibid.

101. V. O. Williams, "Fabulous Vizcaya Finds It Tough to Make Culture Pay." The article reports that Vizcaya earned around $180,000 in 1955, which was considerably less than the more than $288,000 that the county had budgeted.

102. Ibid.

103. A. G. Adams, "Vizcaya," 29.

104. McHale, "Restoring Vizcaya Is No Ball for Its Dedicated Supporters."

105. "Gains Told to Vizcayan Membership."

106. Reno, "Dade Has Reason for Pride in Vizcaya." In addition to his background, Bob Hunter is quoted in the article as referring to Vizcaya as "unique among great houses" and "genuine and beautiful, and the things there—tapestries, furniture, rugs and decorations—are good in themselves and worthy of being cherished as community treasures," but makes no mention of the gardens.

107. Heinrich, "Vizcaya Sale Move Stirs Angry Protest."

108. McHale, "Restoring Vizcaya Is No Ball for Its Dedicated Supporters."

109. Barry, "Ol' Sam Bids Vizcaya Farewell."

110. Gallagher, "The Garden's Guardian."

111. Fosse quoted in T. J. King, "Fosse Met Vizcayan Challenge with Genius for Growing Things."

112. There are reports that county commissioners waived entrance and event fees for celebrities and other notable visitors.

113. Meadows, "Vizcaya Is Facing an Inspection by Museum Group." AAM inspectors that year had criticized Vizcaya for permitting the "wear-and-tear that lavish weddings bring," and for not having more interpretive programs in its formal gardens.

114. Ibid. This approach to management was employed at other local facilities, such as the Miami Museum of Science, the Miami Art Museum, the Historical Museum of Southern Florida, and the Fairchild Tropical Garden.

115. "Miami-Dade Legislative Item File Number: 001616: Expanded Authority of Vizcaya Museum and Gardens Trust," miami.gov, accessed August 4, 2021, http://www.miamidade.gov/govaction/matter.asp?matter=001616&file=false&fileAnalysis=false&yearFolder=Y2000.

116. Miami-Dade County, "Vizcaya's Restoration of Main House and Gardens."

117. Simpkins had previously worked at Tyler Arboretum and the Scott Arboretum at Swarthmore College.

118. Ian Simpkins, discussion with the author, March 5, 2019.

119. Ibid.

120. Olin, letter to the editor, *Miami Herald.*

121. Panamerican Consultants, Inc., *Viewshed Impact Assessment of the Proposed 300 Grove Bay Condominium Project on Vizcaya Museum and Gardens,* 18.

122. In May 2008, the Miami-Dade Circuit Court nullified the project's approval on grounds of procedural violations and claimed the commission broke its own zoning ordinance with unlawful spot zoning.

123. "South Florida Album." The quotation came from the Southern Region director of the National Trust for Historic Preservation.

124. Simpkins, interview.

125. American Public Gardens Association, "Vizcaya Museum and Gardens Promotes Ian Simpkins to Deputy Director for Horticulture and Urban Agriculture."

126. M. C. Harry Associates, "Vizcaya Master Plan: Restoring the Village," presentation to Miami-Dade County Commissioners, July 6, 2017, accessed November 22, 2020, http://www.miamidade.gov/govaction/legistarfiles/Matters/Y2017/171366.pdf.

127. A. G. Adams, "Vizcaya," 31.

128. Lopez, "As COVID-19 Cancels Events, Vizcaya Asks for Help."

129. Tester, "Crown Jewel of Miami Attractions, Vizcaya Hit Hard by COVID-19."

130. "Miami," episode 4 of *Sinking Cities: A Four-part Series on the Threat of Climate Change,* PBS, November 21, 2018, https://www.pbs.org/video/sinking-cities-miami-bcdxzj/.

131. In 2019, gardens' staff experimented with tiger dams, which are long inflatable tubes filled with water that function as linear sandbags. Stackable to over thirty feet high and extendable for miles without seams, they offer an intriguing option for temporarily holding water at bay.

2. Cultural Sites

Epigraph: Grady Clay, "Whose Time Is This Place? The Emerging Science of Garden Preservation," *Landscape Architecture,* May 1976, 217.

1. National Park Service (NPS), "Defining Landscape Terminology."

2. Ibid.

3. "A National Park Service."

4. Mayall, "Recording Historic American Landscape Architecture," 1.

5. The gardens around the Lindens, a house in the Province of Massachusetts Bay, were documented because the house was being moved to Washington, D.C.

6. Mayall, "Recording Historic American Landscape Architecture," 4.

7. NPS, "Preservation of Historic Sites Act, 1935."

8. Quoted in Mackintosh, McDonnell, and Sprinkle, "The National Parks," 43.

9. NPS, "National Historic Preservation Act, 1966."

10. Mackintosh, McDonnell, and Sprinkle, "The National Parks," 43.

11. Richard Nixon Foundation, "President Nixon's Legacy of Parks."

12. NPS, "America's National Park System."

13. Von Eckardt, "Landmark Decision."

14. Buggey, "On the Founding and Early Years of the Alliance."

15. Fitch, "Preservation Requires Tact, Modesty and Honesty among Designers," 278.

16. U.S. Congress, *H.R.5496—National Historic Preservation Act Amendments of 1980,* 96th Congress, December 12, 1980, 2988.

17. The four general options for the treatment of historic buildings and landscapes as identified by the Department of the Interior are *preservation,* which focuses on the maintenance and repair of existing historic materials and retention of a property's form as it has evolved over time; *rehabilitation,* which acknowledges the need to alter or add to a historic property to meet continuing or changing uses while retaining the property's historic character; *restoration,* which depicts a property at a particular period of time in its history, while removing evidence of other periods; and *reconstruction,* which re-creates vanished or nonsurviving portions of a property for interpretive purposes.

18. In 1988, NPS management policies formally identified landscapes as a type of cultural resource. A policy was established to recognize and protect landscapes with significant historic, design, archaeological, and ethnographic values. This policy also recognized the importance of considering both built and natural features, and the dynamics inherent in natural processes and continued use.

19. O'Donnell quoted in McCormick, "Vaulting the Garden Wall," 75. O'Donnell founded the influential firm Heritage Landscapes in 1987. In his letter nominating the firm for ASLA's Firm of the Year in 2019 (https://www.asla.org/uploadedFiles/CMS/About__Us/Honors_and_Awards_New/2019/HeritageLandscapesLLCNomination.pdf), the landscape architect Laurie Olin wrote: "No other firm or group of individuals in the country or world has done as much for the recognition and active rescue, rehabilitation, restoration, and management for ongoing use of historic and cultural landscapes than Patricia O'Donnell and her colleagues at Heritage Landscapes."

20. Birnbaum and Peters, eds., "Secretary of the Interior's Standards for the Treatment of Historic Properties with Guidelines for the Treatment of Cultural Landscapes."

21. NPS, "HALS: Historic American Landscapes Survey."

22. Mount Auburn Cemetery, "Our Mission."

23. Abraham Lincoln eliminated the writ of habeas corpus during this time and instructed people deemed disloyal to the Union to be confined at Alcatraz.

24. Randall, "The Historic Gardens of Alcatraz," 26. Randall writes that in addition to the poppy, coyote bush, blue elderberry, and blackberry probably came over with the soil.

25. In 1940, a massive, 250,000-gallon water tank was built and refilled twice a week to serve the needs of the prison.

26. This unattributed quotation is taken from an undated brochure, "The Gardens of Alcatraz," which is attributed to the Garden Conservancy, the Golden Gate National Parks Conservancy, and the National Park Service.

27. P. L. Brown, "The Plant Men of Alcatraz," 8. Captain Philip R. Bergen said this to the *New York Times* writer when he ninety-six. He had been an associate warden and had lived on the island with his family between 1939 and 1955.

28. Ibid.

29. Beatty, "Gardens on Alcatraz?," 11.

30. Reichel quoted in Golden Gate National Parks Conservancy, "Gardens of Alcatraz."

31. Beatty, "Gardens on Alcatraz?," 12. Beatty quotes Michener as saying he also planted chrysanthemum and delphinium.

32. P. L. Brown, "The Plant Men of Alcatraz."

33. Fritz, *Alcatraz Gardens.*

34. Following a series of subsequent acquisitions, the Recreation Area expanded to more than 80,000 acres.

35. Kushins, "Alcatraz's Gardens Are Flourishing, against All Odds." A student leader of the group later recalled, "I did not see any gardens."

36. NPS, Save America's Treasures Grants. The program started in 1988 and is a joint effort of the NPS, the National Endowment for the Arts, the National Endowment for the Humanities, and the Institute of Museum and Library Services.

37. Adezio quoted in Garden Conservancy, "Save America's Treasures Grant Funds Alcatraz Historic Gardens Project," 1.

38. NPS, Golden Gate National Parks Conservancy, and Garden Conservancy, "The Gardens of Alcatraz," project prospectus, accessed May 21, 2022, https://www.nps.gov/goga/getinvolved/supportyourpark/upload/garden_prospectus.pdf.

39. NPS, "Cultural Landscape Inventory: Alcatraz Island," (2005) Part 4, 10.

40. NPS, Golden Gate National Parks Conservancy, and Garden Conservancy, "West Side Gardens Treatment Plan," October 2008, 37.

41. Gardens of Alcatraz, "Remembering Carola Ashford."

42. Bremier, "Rediscovering Isolated Beauty."

43. Syon Park website, accessed June 2, 2020, https://www.syonpark.co.uk/.

44. Shelagh Fritz (program manager, Alcatraz Gardens), in discussion with author, June 3, 2020.

45. Gardens of Alcatraz, "Welcome to the Gardens of Alcatraz Blog," accessed June 6, 2020, http://alcatrazgardens.org/blog/index.php/2010/11/. As reported by Shelagh Fritz in her 11/29/10 blog post: "I have been lucky enough to speak with many visitors who came to the island when the gardens were overrun with blackberries and are amazed upon returning years later to an island that is blooming with tended gardens."

46. Golden Gate National Parks Conservancy, Garden Conservancy, and NPS, "Golden Gate National Parks Conservancy, Garden Conservancy, and National Park Service Celebrate Completion of Historic Gardens of Alcatraz Project."

47. Gardens of Alcatraz, "A Milestone Has Been Reached," blog post, accessed June 6, 2020, http://alcatrazgardens.org/blog/index.php/2014/08/a-milestone-has-been-reached/

48. Golden Gate National Parks Conservancy, "Alcatraz Historic Gardens," accessed June 6, 2020, https://parksconservancy.org/programs/npn/alcatraz-historic-gardens.

49. Shelagh Fritz, in discussion with author, June 3, 2020.

50. Ibid.

51. Barreira, "Golden Gate National Parks Conservancy to Permanently Cut Workforce by 27%."

52. Russolillo, "California Has Driest Year Since 1924."

53. California Energy Commission, *San Francisco Bay Area Region Report.* The August 2018 report states: "Several studies suggest that coastal fog along the California coast, so critical to our Bay Area climate, is less frequent than before. However, the causes of this decline and implications of climate change are complicated because coastal fog formation is the result of a delicate moving balance between heat and humidity from three different sources: ocean, air, and land."

54. This was an agency-specific iteration of a broader strategy created by the secretary of the interior in 2009 as the Department of the Interior's "first-ever coordinated strategy to address current and future impacts of climate change on America's land, water, ocean, fish, wildlife, and cultural resources."

55. NPS, "Climate Change Response Strategy."

56. In Philadelphia, an epidemic of yellow fever killed five thousand people in 1793. New York City, which by 1800 packed 60,000 people into one and a half square miles, had three such outbreaks between 1798 and 1803, prompting the formation of the first New York City Board of Health. People kept dying, but there was significantly less space in which to bury their bodies. In Lower Manhattan, the two-and-a-half-acre site next to Trinity Church became the burial site for as many as 120,000 people by 1822, the same year that yet another epidemic hit the city. Some feared the lingering stench of decaying bodies was making people sick, and burials near the church were banned.

57. Bigelow, *A History of the Cemetery of Mount Auburn,* 2.

58. Linden, *Silent City on a Hill,* 147.

59. Bigelow, *A History of the Cemetery of Mount Auburn,* 135.

60. Dearborn was influenced by Cimitière du Père Lachaise, the over one-hundred-acre cemetery near the heart of Paris. He also respected the English landscape designer Humphrey Repton and is known to have read Repton's three famous books on landscape gardening. Dearborn praised Repton and other famous English gardeners as being "champions of true taste." The qualities of the picturesque landscapes they created are evident in Mount Auburn's lush woodlands, rolling hills, and inherent drama, capped by a central mound standing 125 above the nearby Charles River.

61. A committee of nine men—including Bigelow, Dearborn, Brimmer, and Story—was formed to oversee the construction of Mount Auburn and to manage the cemetery, the garden, and the more natural landscape. Three-hundred-square-foot lots sold for $60, and, after two years, $30,000 had been spent on grading, laying out roads and pedestrian paths, and other improvements needed to open the cemetery. Dearborn provided most of the detailed layout work and even transplanted more than one thousand trees from his nursery to create the public front of the cemetery. His vision for Mount Auburn's landscape extended beyond cemetery grounds, and he continued to push the experimental garden as integral to the mission.

62. "Cemetery at Mount Auburn."

63. Story, *An Address Delivered on the Dedication of the Cemetery at Mount Auburn,* 19.

64. "No Place Will Possess Stronger Attractions to the Visitor."

65. Reed, "Sweet Auburn."

66. Cothran and Danylchak, *Grave Landscapes.*

67. Dearborn, *A Concise History of, and Guide through Mount Auburn,* 3.

68. Bigelow, *A History of the Cemetery of Mount Auburn,* 108.

69. Ibid., 118.

70. Downing quoted in Linden, *Silent City on A Hill,* 256.

71. Bigelow, *A History of the Cemetery of Mount Auburn,* 95.

72. Ibid., 108.

73. Ames, "Mount Auburn's Sixscore Years," 92.

74. Bigelow, *A History of the Cemetery of Mount Auburn,* 96.

75. Ibid., 121.

76. Ibid., 90–91.

77. Merrifield, "From Necessity to Honor."

78. Linden, *Silent City on a Hill,* 13.

79. Giguere, "'The Americanized Sphinx,'" 62. Quotation is attributed to Jacob Bigelow.

80. Ibid., 66.

81. The first lawn mower was invented and patented near Gloucestershire, England, by Edwin Budding in 1830. It was made of wrought iron and was designed to be pushed from behind to cut grass primarily on sports grounds, cemeteries, and large gardens.

82. D. Collins, "Balancing Historical Integrity with Horticultural Innovation."

83. Strauch worked at Spring Grove for nearly thirty years and designed landscapes for numerous private estates.

84. Linden, "Nineteenth-Century German-American Landscape Designers."

85. Berg, *Mount Auburn Cemetery Master Plan,* 32.

86. *Boston Globe,* February 6, 1883. An average annual planting around that time was 250 trees and 400 shrubs.

87. "Mount Auburn Cemetery: The Corporators Elect Officers and Hear Reports."

88. Ames, "Mount Auburn's Sixscore Years," 94.

89. National Historic Landmark Nomination: Mount Auburn Cemetery, National Park Service, undated. Accessed September 7, 2020, https://npgallery.nps.gov/NRHP/GetAsset/NHLS/75000254_text.

90. Caldwell quoted in Bussman, "Memorial Park Concept at Willow Pond Designed to Meet 20th-Century Needs."

91. National Historic Landmark Nomination.

92. Ames quoted in Linden, *Silent City on a Hill,* 283.

93. Berg, *Mount Auburn Cemetery Master Plan,* 45.

94. Cappucci, "A Look Back at The Disastrous Great New England Hurricane of 1938."

95. D. Collins, "The Hurricane of 1938."

96. S. Jackson, "A Living Museum of Trees." The storm destroyed 811 trees and inflicted substantial damage on 777 others, as reported by Mount Auburn's plant records and GIS manager.

97. Ames, "Mount Auburn's Sixscore Years," 94.

98. Ibid., 94.

99. National Historic Landmark Nomination.

100. Ames, "Mount Auburn's Sixscore Years," 93.

101. Ibid., 94.

102. Reed, "Sweet Auburn," 21.

103. National Historic Landmark Nomination.

104. Chesney, "Announcing: The Friends of Mount Auburn."

105. Ibid.

106. Wilson, "Sites and Insights: Mount Auburn Cemetery."

107. Chesney, "Retirement of Superintendent Duncan W. Munro."

108. Sands, "I.P.M.: For the Care of Mount Auburn's Trees and Shrubs."

109. "Cemetery Fence."

110. Trustees of the Reservation, "Our Work."

111. Kendrick, *The Lively Place,* 216. The quotation is attributed to Dennis Collins, Mount Auburn's horticultural curator, and refers to years in the tenures of Ames and Chesney.

112. Anable, "A Conversation with Bill Clendaniel," 2.

113. Ibid.,2.

114. David Barnett, conversation with the author, June 14, 2019.

115. Clendaniel, "The Friends Becomes a Charitable Trust and Receives Its First Grant."

116. Halvorson Company, "*Mount Auburn Master Plan,*" 1993, 1:3.

117. Anable, "175 Years of Stewardship." This is an interpretation of a quotation by Mount Auburn's archivist, Meg Winslow: "When we have a site that we want to develop for burials, for instance, the staff goes to the archives and looks through historic images and records to learn what the site used to be like and refers to the Master Plan. The staff then incorporates that information into decisions about the new. So we're not copying what used to be—we're using it as a reference to create new landscapes that reflect our contemporary feelings about memorialization and commemoration."

118. Barnett conversation.

119. Ibid.

120. Barnett, "President's Corner."

121. Story, *An Address Delivered at the Dedication of the Cemetery at Mount Auburn,* 17.

122. Mount Auburn, "Mission & Vision," accessed October 20, 2020, https://mountauburn.org/mission/.

123. The Leadership in Energy and Environmental Design (LEED) started in 1994 and by 2006 had expanded from one general standard for "green" building to a comprehensive system for all aspects of design, development, and construction. Although it never adequately addressed issues and conditions specific to the landscape, many university campuses and other designed landscapes were pushed toward more sustainable practices.

124. Anable, "David P. Barnett, Ph.D."

125. Barnett conversation.

126. Mount Auburn Cemetery, *186th Annual Report.* The annual report states that between April 1, 2017, and March 31, 2018, "the Cemetery generated $4.1 million in revenue from the sale of new burial space, exceeding its annual goal by 17%."

127. Mount Auburn Cemetery, "A Message from the President & CEO."

128. Mount Auburn Cemetery, "Celebrating Dave Barnett."

129. Stephens, "President's Corner," 1.

130. Stephens, "Cemeteries Exploring Climate-Friendly Burials." As he explains in the piece, "Cremation, fueled by natural gas relies on a fuel source that is not sustainable. For body burials, most cemeteries require concrete liners or burial vaults, which prevent graves from collapsing under the weight of heavy equipment like backhoes or lawn mowers. Raw materials and energy to manufacture and ship these liners have significant carbon footprints." As alternatives, he recommends: "Hydrolysis uses heat, pressure, water, and base chemical agents to reduce human remains to bone fragments and essential elements. Natural organic reduction accelerates the conversion of remains to soil. As with other methods, both are conducted by professionals in controlled facilities."

3. Campuses

Epigraph: Laurie Olin, "Exceptional Grounds: The Landscapes of Institutions," *Sitelines,* Spring 2013, 3.

1. Stanford and his business partners and fellow opportunists Collis Potter Huntington, Mark Hopkins, and Charles Crocker became known as the Big Four. Stanford provided the political connections and helped secure public investment and land grants and subsidies from the state to make the transcontinental railroad a reality.

2. Jane Stanford transformed her son's room on the upper floor of the family's Nob Hill mansion into The Leland Stanford, Jr. Museum, displaying artifacts and curiosities he had collected on his many travels. She wanted to build a more expansive museum to honor his memory, while his father believed a university would be a more suitable monument. They ended up doing both, building a museum named for him on a campus named for him.

3. They envisioned as the school's fundamental objective "to promote the public welfare by exercising an influence in behalf of humanity and civilization, teaching the blessings of liberty regulated by law, and inculcating love and reverence for the great principles of government as derived from the inalienable rights of man to life, liberty, and the pursuit of happiness." It was to be a university open to the poor, with no tuition and the cost of room and board low enough to accommodate students of modest means.

4. Coolidge was a founding partner of Shepley, Rutan & Coolidge, one of the longest-lasting and most heralded architecture firms in American history. He designed numerous academic buildings, libraries, museums, and medical buildings at Stanford, Yale, Brown, Vanderbilt, Harvard, and the University of Chicago. He also designed the Art Institute of Chicago and Chicago's public library.

5. The founding grant assigns Leland and Jane Stanford the "right to improve, manage, and control the trust property, as if this Grant had not been made."

6. Frederick Law Olmsted to Dr. Charles Eliot, March 2, 1886. LoC FLOP: Subject File, 1857–1952; School buildings and grounds; Stanford University, Stanford, Calif., 1881–1892.

7. Frederick Law Olmsted to John Charles Olmsted, September 29, 1886. LoC FLOP: Subject File, 1857–1952; School buildings and grounds; Stanford University, Stanford, Calif., 1881–1892.

8. Frederick Law Olmsted, Sr. to The Hon. Leland Stanford, November 27, 1886. LoC FLOP: Subject File, 1857–1952; School buildings and grounds; Stanford University, Stanford, Calif., 1881–1892.

9. The Beaux-Arts style, with its opulence and adherence to symmetry and formality, had yet to permeate American culture, but its influence had guided European architecture and design for decades. The Stanfords surely saw examples of this on their travels, and Leland likely considered it a viable precedent for the university. Olmsted, on the other hand, was years away from his celebrated commission for the Beaux-Arts–inspired World Columbian Exposition of 1893 in Chicago and still favored the picturesque style that had made him the most famous landscape architect in the world. His preferred campus plan involved tucking a system of small "cottages" into the landscape in harmony with its topography, circulation system, and immediate site conditions. Such was the approach he took for his plan for the University of California, Berkeley, and the campus of Smith College, among others.

10. Turner, *Campus,* 172.

11. Olmsted, letter to Stanford, June 20, 1889.

12. Frederick Law Olmsted, Sr. to The Hon. Leland Stanford, June 20, 1889. LoC FLOP: Subject File, 1857–1952; School buildings and grounds; Stanford University, Stanford, Calif., 1881–1892.

13. Leland Stanford to F. L. Olmsted, November 9, 1891. LoC FLOP: Subject File, 1857–1952; School buildings and grounds; Stanford University, Stanford, Calif., 1881–1892.

14. "Opening of the Great Leland Stanford, Jr. University."

15. Johnston, "Meet President Jordan."

16. Wolfe, "Who Killed Jane Stanford?"

17. In a display of how her tastes differed from her husband's, Jane Stanford engaged a new set of architects to design the buildings in the popular neoclassical style, thus rejecting the campus's established mission-style architecture.

18. Founding grant, 2. The text regarding the amendment states: "One more safeguard was provided. An act of the legislature, passed February 10, 1903, authorized the Trustees to petition the courts for 'the ascertainment of the existence and terms of, and for the determination of the validity and legal effect of, grants and other instruments creating, changing, or affecting trusts and estates for the founding, endowment, and maintenance of, the Leland Stanford Junior University.' After due process a judicial decree was obtained, July 3,1903, from the Superior Court of Santa Clara County, which gave confirmation to all claims and statements set forth in the Petition to the Court and vested irrevocably in the Trustees the titles to the Stanford properties already conveyed to the University."

19. Founding grant, 6.

20. Elliott, *Stanford University: The First Twenty-five Years,* 147.

21. Ibid., 147.

22. Olmsted Brothers to Stanford Board of Trustees, May 8, 1914. LoC FLOP: Subject File, 1857–1952; School buildings and grounds; Stanford University, Stanford, Calif., 1881–1892.

23. Ibid., 8.

24. Ibid., 13.

25. Price, "The Stanford Arboretum." There is not much to be found about Price's credibility or standing as a historian. He is a dual Stanford graduate who researched at least two aspects of the university's history. He does not specifically cite the sources he used for this paper on the arboretum.

26. Ibid. Price quotes the *Palo Alto Time* (likely *Times*) from March 23 of that year.

27. "$30,000 Worth?"

28. "Stanford Stories from the Archives: 1940s," Stanford Libraries, accessed January 12, 2021, https://exhibits.stanford.edu/stanford-stories/feature/1940s.

29. Jahnke, "Planning Office."

30. Ibid.

31. Eldridge T. Spencer, "A Coordinated Plan for the University," Stanford University, Planning Office, Records (SC0486), 2 (SUL).

32. Lewis Mumford, "Memorandum on Planning," March 6, 1947, Stanford University, Planning Office, Records (SC0486), 1 (SUL).

33. Ibid., 1.

34. D. L. Miller, *Lewis Mumford, a Life,* 480.

35. An article in the *Stanford Daily* dated September 22, 1941, describes the widening of "five main arteries" and diagonal parking lining Serra Street, which is the primary corridor that intersects with Palm Drive at the front of the original campus.

36. Mumford, "Memorandum on Planning," 1.

37. Harmsen, letter to the editor, *Stanford Daily.*

38. Moore, "Stop Those Monsters!"

39. "Stanford Plans Expansion of Academic, Scientific Facilities in Areas Adjacent to the Present Quad," *Stanford* Daily, May 27, 1949. As quoted in the article: "There is no immediate plan to pave the Inner Quad, and plans to reconstruct the Quad islands have been abandoned at present because of lack of funds."

40. "Farm Landscaping Emphasizes Trees."

41. Harry Sanders, "A Stanford Planner Charts the History of Campus Planning, and Church's Contribution." Interview by Suzanne B. Riess. Stanford University, Planning Office, Records (SC0486),1978: 637 (SUL).

42. "$25 Million Ford Grant Given Stanford."

43. Rankin, "$100 Million SU Drive."

44. Palmer, "Harry Sanders."

45. Palmer, "Harry Sanders."

46. Sanders, "A Stanford Planner Charts the History of Campus Planning, and Church's Contribution," 638. Sanders says the trustees and administrators of the university shared this opinion.

47. Hardie, "He Changed the Landscape."

48. Sanders, "A Stanford Planner Charts the History of Campus Planning, and Church's Contribution," 638.

49. Ibid., 650.

50. "Oscar Nelson," *San Francisco Chronicle,* March 27, 1998.

51. Ibid.

52. Sanders, "A Stanford Planner Charts the History of Campus Planning, and Church's Contribution," 665.

53. Broadwin, "Farm Wheels on Way Out."

54. M. Collins, letter to the editor.

55. Scott, "Closing the Books."

56. Weinstein, "Stanford Starts $83 Million Building Program."

57. Sanders, "A Stanford Planner Charts the History of Campus Planning, and Church's Contribution," 640.

58. Weinstein, "Stanford Starts $83 Million Building Program."

59. "Visions of Red Tile & Sandstone."

60. Fong. "Stanford Grounds and Plants," 5.

61. "Bug Me."

62. Ibid.

63. "Renovation: Yes, Asphalt: No."

64. Trousdale, "Quad Revamping Will Soon Add Splash of Color."

65. Ibid.

66. P. Williams, "Quad Plans Are Not Finalized."

67. Callon, "Buildings Grow on Farm,"

68. Johnston, "All Right Now." More than one hundred buildings were structurally strengthened and restorations were made to cherished architectural works such as Main Quad buildings, Memorial Church, and the Stanford Museum.

69. Joncas, Neuman, and Turner, *Stanford University,* 143.

70. Palmer, "Neuman Leaves Landscape Built on Clarity, 'Sense of Place.'"

71. Neuman, "Success and Failure in Campus Design in the Post–World War II Era."

72. This quotation can be found in a white paper titled "Learning in Place: Shaping the Future of Campus and Facility Planning" and produced by Smithgroup JJR, which hosted a gathering of planners, designers, and university administrators at the Society of College and University Planning conference in Chicago in 2012. The author was one of the invited participants. The paper can be found at https://www.smithgroup.com/sites/default/files/2018-06/LearninginPlace.pdf.

73. Bajoria, "Bike Program Nears Its End." According to John Ciccarelli, Stanford's bike coordinator, the program was established to "support the bike as an effective mode of transportation across the campus."

74. Frederick Law Olmsted to Dr. Charles Eliot, July 20, 1886. LoC FLOP: Subject File, 1857–1952; School buildings and grounds; Stanford University, Stanford, Calif., 1881–1892.

75. Stanford University, "Water Efficiency at Stanford."

76. Stanford University, "Capital Plan and Capital Budget." This figure represents money spent on facilities, infrastructure, and commercial real estate on university lands.

77. "Stanford University Board of Trustees' Commendation for David Neuman," Neu Campus

Planning, accessed January 23, 2021, http://www.neucampusplanning.com/wp-content/uploads/2016/08/Stanford-Commendation.pdf.

78. Peña, "Ohio Building Designer David Lenox Named Stanford University Architect."

79. Cathy Blake, interview by author, January 30, 2021.

80. Stanford University, "Stanford University Grounds Maintenance Guidelines."

81. Stanford University, "Budget Plan 2019/2020, 73"

82. The act, which was signed on May 28,1830 by President Andrew Jackson, authorized the president of the United States to grant property on land west of the Mississippi to Indigenous people in exchange for land they held within existing state borders. In 1838 and 1839, during one of the country's darkest moments, four thousand Cherokees forced to leave their homes died on a westward march that became known as the Trail of Tears.

83. Grandison, "From Plantation to Campus." 9.

84. The loan came from James Fowler Baldwin Marshall (1818–1891), who was an incorporator and trustee for the Hampton Normal and Agricultural Institute for the education of Black teachers, from which Washington graduated and where he worked prior to moving to Tuskegee. Marshall thought it would be inappropriate for the Hampton Institute to loan the money, so gave it to Washington from his personal account.

85. Harlan et al., *Booker T. Washington Papers: Vol. 2,* 34–35.

86. Grandison, "From Plantation to Campus," 9.

87. Harlan et al., *Booker T. Washington Papers: Vol. 2,* 392.

88. Ibid., 392.

89. Grandison, "Negotiated Space," 568.

90. NPS, *Tuskegee Institute National Historic Site, General Management Plan,* U.S. Department of the Interior, 1978. 7. Carver was awarded the Roosevelt Medal in 1939.

91. NPS, Southeast Regional Cultural Resources Division, "Tuskegee Institute National Historic Site Historic Resource Study," 2019, 3–24. The NPS credits the quotation to Christina Vella's biography of Carver: *George Washington Carver, A Life,* published by Louisiana State University Press in 2015.

92. Thrasher, *Tuskegee,* 41.

93. Harlan et al., *Booker T. Washington Papers: Vol. 3,* 469.

94. Ibid., 470.

95. Weiss, *Robert R. Taylor and Tuskegee,* 105.

96. Weiss, "Tuskegee," 32.

97. WLA Studio, *The Oaks,* 22.

98. Ibid., 24.

99. Ibid., 21.

100. Washington quoted in Clement & Wynn, "Campus Heritage Plan: Tuskegee University," 2009, 13. The plan credits this quotation to a doctoral dissertation by Richard Kevin Dozier at the University of Michigan. I was not able to find a copy of the dissertation to confirm the quotation so am relying on the Heritage Plan.

101. Robert R. Moton, "Special Report of the Principal to the Board of Trustees, April 1930," in *Julius Rosenwald Fund Collection* (Amistad Research Center, Tulane University).

102. Weiss, *Robert R. Taylor and Tuskegee,* 30. After his time at Tuskegee and subsequent studies at Meharry Medical College, Eneas returned to Nassau, Bahamas, where he became an educator and successful dentist. He wrote a book, *Bain Town,* about his family's African heritage and subsequent capture into slavery.

103. C. W. Kelly, "News among the Colored People."

104. Ibid.

105. F. Patterson, "Southern Viewpoint."

106. Ibid.

107. Kellum, "Designing His Own Path," *Landscape Architecture Magazine,* May 1999, 138.

108. Ibid., 139.

109. E. V. Brown, "Designing Beauty."

110. Alabama Chapter of the American Society of Landscape Architects, "Acclaim: Alabama Chapter Newsletter," January 1993.

111. Kellum, "Designing His Own Path," 138.

112. "Dr. Foster Announces Million Dollar Expansion at Tuskegee."

113. Tuskegee University, "History of the Chapel.".

114. Cotton, "Stewarding an Educational Legacy." Cotton cites the "Campus Heritage Plan: Tuskegee University" by Clement & Winn Architects for this information. She references the plan several times in her paper. I could not locate a copy of the plan to cite directly.

115. "Allen Introduces Bill to Preserve Historical Buildings at Tuskegee."

116. The quotation is attributed to Jack O. Horton from a letter to Senator Henry M. Jackson, chairman of the Senate Interior Committee, and cited in "Legislation Would Make Historic Site in Area of Tuskegee Institute."

117. NPS, "General Management Plan," 29.

118. Hirsley, "Debt Puts Stranglehold on Many Black Colleges Nationwide."

119. Ibid.

120. "Tuskegee Is a Model for Black Universities." This was a reprint of an article first published in the *Chicago Tribune.* Hampton University had an endowment of $40 million at the time.

121. NPS, *General Management Plan,* 12.

122. Ibid. *General Management Plan,* 15. Visitor statistics by the NPS document that 54,562 people visited Tuskegee in 1999.

123. NPS, "Long Range Interpretive Plan: Tuskegee Institute National Historic Site," 2003, 15.

124. NPS, *General Management Plan,* 19.

125. NPS, "Long Range Interpretive Plan," 48.

126. NPS, *General Management Plan,* 39.

127. Ehrenkrantz Group and Edward Pryce, "Master Plan for Tuskegee University: Toward the Year 2000," April 10, 1987, 20.

128. Ehrenkrantz Group and Price, "Master Plan for Tuskegee University," 21.

129. Ibid. 21.

130. Ibid., 41.

131. "About Us," White House Initiative on Historically Black Colleges and Universities, accessed February 23, 2021, https://sites.ed.gov/whhbcu/about-us/#:~:text=In%20 1980%2C%20President%20Jimmy%20Carter,universities%20to%20provide%20quality%20 education.%E2%80%9D.

132. U.S. General Accounting Office, "Historic Preservation: Cost to Restore Historic Properties at Historically Black Colleges and Universities."

133. National Trust for Historic Preservation, "HBCU Cultural Heritage Stewardship Initiative." The partnership includes the National Endowment for the Humanities, Ford Foundation, the JPB Foundation, the J. M. Kaplan Fund, the Executive Leadership Council, the Chipstone Foundation, the Wunsch Americana Foundation, and the James Marston Fitch Charitable Foundation. The $1 million program is part of the larger African American Cultural Heritage Action Fund, a $25 million campaign to support 150 historic places that have been overlooked in American history and represent centuries of African American activism, achievement, and resilience.

134. F. E. Martin, "Every Campus Tells a Story."

135. In 1993, Harvard embarked on a restoration of Harvard Yard following the decline of its stand of American Elms. When I started at Duke in 2000, the historic quadrangles were severely denuded and undermaintained. Reading through the synopses of Getty grant recipients, it is clear than many of those campuses had landscapes in need of restoration.

136. Getty Center, "Campus Heritage Grants."

137. Dan Levine, "Miami to Survey Landmarks," *Journal News,* July 6, 2007.

138. Cotton, "Stewarding an Educational Legacy," 66. The plan, titled "Campus Heritage Plan: Tuskegee University," was produced by Clement & Winn Architects.

139. Cotton, "Stewarding an Educational Legacy," 72, citing the Clement & Winn "Campus Heritage Plan."

140. "Tuskegee University Students Participate in Historic Preservation Training Project."

141. Tuskegee University, "Campus Master Plan 2017."

142. McNair quoted in Koplowitz, "Tuskegee University Receives $20 Million Donation from MacKenzie Scott."

143. "Tuskegee University Students."

4. Communities

Epigraph: This quotation is widely published. This citing was confirmed here: Henry H. Saylor, "Make No Little Plans," *Journal of the American Institute of Architects,* March 1957, 95–99.

1. Mundigo and Crouch. "The City Planning Ordinances of the Laws of the Indies Revisited," 252.

2. Ibid., 254.

3. John W. Reps, "William Penn and the Planning of Philadelphia," *The Town Planning Review* 27, no. 1 (1956): 29. Part of the reason for the generously sized lots was to keep spaces between houses in case of a fire similar to the one that struck London in 1666, through which Penn survived.

4. Reinberger, "Oglethorpe's Plan of Savannah," 847.

5. Ibid., 845.

6. AP, "Farm Population Lowest Since 1850's," *New York Times,* July 20, 1988.

7. American Scenic and Historic Preservation Society, *Twenty-Fourth Annual Report of the American Scenic and Historic Preservation Society, 1919,* 99.

8. Ibid., 99–100.

9. "Valuable Property for Sale."

10. "Gramercy Park Not Disturbed by Loft Buildings."

11. "Declaration of Restriction No. 1 Affecting Radburn, Property of City Housing Corporation in the Borough of Fair Lawn, Bergen County, New Jersey," March 15, 1929. The lawyer credited with writing the restrictions was Charles Asher, who was part of the planning team that created Radburn.

12. In the late 1930s, following the Great Depression, the Home Owners' Loan Corporation, with the aid of realtors and appraisers, graded neighborhoods into four categories, based in large part on their racial makeup. Neighborhoods with minority occupants were marked in red (the origin of the term "redlining") and considered high risk for mortgage lenders. While white families could easily get mortgages and buy homes in the newly sprawling suburbs, Black families found it difficult or impossible to acquire federally backed mortgages. This facilitated the decline of many urban communities, while places such as Riverside thrived.

13. Riverside Improvement Company, "Riverside: Progress Prospectus," August 1869, as found in Hubbard et al., "Riverside, Illinois," 285.

14. Numerous newspaper accounts present widely differing numbers.

15. "Schools Are the Issue." The article explains the estimate is based on the 350 votes placed by people in the village in the presidential election.

16. "Chicago's Suburbs Stirred by Oak Park's Landscape 'Color Scheme.'"

17. Perce, "By the River's Brink."

18. Frederick Law Olmsted, as found in Hubbard et al., "Riverside, Illinois," 260.

19. "Schools Are the Issue."

20. "Riverside All Set for Boom in House Building."

21. Hubbard et al., "Riverside, Illinois," 274.

22. "Riverside All Set for Boom in House Building."

23. McCormack, Black and Co., "Riverside."

24. Gray, "An Everyday Wilderness," 12.

25. "Mrs. McCormick Gives Huge Zoo Site to County."

26. J. O'D. Bennett, "Chicago's New Zoo Paradise for Animals."

27. "Riverside Okays 9 New Project."

28. "Set Meeting in Riverside on Land Use."

29. Cohen, "Navy Pier Students Object to New Site."

30. Ibid.

31. Ibid.

32. Clay, "Landscape Architects Honor Kentucky Pioneer Olmsted." It is unclear exactly which 1864 date the two groups were recognizing. Olmsted had reportedly coined the name "landscape architecture" several years earlier when he and Vaux opened their shop to work on Central Park. Clay writes, "Honoring Frederick Law Olmsted, whose work in Louisville and town plan for Anchorage, Ky., as well as work in dozens of other cities, marked the beginning of professional landscape architecture in the United States."

33. Maersch, "Charm of Riverside Is Due to Design."

34. "Move to Determine If Riverside Is a National Landmark." I was not able to determine the outcome of the lawsuit.

35. Snell, "National Register of Historic Places Inventory—Nomination Form." Snell was the survey historian with the NPS who was invited to Riverside to evaluate its potential status on the register.

36. Bassman, letter to the editor.

37. "Budding Controversy Blooms as Petunia Wars in Riverside," *Chicago Tribune,* April 3, 1998.

38. Ibid. Brackets in the original.

39. "Care of Elms Told by Expert." Because of its harmful effects on people and the environment, DDT was formally banned from agricultural use in the United States in 1972.

40. "Tree Future Grim," *Suburban Life/Citizen.*

41. "Can the American Elm Trees Be Saved?"

42. Phillips, "The Story of Riverside Is a Tale of Two Men and a City."

43. Cairns and Kesler, "Protecting a Prototype," 63. The work was led by landscape architecture faculty from the University of Illinois at Urbana-Champaign.

44. Cairns and Kesler, "Protecting a Prototype," 65.

45. Cairns and Kesler, "Protecting a Prototype," 100.

46. Uphues, "Once a Wall of Buckthorn, Now a Haven for Native Species."

47. Enstad, "It's Hard to Pin Blame in Case of Des Plaines."

48. Uphues, "Stemming the Tide." Riverside residents Ladd and Claudia Kulhanek contacted the Illinois Department of Natural Resources, the U.S. Army Corps of Engineers, and the Cook County Metropolitan Water Reclamation District.

49. Ibid.

50. Uphues, "Board Vote Sets Stage for Hoffman Dam Removal."

51. James D. Reynolds, "Letter from the President," *Frederick Law Olmsted Society of Riverside Newsletter,* Fall 2008, 2. The Village economy was suffering in conjunction with the global economic crisis, which was considered the worst economic disaster since the Great Depression.

52. Landscape Advisory Commission, Village of Riverside, "Master Landscape Plan," November 2015, 4.

53. Village of Riverside, "ArbNet Accreditation."

54. Landscape Advisory Commission, Village of Riverside, "Master Landscape Plan," 4.

55. Bowers Realty Group, "How Is the Real Estate Market Doing in Riverside, IL?"

56. Ho, "Des Plaines River Flood Proposals Described."

57. Ibid.

58. Uphues, "National Olmsted Group Objects to Riverside Floodwall."

59. Cappucci, "More than 10 Inches of Rain Falls in Illinois."

60. "Riverside, Illinois," Data USA, accessed April 18, 2021, https://datausa.io/profile/geo/riverside-il#demographics. This data was corroborated on several websites. The 90 percent reflects a combination of 81.2 percent identifying as white (non-Hispanic) and 10.8 percent identifying as white (Hispanic) for the year 2018.

61. Holmgren, *Hilton Head,* 35.

62. Danielson, *Profits and Politics in Paradise,* 11.

63. Halfacre, *A Delicate Balance,* 171.

64. Oatis, "Hilton Head Opened New Era in Development."

65. McPhee, *Encounters with the Archdruid,* 45.

66. Horn, "Nothing to Do—but Enjoy Yourself."

67. In its first year of service, 48,000 cars paid the $2.50 toll to cross the Intracoastal Waterway onto Hilton Head.

68. Charles Fraser, interview by Patrick Grafton, in the Twenty-Fifth Anniversary Issue of *Islander* magazine, June 1982.

69. Town of Hilton Head Island Town Council. *Town of Hilton Head Island Design Guide.*

70. C. Ryan, *My Life with Charles Fraser,* 75. The quotation comes from a recollection of Fraser by the landscape architect, Ed Pinckney, in ibid.

71. Rhodes, "Alchemy Island," 31.

72. Ibid.

73. C. Ryan, *My Life with Charles Fraser,* 184. Fraser's daughter, Laura Lawton Fraser, recalls the firm was recommended to Fraser, in ibid.

74. Ibid., 1. The landscape architect, Stuart Dawson, recounts this from personal experience.

75. Ibid., 1.

76. Ibid., 4.

77. MacNabb, "Hilton Head Island Growing Fast."

78. "Care to Invest as You Enjoy Your Vacation?"

79. Danielson, *Profits and Politics in Paradise,* 41.

80. For instance, a large landscaped traffic circle was created to save what later became known as the Fraser Oak (dated to circa 1760).

81. Sea Pines Plantation Company, "Declaration of Rights, Restrictions, Conditions, Etc.," August 19, 1957.

82. Sea Pines Plantation Company, "Sea Pines Land Use Covenants," April 4, 1970, 4.

83. McPhee, *Encounters with the Archdruid,* 47.

84. Ibid.

85. Danielson, *Profits and Politics in Paradise,* 48.

86. Horn, "Nothing to Do—but Enjoy Yourself."

87. "Announcing the First Villas in Our Exciting New Resort Community." The quotation is in the body of an advertisement for Harbour Town. There are other references to Fraser wanting the yacht basin and shopping area to resemble Portofino or other Italian port towns.

88. David Pardue, letter to "Sea Pines Friends," November 2018 (accessed from Sea Pines Homeowners private group on Facebook). Pardue is a lawyer and permanent Sea Pines resident since 1974. In the letter, he details the history of Fraser's covenants in a concise, clear, and, according to all my other research, very accurate manner.

89. Sea Pines Plantation Company, "Land Use Restrictions, Covenants, and Affirmative Obligations Affecting the Sea Pines Forest Preserve," June 29, 1970, 2.

90. Town of Hilton Head website, "Hilton Head Island History," accessed April 25, 2020, https://hiltonheadislandsc.gov/ourisland/history.cfm?menuheader=2.

91. *New York Times,* "Hilton Head Islanders Split Racially over Incorporation Vote," May 10, 1983.

92. Sawyer, "Almost Forgotten Island Awakes to Tourists."

93. Some of my earliest memories of driving across Hilton Head in the early 1970s are seeing these families outside their small homes—some likely not more than 150 square feet—that lined the main road. These gradually faded from the main view of most tourists over time.

94. Scardino, "A Gust of Bankruptcy and Scandal Rattles Elegant Hilton Head Island."

95. Sally Mahan, "Back to Basics: Could a Resurgence of the 'Old Hilton Head' Architectural Style Be the Next Big Thing?," *Hilton Head Monthly,* November 29, 2011.

96. Pardue, letter to "Sea Pines Friends."

97. Danielson, *Profits and Politics in Paradise,* 109.

98. Rhodes, "Alchemy Island," 33. Scardino's piece in the *New York Times* ("A Gust of Bankruptcy") reports that the company paid another $10 million for the real estate owned by the Sea Pines Company.

99. Scardino, "A Gust of Bankruptcy and Scandal Rattles Elegant Hilton Head Island." Fraser also worked as chair of a committee that sought to create sixty public parks on the island.

100. Stertz, "Failed Ambitions and Lawsuits on Hilton Head Island."

101. Kienzl, "Golfer Sinks Self into Site." The quotation is attributed to Bill Cullison, a Sea Pines resident and member of the original Sea Pines Associates group.

102. Sea Pines Community Services Associates, accessed August 10, 2019, https://www.seapinesliving.com/property-owners/services/community-services-associates-board/.

103. Leland, "2 Faces of Hilton Head."

104. Ibid. The quotation is attributed to Walter Mack, who worked at the Penn Center.

105. C. Ryan, *My Life with Charles Fraser,* 75.

106. Ibid., 75. Ed Pinckney recalls Fraser's mother "sending truckloads of azaleas" from Hinesville, Georgia, to be planted all over Sea Pines.

107. Sea Pines Company, "Sea Pines Land Use Covenants," April 4, 1970, 5. One of the only direct references to general planting for residences states: "There shall not be maintained any plants or animals, or device or thing of any sort whose normal activities or existence is in anyway noxious, dangerous, unsightly, unpleasant or of a nature as may diminish or destroy the enjoyment of other property in the neighborhood by the owners thereof."

108. C. Ryan, *My Life with Charles Fraser,* 9. This account comes from David Pearson, Sea Pines' first director of public relations.

109. Mandy Matney, "More than Half of SC Alligator Attacks were in Beaufort County. Why So Many on Hilton Head?," *Island Packet,* August 30, 2019.

110. McPhee, *Encounters with the Archdruid,* 45.

111. Matney, "More Than Half of SC Alligator Attacks Were in Beaufort County."

112. Ibid.

113. "Judge OKs Hilton Head Deer Kill."

114. Pressley, "In America's Suburbs, the Deer Next Door."

115. Ibid.

116. Kokal, "Hilton Head Gated Community to Shoot Deer to Curb Crashes, 'Vegetative Destruction.'"

117. Rembert, "Sewage . . . Made Beautiful."

118. David Henderson (director of Special Projects and Operations at CSA), discussion with the author, August 28, 2019.

119. Ibid.

120. Ibid.

121. AP, "Developer Seeks to Buy Resort of Sea Pines."

122. "Sea Pines Associates, Inc." United States Securities and Exchange Commission, accessed March 26, 2021, https://www.sec.gov/Archives/edgar/data/846926/000095014404007406/g90162defa14a.htm.

123. The resort owns a portion of Forest Preserve but has committed to transfer the land to CSA. The property is 181 acres, about the size needed for an eighteen-hole golf course, which was identified as a possible use.

124. Heffernan, "Sea Pines Hardest Hit by Hurricane Matthew on Hilton Head."

125. Angst, "How Trees Lost to Matthew Affected Hilton Head Island Residents—for Good and Bad."

126. Professional Grounds Management Society, "Green Star Awards."

127. Moss, "Sea Pines Wins National Landscaping Award."

128. Sea Pines Community Services Associates, "Sea Pines Critical Needs Referendum," 27.

129. Ibid., 27.

130. Ibid., 25.

131. The author lived in Sea Pines for almost six months and visited regularly during this period, so this is based on personal experience.

132. Ogozalek, "'It's beyond Crazy.'"

133. Hilton Head Area Association of Realtors, "Local Market Update—June 2021: Sea Pines."

5. Parks

Epigraph: Robert Smithson, *Collected Writings* (Berkeley: University of California Press, 1996), 160.

1. In the 1960s, officials in Portland hired Halprin's firm to help energize their public spaces by redesigning an eight-block sequence of open spaces aptly called the Open Space Sequences. Famous among these spaces are Lovejoy Plaza, Pettygrove Park, and the Ira Keller Forecourt Fountain.

2. Alanen and Bjorkman, "Plats, Parks, Playgrounds, and Plants," 54.

3. Clurman quoted in McKenna, "Central Park."

4. The Task Force raised private money for improvements while also assisting the parks department by coordinating volunteer clean-ups, overseeing projects, and organizing youth programs. The Community Fund also raised money, as well as funding a report of the park's condition titled the *Savas Report,* which recommended consolidating management of Central Park and creating a citizen-led organization to develop a strategy for restoring for the park and raising money.

5. These cities did not all use the term conservancy. Houston, for instance, created the Buffalo Bayou Partnership and St. Louis called its organization Forest Park Forever.

6. Harnik and Martin, *Public Spaces/Private Money.*

7. Whitehall, *Dumbarton Oaks.*

8. Joseph, Fanning, and Davison, *Cultural Landscape Report,* 40.

9. Ibid., 12.

10. Whitehall, *Dumbarton Oaks,* 58.

11. Farrand quoted in Tankard, "Dumbarton Oaks Garden and Park," 3.

12. Beatrix Farrand to Mildred Bliss, June 24 and 25, 1922, B:BF 1922.06.24, Dumbarton Oaks RBC.

13. Ibid.

14. Ibid.

15. Farrand to Harry T. Thompson, March 29, 1943, reprinted in Joseph, Fanning, and Davison, *Cultural Landscape Report,* 287.

16. Joseph, Fanning, and Davison, *Cultural Landscape Report,* 35–36.

17. U.S. Congress, "An Act Authorizing the Establishing of a Public Park in the District of Columbia," September 27, 1890.

18. Bliss to Farrand, March 22, 1938. Dumbarton Oaks Garden Archives, B:MB 1938.03.22, Dumbarton Oaks RBC.

19. Farrand to Bliss, September 23, 1942. Dumbarton Oaks Garden Archives, B:BF 1942.09.23, Dumbarton Oaks RBC.

20. Ibid.

21. Ibid.

22. Irving C. Root, U.S. Department of the Interior, press release, July 8, 1941.

23. Ibid.

24. NPS, "Photographs, Reduced Copies of Measured Drawings, Written Historical and Descriptive Data," Historic American Buildings Survey (HABS), undated, p.14.

25. Joseph, Fanning, and Davison, *Cultural Landscape Report,* 44.

26. Joseph, Fanning, and Davison, *Cultural Landscape Report,* 40.

27. Farrand, "Dumbarton Oaks," 134.

28. NPS, "Photographs, Reduced Copies," 20. A letter from a user in 1956 says he had never seen the park as beautiful as at that time.

29. Joseph, Fanning, and Davison, *Cultural Landscape Report,* 54.

30. Hill quoted in Meyer, "Neglected No More."

31. Donald E. Smith, Assistant Superintendent (1952–1974) and Superintendent (1974–1992) of Gardens and Grounds, Dumbarton Oaks, interview over several days in July, August, and September 1992. Dumbarton Oaks Research Library, Oral History Project.

32. Ibid.

33. Joseph, Fanning, and Davison, *Cultural Landscape Report,* 232.

34. U.S. Environmental Protection Agency, "What Is the National Environmental Protection Act?" EPA website, accessed March 26, 2020: https://www.epa.gov/nepa/what-national-environmental-policy-act.

35. Darwina Neal (retired landscape architect with the NPS) in email to author, April 28, 2020.

36. Joseph, Fanning, and Davison, *Cultural Landscape Report,* 55.

37. Joseph quoted in Heckler, "Out of the Wilderness," 77.

38. NPS, HABS, 2.

39. George Washington University Historic Landscape Preservation Program Studio, "Preservation Needs Assessment, Dumbarton Oaks Park, Washington, DC."

40. Ibid.

41. Fanning and Joseph, "Discovering Dumbarton Oaks Park," 63.

42. Ibid., 64.

43. Joseph, Fanning, and Davison, *Cultural Landscape Report,* iii. It is difficult to ascertain

precisely the role Harvard staff in the garden played in the CLR. Access to the archives and library would surely have been important for documenting history. The full quotation from the superintendent of Rock Creek Park is: "Over the past three years of close collaboration, the National Park Service and Dumbarton Oaks Gardens have come to share a commitment to revealing, once again, the vision of Mildred and Robert Bliss and Beatrix Farrand for the gardens of Dumbarton Oaks."

44. Joseph, Fanning, and Davison, *Cultural Landscape Report,* 2.

45. Ibid., 247.

46. This can be seen clearly in the Pebble Garden, which shows Havey's preference for flourish elements and embedded images of scrolls.

47. Joseph, Fanning, and Davison, *Cultural Landscape Report,* 81.

48. Architrave Architects, Rhodeside & Harwell Landscape Architecture & Planning, and Robinson & Associates Architectural & Landscape Historians, *Montrose Park Cultural Landscape Report* (Washington, DC: U.S. Department of the Interior, National Park Service, 2004), 70.

49. NPS, "National Register of Historic Places Registration Form," signed March 12, 2004.

50. Ibid.

51. Cultural Landscape Foundation, "Dumbarton Oaks Park."

52. Trafton quoted in Higgins, "Conservancy Formed to Rescue Dumbarton Oaks Park."

53. Elizabeth Gilbert, in-person interview by author, March 10, 2020.

54. Joseph, Fanning, and Davison, *Cultural Landscape Report: Part 2,* ix. This comment was made in a statement on the acknowledgments page, which was added in 2015. Joseph also thanked the DOPC for its partnership, especially for its "drive to seek other funding opportunities" for the treatment proposed in the *Cultural Landscape Report.*

55. Ibid.

56. Gilbert, interview.

57. Higgins, "Revered, Then Reviled." The author quotes Ann Aldrich, the DOPC's invasive plant management specialist.

58. Farrand to John Thacher, March 30, 1942, D:BF 1942.03.30B, Dumbarton Oaks OHP.

59. Farrand to Harry T. Thompson, March 29, 1943, Dumbarton Oaks OHP. *Cultural Landscape Report,* 287.

60. Farrand to Thompson, March 29, 1943,

61. Olmsted Center for Landscape Preservation, *Landscape Preservation Maintenance Plan, Dumbarton Oaks Park.*

62. Kulakowski, "Dumbarton Oaks Park Hydrologic Analysis Update, September–November 2011.".

63. Ibid.

64. Gilbert, interview.

65. Dumbarton Oaks Park Conservancy and National Parks Service, "Masterplanning Workshop," 2.

66. Ibid., 2.

67. Ibid., 15.

68. Nick Bartolomeo, chief of resource management of Rock Creek Park, expressed this sentiment at the Dumbarton Oaks Conservancy Masterplanning Workshop in November 2017.

69. Olmsted quoted in Lancaster, *Prospect Park Handbook.* This description was contained in a summarized version of a lengthy essay by Olmsted and Vaux that accompanied their plan in 1866.

70. Beveridge, *Frederick Law Olmsted,* 436.

71. "Our Village."

72. "A Glance Ahead!," 2.

73. Ibid.

74. Lancaster, *Prospect Park Handbook,* 21.

75. Brooklyn Park Commission, "Report of the Brooklyn Park Commissioners for the Year 1868," 1868, 3.

76. "An Account of the Dinner by the Hamilton Club."

77. "Prospect Park and J. S. T. Stranahan."

78. "Mr. Stranahan Dead."

79. "Prospect Park."

80. Ibid.

81. Brooklyn Park Commission, "Report of the Brooklyn Park Commissioners from January 1874, to December 31, 1879," 1880, 30.

82. Olmsted, Vaux & Co., *Preliminary Report to the Commissioners for Laying Out a Park,* 9.

83. "Brooklyn Parks and Boulevards."

84. Brooklyn Park Commission, *Twenty-Sixth Annual Report of the Brooklyn Park Commissioners for the Year 1886,* 6.

85. McLoughlin, "Prospect Park Lasting Memorial to Culyer."

86. "Prospect Park's Decay."

87. "To Improve Prospect Park," *New York Times,* May 30, 1888.

88. Stranahan quoted in Thomas C. Campanella, *Brooklyn: The Once and Future City* (Princeton, NJ: Princeton University Press, 2019), 91.

89. Burrows and Wallace, *Gotham,* 1228.

90. Ibid., 1222.

91. Culter quoted in "New Park Entrances, How the Two Will Look When Completed," *Brooklyn Daily Eagle,* August 3, 1890.

92. Brooklyn Park Commission, "Thirty-Third Annual Report of the Brooklyn Park Commissioners for the Year 1893," 1894, 8.

93. "Aneurin Jones' Pull."

94. Schuyler and Kaliss, *The Papers of Frederick Law Olmsted: Vol. 9,* 748. This was in a letter from Olmsted to John Charles Olmsted and Charles Eliot dated February 19, 1894.

95. Ibid., 757.

96. Olmsted quoted in Colley, *Prospect Park,* 128.

97. Brooklyn Park Commission, "Thirty-Third Annual Report," 8.

98. Robert A. Caro, *The Power Broker: Robert Moses and the Fall of New York* (New York: Alfred A. Knopf, 1974), 20.

99. Hylan quoted in Brooklyn Park Commission, *Twenty-First Annual Report of the Department of Parks for the Year 1918,* 5.

100. Ibid., 6.

101. "Kennedy Opposes Soldiers' Monument in Prospect Park."

102. Ibid.

103. "Park Memorial to Cost $35,000."

104. Brooklyn Park Commission, *Annual Report of the Brooklyn Park Commissioners 1926–1927,* 14.

105. "Olmsted and Moses Were the Key Figures in Development of City Parks." Although the quotation refers specifically to Central Park in regard to its abuse and neglect, it makes clear that the issue was pertinent to other landscapes, including Prospect Park.

106. Siegel, *The Future Once Happened Here,* 22.

107. Caro, *Power Broker,* 87.

108. "Commissioner Moses," *New York Times,* January 20, 1934.

109. Rogers, "Robert Moses."

110. Moses quoted in Woolf, "Our Park Planner Talks of the Future."

111. Ibid.

112. Ibid. Moses was speaking specifically about Central Park and the designers' inability to foresee the rise of tenements around the park and their influence on how the park was used.

113. Moses, "Memorandum on Organization of Work Relief Projects."

114. Ibid., 1.

115. The Civil Works Administration was a temporary federal program established to fund public works projects.

116. Hickok, "He's One Reason Why Prospect Park's Beautiful."

117. New York City Department of Parks, press release, May 1, 1941, 2.

118. Zorah W. Gristede to Commissioner Moses, April 17, 1944. Gristede was the daughter-

in-law of Charles Gristede, who started the Gristedes grocery store chain in Manhattan in 1888. NYC.gov, https://www.nyc.gov/html/records/pdf/govpub/42001944_press_releases_part1.pdf.

119. Robert Moses to Zorah W. Gristede, April 18, 1944.

120. "Quiet Milestone in Prospect Park."

121. Helmle., letter to the editor, *New York Times.*

122. Lindsay quoted in Colley, *Prospect Park,* 155. This was likely a thinly veiled critique of Moses.

123. Hoving quoted in "New City Parks Chief:" Olmsted's name is once again misspelled in this article as "Olmstead." The variant spelling did not appear to be an issue while Olmsted was living.

124. Hoving quoted in "Two Curators Named as Parks Go Artistic."

125. "Quiet Milestone in Prospect Park."

126. Makla quoted in Knight, "Prospect Park's Future Is Pondered." The full quotation, by the president of the Park Slope Civic Council, is: "The City is making it sort of beautiful in a concretized and modernized way, when before it was wild and beautiful."

127. Ibid.

128. Ibid. The quotation is attributed to "Charles Monaghan, president of the Central Brooklyn Independent Democrats and a former trustee of the Park Slope Civic Council."

129. Campanella, *Brooklyn,* 431.

130. Clurman quoted in Hagan, "Every Prospect Pleases, and Only Man is Vile."

131. Richard M. Clurman, letter to the editor, *New York Times,* March 4, 1973.

132. Ibid.

133. Clurman quoted in Wade, "In Which Mr. Clurman, the New City Parks Administrator, is Taken, Over the Coals for What he Said."

134. New York City was saved from bankruptcy by the teachers' union, which agreed to purchase city bonds with teachers' pensions to raise needed money.

135. Makla, "Wants National Park Service to Take Over Prospect Park."

136. Koch quoted in Quindlen, "For Prospect Park, $10 Million to Recapture What It Was."

137. Ibid.

138. Fahim, "Returning Prospect Park to the People."

139. Thomas quoted in Copage, "Greening of the Park."

140. Santangelo, "Proud to Call It Home."

141. Carmody, "A Born-Again Prospect Park Is Greener, Safer and Busier."

142. Ibid. Tupper Thomas says the city invested "many hundreds of the thousands of dollars" in creating a master plan for the park and that "there are very few places where people are looking

at their urban parks the way we are as a design[ed] space and as a long-range thing that must be preserved, not only to save the grass and the trees, but also the way the park looks."

143. Thomas quoted in Brown, "Preservation Heroes."

144. Thomas quoted in Schwartz, "An Improved Prospect," 78.

145. Colley, *Prospect Park,* 171.

146. Olmsted, Vaux and Co., *Preliminary Report to the Commissioners,* 13.

147. Schwartz, "An Improved Prospect," 80. Thomas credits "newly hired gardening staff" for changing opinions about the landscape.

148. The Lila Wallace-Reader's Digest Fund emerged from the estate of Lila Wallace, who, along with her husband, DeWitt, amassed a fortune from their publication, *Reader's Digest.* After their deaths, their fortune was split into two foundations and a focus for Lila Wallace's foundation became urban parks. Prospect Park was an early recipient of such a grant.

149. Toth and Sauer, *A Landscape Management Plan for the Natural Areas of Prospect Park,* 9.

150. Ibid., 5.

151. Christian Zimmerman, interview by author, Prospect Park, August 20, 2021.

152. B. Stewart, "Nature Meets Artifice, and a Ravine Is Reborn."

153. Fox quoted in Schwartz, "An Improved Prospect," 85.

154. DiCicco, "Long-Term Urban Park Ecological Restoration," 318.

155. Hamill, "Top Town for Summer Is NYC."

156. Hamill, "Park Rape."

157. Schwartz, "An Improved Prospect," 108.

158. Thomas quoted in D. Martin, "A Lush Park Woodland, the Way It Used to Be."

159. Stewart, "Nature Meets Artifice, and a Ravine Is Reborn."

160. Benepe, "From Playground Tot to Parks Commissioner"; Fahim, "Returning Prospect Park to the People."

161. "A New Leader for Central Park." She is quoted as saying, "We really need to know that we're protecting the Woodlands areas, where for 20 years we've been letting the natural ground cover regenerate itself from the leaves and trees that fall down, and not letting the press of people do damage to the area. We don't have the staff to be watching them all the time."

162. Lloyd quoted in "A New Leader."

163. Rehak, "Heaven's Skate," 114. The quotation is attributed to the architect Billie Tsien, who, along with Tod Williams, was the architect for Lakeside Center.

164. Zimmerman, interview.

165. Lange, "A Skating Rink Grows in Brooklyn."

166. Schumacher, "Tells of Floral Display in Prospect Park."

167. "Prospect Park to Get a Facelift."

168. Desmond, "Lindsay Asks Pressure on TA & Con Ed to Reduce Pollution."

169. White, "Urge Probe of Prospect Park 'Botch' Job." The curator of the park at the time was Donald Simon.

170. James, *In the Vale of Cashmere,* 7. The photographer is Thomas Roma (b. 1950).

171. Bradley-Smith, "A New Vision for Prospect Park's Vale of Cashmere Sparks Controversy at Landmarks Hearing."

172. Wachs, "A Proposed Prospect Park Development Could Eliminate Important LGBT Historic Site." According to its website, the Victorian Society was founded in New York City in 1966 and is dedicated to fostering appreciation for and preservation of our nation's nineteenth-century heritage, as well as that of the early twentieth century (1837–1917). The Victorian Society New York, the oldest of numerous chapters now flourishing throughout the United States, is an independent nonprofit organization affiliated with the national society.

173. Ibid.

174. Ibid.

175. Zimmerman, interview.

176. "Dream Come True."

Conclusion

1. Schuyler and Kaliss, *Papers of Frederick Law Olmsted, Volume 9: The Last Great Projects,* 295.

2. Van Valkenburgh, "Landscapes over Time: The Maintenance Imperative," 108.

3. Ibid., 114.

4. Kiley and Amidon, *Dan Kiley,* 96.

BIBLIOGRAPHY

"$25 Million Ford Grant Given Stanford." *Stanford Daily,* September 26, 1960.
"$30,000 Worth?" Editorial. *Stanford Daily,* May 1, 1931.
Abbott, Stanley W. "Historic Preservation: Perpetuation of Scenes Where History Becomes Real." *Landscape Architecture* 40, no. 4 (1950): 153–57.
"An Account of the Dinner by the Hamilton Club to Hon. James S. T. Stranahan, Thursday Evening, December 13, 1888." *Eagle Press,* December 31, 1889.
"An Act to Enable Henry Shaw to Convey of Devise to Trustees Certain Lands." General Assembly of the State of Missouri. Accessed December 18, 2020. https://mdh.contentdm.oclc.org/digital/collection/molaws/id/6061.
Adams, Adam G. "Vizcaya." *Tequesta: The Journal of the Historical Association of Southern Florida* 15 (1955): 29–39.
Adams, Carolyn. "New Buildings Complement Existing Campus Architecture." *Stanford Daily,* August 16, 1990.
Adams, Thomas Boylston. "Boston Achieves a Triumph in Architecture." *Boston Globe,* December 21, 1969.
Adezio, Antonia. "The Garden Conservancy Mission, Revisited." *Newsletter of the Garden Conservancy,* Spring/Summer 1991.
———. "A Garden Conservancy Project: How It Works." *Newsletter of the Garden Conservancy,* Spring/Summer 1991.

Alanen, Arnold R., and Lynn Bjorkman. "Plats, Parks, Playgrounds, and Plants: Warren H. Manning's Landscape Designs for the Mining Districts of Michigan's Upper Peninsula, 1899–1932." *IA: The Journal of the Society for Industrial Archeology* 24, no. 1 (1998): 41–60.
"Allen Introduces Bill to Preserve Historical Buildings at Tuskegee." *Union Springs Herald,* January 18, 1973.
American Public Gardens Association. "Vizcaya Museum and Gardens Promotes Ian Simpkins to Deputy Director for Horticulture and Urban Agriculture." Publicgardens.org, February 12, 2016.
American Scenic and Historic Preservation Society. *Twenty-Fourth Annual Report of the American Scenic and Historic Preservation Society, 1919.* Author, 1919.
Ames, Oakes. "Mount Auburn's Sixscore Years." *Proceedings of the Cambridge Historical Society* 34 (1951–52): 92.
Anable, Stephen H. "175 Years of Stewardship: Looking Back and Moving Forward." *Sweet Auburn,* Spring 2007.
———. "A Conversation with Bill Clendaniel." *Sweet Auburn,* Summer 2008.
———. "David P. Barnett, Ph.D.: A Connecticut Yankee in Mount Auburn's Court." *Sweet Auburn,* Spring 2009.
"Aneurin Jones' Pull." *Brooklyn Daily Eagle,* August 17, 1890.
Angst, Maggie. "How Trees Lost to Matthew Affected Hilton Head Island Residents—for Good and Bad." *Island Packet,* October 7, 2017.
"Announcing the First Villas in Our Exciting New Resort Community: Harbour Town, on Hilton Head Island." *Atlanta Constitution,* August 2, 1970.
"Another Central Park Danger." *Herald* (New York), April 10, 1919.
AP. "Developer Seeks to Buy Resort of Sea Pines." *Greenville News,* July 31, 2004.
Bajoria, Abhishek. "Bike Program Nears Its End." *Stanford Daily,* January 22, 1999.
Baker, Ray Palmer. "Rensselaer Polytechnic Institute and the Beginnings of Science in the United States." *Scientific Monthly* 19, no. 4 (1924): 337–56.
Barnett, David. "President's Corner." *Sweet Auburn,* Spring 2009.
Barreira, Alex. "Golden Gate National Parks Conservancy to Permanently Cut Workforce by 27%." *San Francisco Business Times,* November 9, 2020.
Barry, Bill. "Ol' Sam Bids Vizcaya Farewell." *Miami News,* March 13, 1966.
Bassman, Herbert J. Letter to the editor. *Forest Park Review,* May 31, 1972.
Beatty, Russell A. "Gardens on Alcatraz?" *Sitelines,* Fall 2012.
Bell, Jefferson. "Told in Miami Homes: A Review of Topics of Daily Interest." *Miami Herald,* February 6, 1926.

Benepe, Adrian. "From Playground Tot to Parks Commissioner: My Life with Robert Moses." *SiteLINES: A Journal of Place* 3, no. 1 (2007): 12–14.
Bennett, James O'Donnell. "Chicago's New Zoo Paradise for Animals." *Chicago Tribune*, September 13, 1931.
Beveridge, Charles E., ed. *Frederick Law Olmsted: Writings on Landscape, Culture, and Society.* New York: Library of America, 2015.
Berg, Shary Page. *Mount Auburn Cemetery Master Plan: Volume II. Historic Landscape Report.* Mount Auburn Cemetery, 1993.
Bigelow, Jacob. *A History of the Cemetery of Mount Auburn.* Boston: James Munroe, 1860.
Birnbaum, Charles, ed. *Preserving Modern Landscape Architecture II.* Washington, DC: Spacemaker Press, 2004.
———. "Preservation Brief 36: Protecting Cultural Landscapes: Planning, Treatment and Management of Historic Landscapes." U.S. Department of the Interior and the National Park Service, 1994.
Birnbaum, Charles, and Christine Capella Peters, eds. "Secretary of the Interior's Standards for the Treatment of Historic Properties with Guidelines for the Treatment of Cultural Landscapes." U.S. Department of the Interior and the National Park Service, 1996.
Black, Chris. "A Floral Facelift for City Hall Plaza." *Boston Globe*, June 7, 1994.
Bowers Realty Group. "How Is the Real Estate Market Doing in Riverside, IL?" Team bowersrealtors.com. Accessed April 18, 2021. https://www.teambowersrealtors.com/reports/Riverside/how-is-the-real-estate-market-doing/.
Bradley-Smith, Anna. "A New Vision for Prospect Park's Vale of Cashmere Sparks Controversy at Landmarks Hearing." *Brownstoner*, March 2, 2023.
Bremier, P. J. "Rediscovering Isolated Beauty: Marin Volunteers Help Restore Prison Gardens on Alcatraz Island." *Marin Independent Journal (California)*, July 15, 2006.
Broadwin, John. "Farm Wheels on Way Out." *Stanford Daily*, December 1, 1964.
"Brooklyn Parks and Boulevards." *New York Times*, January 25, 1875.
Brooklyn Park Commission. *Annual Report of the Brooklyn Park Commissioners 1926–1927.* 1928.
———. *Thirty-Third Annual Report of the Brooklyn Park Commissioners for the Year 1893.* 1894.
———. *Twenty-Sixth Annual Report of the Brooklyn Park Commissioners for the Year 1886.* 1887.
———. *Twenty-First Annual Report of the Department of Parks for the Year 1918.* 1918.
Brown, Elizabeth Via. "Designing Beauty: Retired Architect Sculpts Landscapes into Pieces of Art." *Montgomery Advertiser*, February 22, 1997.
Brown, Jane Roy. "Preservation Heroes: Tupper Thomas, Prospect Park, Brooklyn, NY (2011)." Library of American Landscape History, 2011.

Brown, Patricia Leigh. "The Plant Men of Alcatraz." *New York Times,* June 21, 2001.
"Bug Me." *Stanford Daily,* November 26, 1980.
Buggey, Susan. "On the Founding and Early Years of the Alliance." Paper presented at the Alliance for Historic Landscape Preservation 30th Anniversary Meeting, Montreal, May 31, 2008.
Burrows, Edwin G., and Mike Wallace. *Gotham: A History of New York City to 1898.* Oxford University Press, 1999.
Bussman, Jessica. "Memorial Park Concept at Willow Pond Designed to Meet 20th-Century Needs." *Mount Auburn Cemetery,* December 10, 2014.
Cairns, Malcolm, and Gary Kesler. "Protecting a Prototype." *Landscape Architecture Magazine,* July/August 1987.
California Energy Commission. *San Francisco Bay Area Region Report.* California's Fourth Climate Change Assessment, August 2018.
Callon, John. "Buildings Grow on Farm." *Stanford Daily,* April 21, 1986.
Campbell, Robert. "The Trouble at City Hall Plaza." *Boston Globe,* August 5, 1973.
———. "Giving Purpose to City Hall Plaza." *Boston Globe,* December 12, 1994.
"Can the American Elm Trees Be Saved?" *Cicero Life,* July 5, 1978.
Cappucci, Matthew. "A Look Back at the Disastrous Great New England Hurricane of 1938." *Boston Globe,* September 21, 2018.
———. "More than 10 Inches of Rain Falls in Illinois, Another Exceptional Deluge." *Washington Post,* August 2, 2022.
"Care of Elms Told by Expert." *Berwyn Life,* May 4, 1949.
"Care to Invest as You Enjoy Your Vacation?" *National Post,* December 20, 1958.
Carmody, Deirdre. "A Born-Again Prospect Park Is Greener, Safer and Busier." *New York Times,* August 4, 1984.
"Cemetery at Mount Auburn." *New England Farmer,* September 28, 1831.
"Cemetery Fence." *Harvard Crimson,* October 25, 1980.
Chesney, Alan. "Announcing: The Friends of Mount Auburn." *Sweet Auburn,* Winter 1986.
———. "Retirement of Superintendent Duncan W. Munro." *Sweet Auburn,* Winter 1990.
"Chicago's Suburbs Stirred by Oak Park's Landscape 'Color Scheme.'" *Chicago Tribune,* February 4, 1900.
City of Boston. *Rethink City Hall: Boston City Hall and Plaza Study.* Public Facilities Department, October 2017.
Clay, Grady. "Landscape Architects Honor Kentucky Pioneer Olmsted." *Courier-Journal,* July 19, 1964

———. "Whose Time Is This Place? The Emerging Science of Garden Preservation." *Landscape Architecture,* May 1976.
———. "The Friends Becomes a Charitable Trust and Receives Its First Grant." *Sweet Auburn,* Spring 1991.
Cohen, George. "Navy Pier Students Object to New Site." *Chicago Tribune,* February 26, 1959.
Coleman, Sandy. "Wanted: Ideas for City Hall Plaza." *Boston Globe,* September 25, 1994.
Colley, David P. *Prospect Park: Olmsted & Vaux's Brooklyn Masterpiece.* New York: Princeton Architectural Press, 2013.
Collins, Dennis. "Balancing Historical Integrity with Horticultural Innovation: The Dynamic Mount Auburn Mix." *Sweet Auburn,* Summer 2006.
———. "The Hurricane of 1938: A Lasting Impact on Horticulture at Mount Auburn." *Sweet Auburn,* Summer 2001.
Collins, Mike. Letter to the editor. *Stanford Daily,* January 11, 1966.
"Commissioner Moses." *New York Times,* January 20, 1934.
Cothran, James R.., and Danylchak, Erica. *Grave Landscapes: The Nineteenth-Century Rural Cemetery Movement.* Columbia: University of South Carolina Press, 2018.
Community Services Associates. "Sea Pines Critical Needs Referendum: Key Questions Answered." Brochure, n.d.
Copage, Eric V. "Greening of the Park." *Daily News,* July 6, 1980.
Cotton, Katlyn E. "Stewarding an Educational Legacy: Historic Preservation at Historically Black Colleges and Universities." PhD diss., University of Pennsylvania, 2017.
Crosby, Marshall R. Letter to the editor. *St. Louis Dispatch,* August 30, 1978.
Cultural Landscape Foundation. "Landslide 2006." Dumbarton Oaks Park: Farrand's Garden in the Valley. Cultural Landscape Foundation, 2006.
———. "Landslide 2008." Marvels of Modernism: Boston City Hall Plaza. Cultural Landscape Foundation, 2008.
Danielson, Michael N. *Profits and Politics in Paradise: The Development of Hilton Head Island.* Columbia: University of South Carolina Press, 1995.
Dearborn, Nathaniel. *A Concise History of, and Guide Through Mount Auburn, with a Catalogue of Lots Laid Out in the Cemetery; a Map of the Grounds, and Terms of Subscription, Regulations Concerning Visitors, Interments, &c. &c,.* Boston: Nathaniel Dearborn, 1843.
Desmond, James. "Lindsay Asks Pressure on TA & Con Ed to Reduce Pollution." *Daily News,* September 1, 1965.
DiCicco, Jessica M. "Long-Term Urban Park Ecological Restoration: A Case Study of Prospect Park, Brooklyn, New York." *Ecological Restoration* 32, no. 3 (September 2014): 314–26.

Dickey, Page. *Outstanding American Gardens: A Celebration, 25 Years of the Garden Conservancy.* New York: Stewart, Tabori & Page, 2015.

Dimmock, Thomas. "Henry Shaw: A Biographical Sketch." *Missouri Botanical Garden Annual Report 1890* (1890): 7–25.

Donald, Arnold W. "From the Chairman." *Missouri Botanical Garden Annual Report 2010.* 2011.

Downing, Andrew Jackson. *A Treatise on the Theory and Practice of Landscape Gardening.* New York: Orange Judd Co., 1875.

"Dr. Foster Announces Million Dollar Expansion at Tuskegee." *Huntsville Mirror,* September 5, 1953.

"'Dream Come True': Morgan Monaco Tapped as New Head of Prospect Park Alliance." *Brooklyn Magazine,* October 18, 2022.

Duffy, Robert W. "Sculpture in the Parks—Who Decides?" *St. Louis Post-Dispatch,* May 8, 1977.

Dumbarton Oaks Park Conservancy and National Park Service. "Masterplanning Workshop," November 8–9, 2017.

Elliott, Orin Leslie. *Stanford University: The First Twenty-Five Years.* Palo Alto: Stanford University Press, 1937.

Enstad, Robert. "It's Hard to Pin Blame in Case of Des Plaines." *Chicago Tribune,* September 6, 1987.

Faherty, William Barnaby. *Henry Shaw: His Life and Legacies.* Columbia: University of Missouri Press, 1978.

Fahim, Kareem. "Returning Prospect Park to the People." *New York Times,* April 5, 2010.

Fanning, Kay, and Maureen Joseph. "Dumbarton Oaks Park: Restoring a Masterwork for Modern Needs." *ATP Bulletin: The Journal of Preservation Technology* 30, no 2/3 (1999): 61–66. JSTOR.

"Farm Landscaping Emphasizes Trees." *Stanford Daily,* May 27, 1949.

Farrand, Beatrix. "Dumbarton Oaks: An Historical Setting for the Making of History." *Landscape Architecture Magazine* 34, no. 4 (July 1944): 131–35.

Findlen, Paula. "Anatomy Theaters, Botanical Gardens, and Natural History Collections." In *The Cambridge History of Science, Part 2: Personae and Sites of Natural Knowledge,* ed. Katherine Park and Lorraine Daston. Cambridge University Press, 2006, chap. 12, 272–89.

Fitch, James Marston. "Preservation Requires Tact, Modesty and Honesty among Designers." *Landscape Architecture* 66, no. 3 (1976): 276–80.

Flaherty, William Barnaby. *Henry Shaw: His Life and Legacies.* Columbia: University of Missouri Press, 1987.

Fong, Herb. "Stanford Grounds and Plants: Thirty Years of Change and Beyond." *Sandstone & Tile,* Winter 2008: 3–6.

Fritz, Shelagh. *Alcatraz Gardens.* Video. San Francisco Public Library. https://www.youtube.com/watch?v=FPAyOfeEY6o.
"Gains Told to Vizcayan Membership." *Miami Daily News,* May 21, 1957.
Gallagher, Colleen. "The Garden's Guardian." *Miami Herald,* June 14, 1988.
Garden Club of America. "National Affairs & Legislation and Conservation Committees." Website of the Garden Club of America. Accessed September 12, 2021. https://www.gcamerica.org/what-we-do-NAL-Conservation.
Garden Conservancy. "Garden Preservation Comes of Age." *Newsletter of the Garden Conservancy,* Fall 2009.
———. "#OpenDays25." *Garden Conservancy News,* April 2020.
———. "Save America's Treasures Grant Funds Alcatraz Historic Gardens Project." *Newsletter of the Garden Conservancy,* Winter 2007.
Gardens of Alcatraz. "Remembering Carola Ashford." AlcatrazGardens.org, February 2011.
Gates, David M. "From the Director." *Missouri Botanical Garden Bulletin* 56, no. 5 (September–October 1968).
George Washington University Historic Landscape Preservation Program Studio. "Preservation Needs Assessment, Dumbarton Oaks Park, Washington, DC." August 1993.
Getty Center. "Campus Heritage Grants / Funding Areas." Getty.edu. Undated.
Giguere, Joy M. "'The Americanized Sphinx': Civil War Commemoration, Jacob Bigelow, and the Sphinx at Mount Auburn Cemetery." *Journal of the Civil War Era* 3, no. 1 (March 2013).
"A Glance Ahead!" *Brooklyn Daily Eagle,* June 16, 1846.
Golden Gate National Parks Conservancy. "Alcatraz Historic Gardens." ParksConservancy.org, viewed June 6, 2020.
———. "Gardens of Alcatraz." Brochure. AlcatrazGardens.org, viewed January 21, 2021.
Golden Gate National Parks Conservancy, Garden Conservancy, and National Park Service. "Golden Gate National Parks Conservancy, Garden Conservancy, and National Park Service Celebrate Completion of Historic Gardens of Alcatraz Project; Parks Conservancy to Assume Management of Restored Gardens." Joint press release, October 30, 2014.
"Gramercy Park Not Disturbed by Loft Buildings." *New York Times,* February 25, 1912.
Grandison, Kenrick Ian. "From Plantation to Campus: Progress, Community, and the Lay of the Land in Shaping the Early Tuskegee Campus." *Landscape Journal* 15, no. 1 (1996): 6–22.
———. "Negotiated Space: The Black College Campus as a Cultural Record of Postbellum America." *American Quarterly* 51, no. 3 (1999): 529–79.
Gray, Jennifer. "An Everyday Wilderness: Dwight Perkins and the Cook County Forest Preserve." *Future Anterior: Journal of Historic Preservation, History, Theory, and Criticism* 10, no. 1 (2013): 1–21.

Grizzle, Ralph. "How William Cecil Made the Biltmore Estate into Asheville's Biggest Tourist Attraction." *Our State,* November 1, 2017.

Hagan, Patti. "Every Prospect Pleases, and Only Man Is Vile." *New York Times,* March 4, 1973.

Halfacre, Angela C. *A Delicate Balance: Constructing a Conservation Culture in the South Carolina Lowcountry.* Columbia: University of South Carolina Press, 2012.

Hamill, Denis. "Park Rape: Blame Thug, but Learn from the Victim." *Daily News,* November 20, 1996.

———. "Top Town for Summer Is NYC." *Daily News,* June 19, 2011.

Hardie, Raymond. "He Changed the Landscape." *Stanford Magazine,* January/February, 2003.

Harlan, Louis R., Pete Daniel, Stuart B. Kaufman, Raymond W. Smock, and William M. Welty, eds. *Booker T. Washington Papers: Vol. 2, 1860–89.* Champaign: University of Illinois Press, 1972.

———, eds. *Booker T. Washington Papers: Vol. 3, 1889–95.* Champaign: University of Illinois Press, 1972.,

Harmsen, Tyrus G. Letter to the editor. *Stanford Daily,* February 4, 1947.

Harnik, Peter, and Abby Martin. *Public Spaces/Private Money: The Triumphs and Pitfalls of Urban Park Conservancies.* Trust for Public Land, 2015.

Heckler, Sheryl. "Out of the Wilderness." *Preservation.* September 2000.

Heffernan, Erin. "Sea Pines Hardest Hit by Hurricane Matthew on Hilton Head." *Island Packet,* October 8, 2016.

Heinrich, Ken. "Vizcaya Sale Move Stirs Angry Protest." *Miami News,* January 27, 1957.

Helmle, Frank J. Jr. Letter to the editor. *New York Times,* October 8, 1964.

"Henry Shaw Is No More." *Chicago Tribune,* August 26, 1889.

Hickok, Guy. "He's One Reason Why Prospect Park's Beautiful." *Brooklyn Daily Eagle,* June 28, 1934.

Higgins, Adrian. "Conservancy Formed to Rescue Dumbarton Oaks Park." *Washington Post,* April 13, 2011.

———. "Revered, Then Reviled: Tracking the Rise and Fall of Ivy." *Washington Post,* November 29, 2017.

Hilton Head Area Association of Realtors. "Local Market Update—June 2021: Sea Pines." HHRealtor.com, June 2021.

Hirsley, Michael. "Debt Puts Stranglehold on Many Black Colleges Nationwide." *Chicago Tribune,* June 3, 1985.

Ho, Sally. "Des Plaines River Flood Proposals Described." *Chicago Tribune,* September 26, 2013.

Hollo, David M. "From the Chair." *Missouri Botanical Garden Annual Report 2020.* Accessed

November 12, 2021. https://www.missouribotanicalgarden.org/Portals/0/Annual%20and%20Financial%20Reports/PDFs/IA_AnnualReport_2020_FINAL_WEB.pdf.
Holmgren, Virginia C. *Hilton Head: A Sea Island Chronicle.* Hilton Head Island: Hilton Head Island Publishing Co., 1959.
Homan, Becky. "Go to the Top." *St. Louis Post-Dispatch,* March 8, 2003.
Horn, Huston. "Nothing to Do—but Enjoy Yourself." *Sports Illustrated,* December 17, 1962.
Hubbard, Theodora Kimball, Olmsted, Vaux & Co, F. L. Olmsted, and E. E. Childs. "Riverside, Illinois: A Residential Neighborhood Designed Over Sixty Years Ago: Selections from the Papers of Frederick Law Olmsted, Senior." *Landscape Architecture Magazine* 21, no. 4 (1931): 257–91.
Jackson, Stephen, "A Living Museum of Trees." *Sweet Auburn,* 2016, vol. 1.
Jahnke, Ruth. "Planning Office: 9 Guiding Years." *Stanford Daily,* January 8, 1954.
"James Deering Dies on Ocean Coming Home." *Miami News,* September 22, 1925.
James, G. Winston. *In the Vale of Cashmere.* Brooklyn: powerHouse Books, 2015.
"James Deering Gives Hospital $500,000 Fund" *Miami Daily News and Metropolis,* October 20, 1925.
Johnston, Theresa. "All Right Now." *Stanford Magazine,* September/October 1999.
———. "Meet President Jordan." *Stanford Magazine,* January/February 2010.
Joncas, Richard, David Neuman, and Paul V. Turner. *Stanford University: The Campus Guide.* Princeton, NJ: Princeton Architectural Press, 2006.
Joseph, Maureen DeLay, Kay Fanning, and Mark Davison. *Cultural Landscape Report: Dumbarton Oaks Park, Rock Creek Park.* Washington, DC: U.S. Department of the Interior, National Park Service, 2000.
Jost, Daniel, M. Ellen Deming, Meg Calkins, and Fernando Magallanes. "Charlotte Shouldn't Demolish Its Waterfall Park Uptown." *Charlotte Observer,* May 23, 2023.
"Judge OKs Hilton Head Deer Kill." *Atlanta Journal Constitution,* Thursday, May 6, 1999.
Kay, Jane Holst. "Boston Common Sprucing Up at a Snail's Pace—but It Still Shows." *Boston Globe,* August 22, 1971.
———. "Common, Public Garden Are a Mess" *Boston Globe,* October 25, 1970.
Keller, Timothy, and Genevieve P. Keller. "How to Evaluate and Nominate Designed Historic Landscapes." *National Register Bulletin* 18. U.S. Department of the Interior, National Parks Service, and Interagency Resources Division, 1987.
Kellum, Jo. "Designing His Own Path." *Landscape Architecture* 89, no. 5 (1999): 138–39.
Kelly, C. W. "News among the Colored People." *Tuskegee News,* September 10, 1942.
"Kennedy Opposes Soldiers' Monument in Prospect Park." Letter to the editor. *Brooklyn Daily Eagle,* December 1, 1918.

Kendrick, Stephen. *The Lively Place.* Boston: Beacon Press, 2016.
Kienzl, Ray. "Golfer Sinks Self into Site." *Pittsburgh Press,* May 9, 1988.
Kiley, Dan, and Jane Amidon, Jane. *Dan Kiley: The Complete Works of America's Master Landscape Architect.* Boston: Little, Brown, 1999.
King, Mary. "Garden Says No to Di Suvero." *St. Louis Post Dispatch,* September 2, 1979.
King, Terry Johnson. "Fosse Met Vizcaya Challenge with Genius for Growing Things." *Miami News,* September 17, 1976.
Knight, Michael. "*Prospect Park's Future Is Pondered: A Place to Play or a Place to Find Tranquility?*" *New York Times,* February 20, 1971.
Kokal, Katherine. "Hilton Head Gated Community to Shoot Deer to Curb Crashes, 'Vegetative Destruction.'" *Island Packet,* January 21, 2020.
Koplowitz, Howard. "Tuskegee University Receives $20 Million Donation from MacKenzie Scott." AL.com, July 28, 2020.
Kulakowski, Zoe. "Dumbarton Oaks Park Hydrologic Analysis Update, September—November 2011." Geological Society of America–GeoCorps. December 2011.
Kushins, Jordan. "Alcatraz's Gardens Are Flourishing, against All Odds." NationalGeographic .com, October 21, 2020.
Lancaster, Clay. *Prospect Park Handbook.* New York: Walton H. Rawls, 1967.
Landscape Advisory Commission, Village of Riverside. "Master Landscape Plan," November 2015.
Lange, Alexandra. "A Skating Rink Grows in Brooklyn." *New Yorker,* January 2, 2014.
Le, Phuong. "Budding Controversy Blooms as Petunia Wars in Riverside." *Chicago Tribune,* April 3, 1998.
"Legislation Would Make Historic Site in Area of Tuskegee Institute." *Montgomery Advertiser,* October 21, 1973.
Leland, Elizabeth. "2 Faces of Hilton Head." *Charlotte Observer,* June 12, 1989.
Linden, Blanche M. G. "Nineteenth-Century German-American Landscape Designers." *Site-LINES: A Journal of Place* 1, no. 2 (2006).
———. *Silent City on a Hill: Picturesque Landscapes of Memory and Boston's Mount Auburn Cemetery.* Amherst: University of Massachusetts Press, 2007.
Lopez, Yadira. "As COVID-19 Cancels Events, Vizcaya Asks for Help." *Miami Herald,* August 13, 2020.
MacAdam, David H. *Tower Grove Park of the City of St. Louis: Review of Its Origin and History, Plan of Improvement, Ornamental Features, Etc.; with Illustrations.* R. P. Studley & Co., Printers, 1883.

Mackintosh, Barry, Janet A. McDonnell, and John H. Sprinkle Jr. "The National Parks: Shaping the System." *George Wright Forum* 35, no. 2 (2018): 43.
MacNabb, Betty L. "Hilton Head Island Growing Fast." *Charlotte Observer,* July 7, 1957.
Maersch, Stephen. "Charm of Riverside Is Due to Design." *Chicago Tribune,* July 30, 1964.
Maher, James T. *The Twilight of Splendor, Chronicles of the Age of American Palaces.* Boston: Little, Brown, 1971.
"Maintenance Management: A Powerful Tool." *Courier, The Newsletter of the National Park Service,* March 1986.
Makla, Bob. "Wants National Park Service to Take Over Prospect Park." *Daily News,* October 5, 1979.
Martin, Douglas. "A Lush Park Woodland, the Way It Used to Be." *New York Times,* November 30, 1997.
Mayall, R. Newton. "Recording Historic American Landscape Architecture." *Landscape Architecture Magazine,* October 1935, 1.
McCormack, Black and Co. "Riverside." *Chicago Tribune,* August 6, 1922.
McCormick, Kathleen. "Vaulting the Garden Wall." *Landscape Architecture* 84, no. 5 (1994): 74–81.
McGuire, Kim. "New Garden President Says He'll Stick to What Works." *St. Louis Dispatch,* March 3, 2010.
McHale, Joan Nielsen. "Restoring Vizcaya Is No Ball for Its Dedicated Supporters." *Miami News,* November 18, 1970.
McKenna, Ken. "Central Park: Devastation in Small Doses." *Daily News,* May 27, 1975.
McLaughlin, Charles C. "Frederick Law Olmsted's Parks: Antiques or Urban Necessities." *Field Notes,* National Association for Olmsted Parks, Fall/Winter 1980/1981.
McLoughlin, Maurice E. "Prospect Park Lasting Memorial to Culyer." *Brooklyn Daily Eagle,* August 10, 1931.
McNulty, Elizabeth, ed. *Missouri Botanical Garden: Green for 150 Years.* St. Louis: Missouri Botanical Garden Press, 2009.
McPhee, John. *Encounters with the Archdruid.* New York: Farrar, Strauss and Giroux, 1971.
Meadows, Gail. "Vizcaya Is Facing an Inspection by Museum Group." *Miami Herald,* August 8, 1997.
Merrifield, Kelly. "From Necessity to Honor: The Evolution of National Cemeteries in the United States." National Park Service. Accessed June 22, 2020. https://www.nps.gov/nr/travel/national_cemeteries/Development.html.
Meyer, Eugene L. "Neglected No More." *Washington Post,* March 12, 1999.

Miami-Dade County, "Vizcaya's Restoration of Main House and Gardens." MiamiDade.gov, July 20, 2004.
Miller, Donald L. *Lewis Mumford, a Life.* New York: Grove Press, 1989.
Missouri Botanical Garden. "Annual Report 1967." *Missouri Botanical Garden Bulletin* 56, no. 2 (March–April 1968).
———. "A Brief History of the Missouri Botanical Garden." Undated. Accessed July 16, 2021. https://www.missouribotanicalgarden.org/media/fact-pages/brief-history.aspx.
———. "Cultural Institution." In *Missouri Botanical Garden Strategic Plan 2016–2020: Discover + Share.* 2015.
———. "Gateway to the Garden." MissouriBotanicalGarden.org, undated.
———. "Twenty-Third Annual Report of the Director." *Missouri Botanical Garden Annual Report 1912.* 1912, 12–23.
Missouri Botanical Garden Board of Trustees. "Executive Summary." *Missouri Botanical Garden Strategic Plan 2008–2014: Conservation and Sustainability.* 2007.
Missouri Botanical Garden Consolidated Financial Statements and Supplementary Information, Years Ending September 30, 2021 and 2020. 2021. Accessed September 30, 2022. https://www.missouribotanicalgarden.org/Portals/0/Annual%20and%20Financial%20Reports/PDFs/2021_Missouri_Botanical_Garden_Cons_Audit.pdf.
Moore, Ken. "'Stop Those Monsters!' Quad Islands to Expand." *Stanford Daily,* May 25, 1948.
Moss, Teresa. "Sea Pines Wins National Landscaping Award." *Island Packet,* January 3, 2017.
Mount Auburn Cemetery. *186th Annual Report.* 2017. https://viewer.joomag.com/annual-report-fy-2018/0406082001539104083.
———"Celebrating Dave Barnett." MountAuburnCemetery.org, April 30, 2021.
———. "A Message from the President & CEO." MountAuburnCemetery.org, June 1, 2020.
———. "Our Mission." MountAuburnCemetery.org, October 1, 2020.
"Mount Auburn Cemetery: The Corporators Elect Officers and Hear Reports." *Boston Globe,* February 6, 1883.
"Move to Determine If Riverside Is a National Landmark." *Chicago Tribune,* February 5, 1970.
"Mrs. McCormick Gives Huge Zoo Site to County." *Chicago Tribune,* December 30, 1919.
"Mr. Stranahan Dead." *Brooklyn Daily Eagle,* September 3, 1898.
Mundigo, Axel I., and Dora P. Crouch. "The City Planning Ordinances of the Laws of the Indies Revisited. Part I. Their Philosophy and Implications." *Town Planning Review* 48, no. 3 (1977): 247–68.
"A National Park Service." *New York Times,* May 30, 1916.
National Park Service. "America's National Park System: The Critical Documents." U.S. Department of the Interior, 1971.

———. "Climate Change Response Strategy." U.S. Department of the Interior, September 2010.
———. "HALS: Historic American Landscapes Survey." Brochure. Undated.
———. *Tuskegee Institute National Historic Site: General Management Plan.* U.S. Department of the Interior, 1978.
———. "Preservation of Historic Sites Act, 1935." U.S. Department of the Interior, August 21, 1935.
———. "Save America's Treasures Grant Program Announces $7.6 Million in Awards." Press release. U.S. Department of the Interior, December 12, 2007.
National Trust for Historic Preservation. "HBCU Cultural Heritage Stewardship Initiative." SavingPlaces.org, undated.
Neuman, David. "Success and Failure in Campus Design in the Post-World War II Era: An Historical View and Focus on California." Lecture, Berkeley Art Museum, February 10, 2000.
Oatis, Jonathan W. "Hilton Head Opened New Era in Development." *Charlotte Observer,* September 6, 1982.
Ogozalek, Sam. "'It's beyond Crazy': Hilton Head's Real Estate Market Is a 'Frenzy.' What That Means." *Island Packet,* June 25, 2021.
Olin, Laurie. Letter to the editor. *Miami Herald,* March 30, 2007.
"Olmsted and Moses Were the Key Figures in Development of City Parks." *New York Times,* October 13, 1980.
Olmsted Center for Landscape Preservation. *Landscape Preservation Maintenance Plan, Dumbarton Oaks Park.* April 1997.
Olmsted, Vaux & Co. *Preliminary Report to the Commissioners for Laying Out a Park.* Brooklyn: I. Van Anden's Print, 1866.
"Opening of the Great Leland Stanford Jr., University." *Scientific American* 65, no. 15 (October 10, 1891).
Palmer, Barbara. "Harry Sanders, Former Director of Campus Planning, Dead at 89." Stanford News Service, October 2004.
———. "Neuman Leaves Landscape Built on Clarity, 'Sense of Place.'" *Stanford Report,* January 28, 2004.
Panamerican Consultants, Inc. *Viewshed Impact Assessment of the Proposed 300 Grove Bay Condominium Project on Vizcaya Museum and Gardens, A National Historic Landmark, Miami-Dade County, Florida.* February 2007.
"Park Memorial to Cost $35,000." *Brooklyn Daily Times,* November 23, 1919.
"Passed Away Quietly." *St. Louis Post-Dispatch,* August 26, 1889.
Peña, Michael. "Ohio Building Designer David Lenox Named Stanford University Architect."

Stanford News Service website, 2005. Accessed February 12, 2021. https://news.stanford.edu/pr/2005/pr-lenox-011905.html.
Perce, Elizabeth. "By the River's Brink: The Shady Driveways and Walks of Beautiful Riverside." *Inter Ocean,* August 7, 1898.
Phillips, Richard. "The Story of Riverside Is a Tale of Two Men and a City." *Chicago Tribune,* May 14, 1982.
Pitts, Carolyn. "National Historic Landmark Nomination: Vizcaya." National Park Service, October 27, 1993.
Pressley, Sue Anne. "In America's Suburbs, the Deer Next Door," *Washington Post,* August 27, 2001.
Price, Donald. "The Stanford Arboretum." Stanford Libraries, Stanford University, CA. Accessed January 10, 2021. https://web.stanford.edu/~stimby/TOS/PDF/Price/Price.pdf.
Professional Grounds Management Society. "Green Star Awards." Accessed March 28, 2021. https://pgms.org/page/GreenStarAwards.
"Prospect Park: A Tour through Its Beautiful Walks." *Brooklyn Daily Eagle,* October 8, 1874.
"Prospect Park and J. S. T. Stranahan." *Brooklyn Daily Eagle,* October 14, 1917.
"Prospect Park to Get a Facelift." *Daily News,* August 8, 1965.
"Prospect Park's Decay: Mayor Chapin Wants $100,000 Appropriated At Once." *New York Times,* April 20, 1888.
Randall, Jenny. "The Historic Gardens of Alcatraz." *Historic Gardens Review* 29 (October 2013): 24–28.
Rankin, Jerry. "$100 Million SU Drive." *Stanford Daily,* April 19, 1961.
"Rare Trees Lost, Flowers Perish at Shaw's Garden in Sunday Storm." *St. Louis Post-Dispatch,* September 3, 1946.
Raver, Anne. "Garden Notebook; When Gates Open at 100 Private Gardens." *New York Times,* May 18, 1995, C4.
———. "Human Nature; The Keepers of the Garden's Soul." *New York Times,* October 21, 1999, F1.
Reed, Christopher. "Sweet Auburn." *Harvard Magazine,* May 3, 2017.
Rehak, Melanie. "Heaven's Skate." *Landscape Architecture* 104, no. 3 (2014): 104–15.
Rembert, Tracey. "Sewage . . . Made Beautiful." *Times and Democrat,* August 2, 1992.
Reno, Doris. "Dade Has Reason for Pride in Vizcaya, Says New Director." *Miami Herald,* September 22, 1957.
"Renovation: Yes, Asphalt: No." Editorial. *Stanford Daily,* February 9, 1983.
Rhodes, Karl. "Alchemy Island." *Econ Focus,* 19, no.2 (2015).
Richard Nixon Foundation. "President Nixon's Legacy of Parks." Nixon Foundation, April 2020.

Riggs, Trisha. "Reinventing Hilton Head for a New Generation." *Urbanland,* June 25, 2013.
Riley, Laura Date. "America's 10: Who Are They?" *Ladies' Home Journal,* September 1957.
"Riverside All Set for Boom in House Building." *Chicago Tribune,* March 27, 1921.
"Riverside Okays 9 New Projects." *Berwyn Life,* October 19, 1955.
Rogers, Cleveland. "Robert Moses." *Atlantic,* February 1939.
Rogers, Elizabeth Barlow. *Rebuilding Central Park: A Management and Restoration Plan.* Cambridge, MA: MIT Press, 1987.
Rosenblatt, Roger. "The World Is His Garden: Better Tread Carefully." *Time,* April 26, 1999.
Rudolph, Emanuel D. "History of the Botanical Teaching Laboratory in the United States." *American Journal of Botany* 83, no. 5 (1996): 661–71.
———. "One Hundred Years of the Missouri Botanical Garden." *Annals of the Missouri Botanical Garden* 78, no. 1 (1991): 1–18.
Russolillo, Steven. "California Has Driest Year Since 1924." *Wall Street Journal,* October 20, 2021.
Ryan, Andrew. "City Seeks Out Ideas to Repurpose Brick-bound Plaza." *Boston Globe,* March 10, 2015.
Ryan, Charlie. *My Life with Charles Fraser.* Hilton Head Island: Single Star Publishing, 2016.
Rybczynski, Witold, and Laurie Olin. *Vizcaya: An American Villa and Its Makers.* Philadelphia: University of Pennsylvania Press, 2007.
Sands, Philip H. "I.P.M.: For the Care of Mount Auburn's Trees and Shrubs." *Sweet Auburn,* Fall 1986.
Santangelo, Mike. "Proud to Call It Home." *Daily News,* October 17, 1982.
Sawyer, Ann. "Almost Forgotten Island Awakes to Tourists." *Democrat and Chronicle,* July 6, 1958.
Scardino, Albert. "A Gust of Bankruptcy and Scandal Rattles Elegant Hilton Head Island." *New York Times,* March 15, 1987.
"Schools Are the Issue: Riverside Residents Agitated upon Questions of Education." *Chicago Chronicle,* May 30, 1897.
Schumacher, Chas. H. "Tells of Floral Display in Prospect Park." *Brooklyn Daily Eagle,* May 19, 1945.
Schuyler, David, and Gregory Kaliss, eds. *The Papers of Frederick Law Olmsted: Vol. 9. The Last Great Projects, 1890–1895.* Baltimore: Johns Hopkins University Press, 2015.
Schwartz, Anne. "An Improved Prospect." *Landscape Architecture* 93, no. 7 (July 2003).
Scott, Sam. "Closing the Books." *Stanford Magazine,* March/April 2015.
Sea Pines Plantation Company. "Sea Pines Land Use Covenants." April 4, 1970.
———. "Land Use Restrictions, Covenants, and Affirmative Obligations Affecting the Sea Pines Forest Preserve." June 29, 1970.

———. "Declaration of Rights, Restrictions, Conditions, Etc." August 19, 1957.

"Set Meeting in Riverside on Land Use." *Chicago Tribune,* June 30, 1957.

Shaw, Henry. "Henry Shaw's Will." *Missouri Botanical Garden Annual Report 1890.* 1890, 29–55.

"Shaw's Garden Needs Additional Support." *St. Louis Post-Dispatch,* January 14, 1946.

"Shaw's Garden Shuts Station in Move for Economy." *St. Louis Globe–Democrat,* March 17, 1939.

Siegel, Fred. *The Future Once Happened Here: New York, D. C., L. A., and the Fate of America's Big Cities.* San Francisco: Encounter Books, 1997.

Snell, Charles W. "National Register of Historic Places Inventory—Nomination Form: Riverside Landscape Architectural District." February 10, 1970.

"Some Early History of the Garden." *Missouri Botanical Garden Bulletin* 6, no.10 (December 1918): 130.

"Spare the Trees: John Y. Culyer's Protest against the Prospect Park Onslaught." *Brooklyn Daily Eagle,* December 7, 1886.

Stanford University. "Budget Plan 2019/2020." Accessed February 5, 2020. https://budget.stanford.edu/sites/g/files/sbiybj9886/f/budgetplan2019-20.pdf.

———. "Capital Plan and Capital Budget." Accessed February 4, 2021. https://bondholder-information.stanford.edu/pdf/BudBk_2019-20_sec4.pdf.

———. "Stanford University Grounds Maintenance Guidelines." University Architect/Planning Office, July 23, 2010.

———. "Water Efficiency at Stanford." Stanford.edu, accessed December 20, 2023.

"Stanford Plans Expansion of Academic, Scientific Facilities in Areas Adjacent to the Present Quad." *Stanford Daily,* May 27, 1949.

Stephens, Matthew. "Cemeteries Exploring Climate-Friendly Burials." *CommonWealth,* June 11, 2022.

———. "President's Corner." *Sweet Auburn* 2 (2021): 1.

Stertz, Bradley. "Failed Ambitions and Lawsuits on Hilton Head Island." *Tampa Bay Times,* September 27, 1987.

Stewart, Barbara. "Nature Meets Artifice, and a Ravine Is Reborn." *New York Time,.* April 7. 2001.

Stock, Robert W. "Saving the World the Ecologist's Way." *New York Times Magazine,* October 5, 1969.

Story, Joseph. *An Address Delivered on the Dedication of the Cemetery at Mount Auburn.* Boston: Joseph T. & Edward Buckingham, 1831.

"Sulfur Dioxide in Air Menace to Human Life and Vegetation Here." *St. Louis Star,* October 13, 1927.

Tankard, Judith B. "Dumbarton Oaks Garden and Park: A Celebration." *Magnolia,* Fall 2020.

Tester, Hank. "Crown Jewel of Miami Attractions, Vizcaya Hit Hard by COVID-19." CBS4 Miami. Accessed November 21, 2020. https://miami.cbslocal.com/2020/08/27/crown-jewl-of-miami-attractions-vizcaya-hit-hard-by-covid-19/.
Ok Thrasher, Max Bennett. *Tuskegee: Its Story and Its Work.* Boston: Small, Maynard & Co., 1900.
Thompson, J. William. "Land Matters." *Landscape Architecture Magazine,* July 2008.
Tobar, Hector. "Architect Built Bridge between 2 Cultures." *Los Angeles Times,* September 25, 1990.
Town of Hilton Head Island Town Council. *Town of Hilton Head Island Design Guide,* approved October 6, 2015. HiltonHeadIslandSC.gov.
"Tree Future Grim." *Suburban Life/Citizen,* July 24, 1976.
"Two Curators Named as Parks Go Artistic." *Daily News,* January 20, 1966.
———. "Quad Revamping Will Soon Add Splash of Color." *Stanford Daily,* October 4, 1983.
Turner, Paul Venable. *Campus: An American Planning Tradition.* Cambridge, MA: MIT Press, 1984.
———. *The Founders and the Architects: The Design of Stanford University.* Art Department, Stanford University, Palo Alto, 1976.
Tuskegee University. "Campus Master Plan 2017: Executive Summary." Tuskegee.edu, 2017.
"Tuskegee University Students Participate in Historic Preservation Training Project." *Selma Times-Journal,* August 19, 2019.
Stanford University. "Stanford University Grounds Maintenance Guidelines." University Architect/Planning Office, July 23, 2010.
Uphues, Bob. "Board Vote Sets Stage for Hoffman Dam Removal." *Riverside-Brookfield Landmark,* September 27, 2011.
———. "National Olmsted Group Objects to Riverside Floodwall." *Riverside-Brookfield Landmark,* April 13, 2021.
———. "Once a Wall of Buckthorn, Now a Haven for Native Species." *Riverside-Brookfield Landmark,* August 20, 2019.
———. "Stemming the Tide." *Riverside-Brookfield Landmark,* July 7, 2009.
U.S. Congress. H.R.5496—National Historic Preservation Act Amendments of 1980. 96th Congress, December 12, 1980, 2988.
U.S. Department of Education. "White House Initiative on Historically Black Colleges and Universities." Accessed February 23, 2021. https://sites.ed.gov/whhbcu/about-us/#:~:text=In%201980%2C%20President%20Jimmy%20Carter,universities%20to%20provide%20quality%20education.%E2%80%9D.
U.S. General Accounting Office. "Historic Preservation: Cost to Restore Historic Properties

at Historically Black Colleges and Universities." U.S. General Accounting Office, February 1998.

UPI. "Japanese Garden Is Pride of St. Louis." *Daily Journal (Flat River, MO)*, June 10, 1977.

"Valuable Property for Sale." *Evening Post*, November 26, 1833.

Van Valkenburgh, Michael. "Landscapes over Time: The Maintenance Imperative." *Landscape Architecture*, March 2013.

Village of Riverside. "ArbNet Accreditation." Accessed April 18, 2021. https://www.riverside.il.us/375/ArbNet-Accreditation.

"Visions of Red Tile & Sandstone: A Historical Tour of Campus." *Stanford Daily*, August 15, 1980.

Von Eckardt, Wolf. "Landmark Decision." *Washington Post*, October 18, 1980.

———. "Turned to Stone." *Boston Globe*, February 12, 1969.

Wachs, Audrey. "A Proposed Prospect Park Development Could Eliminate Important LGBT Historic Site, Advocates Say." *Architect's Newspaper*, February 28, 2023.

Wade, Richard C. "In Which Mr. Clurman, the New City Parks Administrator, Is Taken, Over the Coals for What He Said. He Then Says It Again." *New York Times*, January 24, 1973.

Walmsley & Co. *The First Historic Landscape Report for the Ravine, Prospect Park, Brooklyn, New York.* New York: City of New York Parks and Recreation Department, 1986.

Weinstein, Bill. "Stanford Starts $83 Million Building Program." *Stanford Daily*, May 2, 1973.

Weiss, Ellen. *Robert R. Taylor and Tuskegee: An African American Architect Designs for Booker T. Washington.* Montgomery, AL: NewSouth Books, 2012.

———. "Tuskegee: Landscape in Black and White." *Winterthur Portfolio* 36, no. 1 (2001): 19–37.

White, Joyce. "Urge Probe of Prospect Park 'Botch' Job." *Daily News*, May 2, 1974.

Whitehall, Walter Muir. *Dumbarton Oaks: The History of a Georgetown House and Garden, 1800–1966.* Cambridge, MA: Belknap Press of Harvard University Press, 1967.

Williams, Philip. "Quad Plans Are Not Finalized." Letter to the editor. *Stanford Daily*, March 3, 1983.

Williams, Verne O. "Fabulous Vizcaya Finds It Tough to Make Culture Pay." *Miami Daily News*, August 8, 1956.

Wilson, Susan. "Sites and Insights: Mount Auburn Cemetery." *Boston Globe*, October 29, 1987.

WLA Studio. *The Oaks: Tuskegee Institute National Historic Site Cultural Landscape Report.* NPS.gov, 2018.

Wolfe, Susan. "Who Killed Jane Stanford?" *Stanford Magazine*, September/October 2003.

"Woodland Restoration Project at Consecration Dell." *Sweet Auburn*, Winter 2004/2005.

INDEX

Page numbers in italics refer to figures.